# Trends in Microextraction Techniques for Sample Preparation

Special Issue Editor
**Victoria F. Samanidou**

MDPI • Basel • Beijing • Wuhan • Barcelona • Belgrade

**MDPI**

*Special Issue Editor*
Victoria F. Samanidou
Aristotle University of Thessaloniki
Greece

*Editorial Office*
MDPI AG
St. Alban-Anlage 66
Basel, Switzerland

This edition is a reprint of the Special Issue published online in the open access journal *Separations* (ISSN 2297-8739) in 2016–2017 (available at: http://www.mdpi.com/journal/separations/special_issues/microextraction-sample-preparation).

For citation purposes, cite each article independently as indicated on the article page online and as indicated below:

Lastname, F.M.; Lastname, F.M. Article title. *Journal Name*. **Year**. *Article number*, page range.

**First Edition 2018**

Image courtesy of Victoria F. Samanidou

ISBN 978-3-03842-705-6 (Pbk)
ISBN 978-3-03842-706-3 (PDF)

# Table of Contents

# About the Special Issue Editor

**Victoria F. Samanidou** is a Professor in the Department of Chemistry, University of Thessaloniki, and obtained her doctorate (Ph.D.) in Chemistry from the Aristotle University of Thessaloniki, Greece. She has published around 150 original research articles in peer reviewed journals and 50 reviews and chapters in scientific books (H-index 28, Scopus December 2017). She is a member of the editorial board of more than 10 scientific journals and has reviewed almost 500 manuscripts in more than 100 scientific journals. She was also the guest editor for three Special Issues. She is listed in the Top 50 Women with high impact in Analytical Chemistry in the 2016 Power List which was created by The Analytical Scientist Magazine by Texere Publishing and she is currently the President of the Division of Central and Western Macedonia of the Greek Chemists Association.

# Preface to "Trends in Microextraction Techniques for Sample Preparation"

Although analytical scientists unequivocally agree that no sample preparation would be the best approach, the fact is that all samples handled in any analytical laboratory need to undergo some treatment to some extent prior to their introduction to the analytical instrument. This step is widely recognized as a major step in the chemical analysis workflow. Therefore, the next best strategy is to find the most adequate methodology that would comply with all current trends in sample preparation, such as speed, automation, operator safety, less solvent consumption, but with no compromise to analytical performance.

Classical methodologies based on solid phase extraction (SPE) and liquidliquid extraction (LLE) tend to have many drawbacks as they include complicated, time-consuming steps, requiring large sample size and large amounts of organic solvent. Therefore, they are being progressively replaced by miniaturized and environment-friendly techniques, such as Micro Extraction by Packed Sorbent (MEPS), Fabric Phase Sorptive Extraction (FPSE), Dispersive LiquidLiquid Micro Extraction (DLLME). Additionally, novel extraction sorbents, due to the evolution of technology and nanotechnology, have been synthesized with improved properties, enhanced selectivity, ease in handling, etc. This combination has opened new potential to extract target analytes from sample matrices with high impact of endogenous interferences. The so-called micro-extraction techniques have sparked excitement in the scientific community and among analytical chemists, as they conform to green analytical chemistry demands and ensure environmental protection and public safety. Savings in cost and time are considered valuable benefits by using novel micro-extraction approaches in sample handling. Selectivity, sensitivity and lower detection limits are also included among the performance characteristics required to meet the legislation criteria.

The target of this Special Issue is to present the state of the art in micro-extraction sample preparation techniques. Modern, simple and efficient methods for preconcentration and separation methods are described for different analytes isolated from various matrices.

Thirteen outstanding contributions are included in this Special Issue.

**Victoria F. Samanidou**
*Special Issue Editor*

*separations*

MDPI

*Editorial*

# Trends in Microextraction Techniques for Sample Preparation

Victoria F. Samanidou

Laboratory of Analytical Chemistry, Department of Chemistry, Aristotle University of Thessaloniki, 541 24 Thessaloniki, Greece; samanidu@chem.auth.gr; Tel.: +30-23-1099-7698

Received: 23 October 2017; Accepted: 18 December 2017; Published: 21 December 2017

Although analytical scientists equivocally agree that "no sample preparation" would be the best approach, the fact is that all samples that are handled in any analytical laboratory need to undergo treatment to some extent prior to their introduction to the analytical instrument. This step has been widely recognized as the major step in the chemical analysis workflow. Therefore, the next best strategy is to find the most adequate methodology that would comply with all of the current trends in sample preparation, such as speed, automation, operator safety, and less solvent consumption, but with no compromise regarding analytical performance.

Classical methodologies based on solid phase extraction (SPE) and liquid-liquid extraction (LLE) tend to have many drawbacks as they include complicated, time-consuming steps that require large sample sizes and large amounts of organic solvent. As a result, they are being progressively replaced by miniaturized, environment-friendly techniques, such as microextraction by packed sorbent (MEPS), fabric phase sorptive extraction (FPSE), and dispersive liquid–liquid microextraction (DLLME). Additionally, due to the evolution of technology and nanotechnology, novel extraction sorbents have been synthesized with improved properties, enhanced selectivity, easiness in handling, etc. This combination has opened up new potentials to extract target analytes from sample matrices with a high impact of endogenous interferences.

The so-called microextraction techniques have sparked the excitement of the scientific community, and already gained interest among analytical chemists, as they conform to green analytical chemistry demands and ensure environmental protection and public safety. Savings in cost and time are considered valuable benefits to using novel microextraction approaches in sample handling. Selectivity, sensitivity, and lower detection limits are also included among the performance characteristics required to meet the legislation criteria.

The target of this Special Issue is to present the state of art microextraction sample preparation techniques. Modern, simple, and efficient methods for preconcentration and separation methods are described for different analytes isolated from various matrices.

Thirteen outstanding contributions are included and briefly presented below.

Abuzar Kabir, Marcello Locatelli, and Halil Ibrahim Ulusoy critically audit the progress of microextraction techniques in recent years in their very comprehensive review, "Recent Trends in Microextraction Techniques Employed in Analytical and Bioanalytical Sample Preparation". Microextraction techniques have indisputably transformed analytical chemistry practices, from biological and therapeutic drugs monitoring to the environmental field, food samples, and phyto-pharmaceutical applications [1].

Soledad Cárdenas and Rafael Lucena present a remarkable review on the recent advances in extraction and stirring integrated techniques. Since microextraction techniques are usually non-exhaustive processes that work under the kinetic range, the improvement of the extraction kinetics necessarily improves the performance. The extraction yield and efficiency is related to how fast the analytes diffuse in samples, therefore, stirring the sample during extraction is crucial. The stirring can be done with an external element, or it can be integrated, with the extraction element in the same device. This article emphasizes the potential of promising approaches rather than their applications [2].

Theodoros Chatzimitakos and Constantine Stalikas provide a snapshot of the most important features and applications of different carbon-based nanomaterials in their excellent review, "Carbon-Based Nanomaterials Functionalized with Ionic Liquids for Microextraction in Sample Preparation". These features include fullerenes, carbon nanotubes, nanofibers, nanohorns, and graphene, all functionalized with ionic liquids for sample preparation. Emphasis is given to the description of the different works that have provided interesting results for the use of graphene and carbon nanotubes in this analytical field [3].

Viktoria Kazantzi and Aristidis Anthemidis focus on the background and sol-gel chemistry for the preparation of new fabric sorbents, as well as applications of fabric phase sorptive extraction (FPSE) for extracting target analytes in their review of fabric sol-gel phase sorptive extraction technique. Some of the new fabric sorbents include various organic and inorganic analytes in different types of environmental and biological samples in high throughput analytical, environmental, and toxicological laboratories [4].

Fabric phase sorptive extraction (FPSE) is a quite recent sample preparation technique that combines the advanced material properties of sol-gel derived microextraction sorbents, and the flexibility and permeability of fabric, to produce a robust, simple, and green device for extracting target analytes directly from various sample matrices. New modes of FPSE, including stir FPSE, stir-bar FPSE, dynamic FPSE, and automated on-line FPSE, are also highlighted and commented upon in detail. Abuzar Kabir, Rodolfo Mesa, Jessica Jurmain, and Kenneth G. Furton in their work "Fabric Phase Sorptive Extraction Explained" present the theory and working principle of fabric phase sorptive extraction (FPSE). As a representative sorbent, sol-gel poly(ethylene glycol) coating was generated on cellulose substrates. Five (cm$^2$) segments of these coated fabrics were used as the FPSE devices for sample preparation using direct immersion mode. An important class of environmental pollutants—substituted phenols—was used as model compounds to evaluate the extraction performance of FPSE. The high primary contact surface area (PCSA) of the FPSE device and porous structure of the sol-gel coatings resulted in very high sample capacities and incredible extraction sensitivities in a relatively short period of time. Different extraction parameters were evaluated and optimized. The new extraction devices demonstrated part per trillion level-detection limits for substitute phenols, a wide range of detection linearity, and good performance reproducibility [5].

Shivender Singh Saini, Abuzar Kabir, Avasarala Lakshmi Jagannadha Rao, Ashok Kumar Malik, and Kenneth G. Furton present, "A Novel Protocol to Monitor Trace Levels of Selected Polycyclic Aromatic Hydrocarbons in Environmental Water Using Fabric Phase Sorptive Extraction Followed by High Performance Liquid Chromatography-Fluorescence Detection". FPSE was applied for the first time, to the trace level determination of selected polycyclic aromatic hydrocarbons (PAHs) in environmental water samples using a non-polar sol-gel C$_{18}$ coated FPSE media. Several extraction parameters were optimized to improve the extraction efficiency and to achieve high detection sensitivity. The developed and validated FPSE-HPLC-FLD protocol is simple, green, fast, and economical, with adequate sensitivity for trace levels of four selected PAHs, and seems to be promising for the routine monitoring of water quality and safety, as proved by application to the analysis of environmental water samples [6].

Natalia Manousi, Georg Raber, and Ioannis Papadoyannis in their work, "Recent Advances in Micro-extraction Techniques of Antipsychotics in Biological Fluids Prior to Liquid Chromatography Analysis", present an overview of microextraction techniques that are used prior to liquid chromatography analyses both for forensic toxicology in different biological matrices as well as therapeutic drug monitoring. Antipsychotic drugs are a class of psychiatric medication worldwide that is used to treat psychotic symptoms, principally bipolar disorder, schizophrenia, and other psycho-organic disorders, and therefore the necessity for sensitive analytical methods for their determination is of utmost importance [7].

Victoria Samanidou, Dimitrios Bitas, Stamatia Charitonos, and Ioannis Papadoyannis, in their review, "On the Extraction of Antibiotics from Shrimps Prior to Chromatographic Analysis", describe the need for sensitive and selective methods of monitoring residue levels in aquaculture species for routine regulatory analysis. It is well known that the widespread use of antibiotics in veterinary practice and aquaculture has led to the increase of antimicrobial resistance in foodborne pathogens that may be transferred to humans [8].

Global concern is reflected in the regulations from different agencies that have set maximum permitted residue limits on antibiotics in different food matrices of animal origin. Since sample preparation is the most important step, several extraction methods have been developed. The review summarizes the extraction trends for several antibiotics classes from shrimps, and compares the performance characteristics of the different approaches. In their work, "Trends in Microextraction-Based Methods for the Determination of Sulfonamides in Milk", Maria Kechagia and Victoria Samanidou describe the state of the art sulfa drugs that are used in the dairy farming industry in several countries to prevent infection. This increases the possibility that residual drugs could pass through milk consumption, even at low levels. These traces of sulfonamides will be detected and quantified in milk. Therefore, microextraction techniques must be developed to quantify antibiotic residues, taking the requirements of green analytical chemistry into consideration as well [9].

Ana Isabel Argente-García, Yolanda Moliner-Martínez, Esther López-García, Pilar Campíns-Falcó, and Rosa Herráez-Hernández, in their research article, "the Application of Carbon Nanotubes Modified Coatings for the Determination of Amphetamines by In-Tube Solid-Phase Microextraction and Capillary Liquid Chromatography", present a study in which polydimethylsiloxane (PDMS)-coated capillary columns (TRB-5 and TRB-35), both unmodified and functionalized with single-wall carbon nanotubes (SWCNTs) or multi-wall carbon nanotubes (MWCNTs), have been tested and compared for the extraction of amphetamine, methamphetamine, and ephedrine by in-tube solid-phase microextraction (IT-SPME). Prior to their extraction, the analytes were derivatized with the fluorogenic reagent 9-fluorenylmethyl chloroformate. The method was applied to the determination of the tested amphetamines in an oral fluid using a TRB-35 capillary column functionalized with MWCNTs [10].

In their work, "Design of a Molecularly Imprinted Stir-Bar for Isolation of Patulin in Apple and LC-MS/MS Detection", Patricia Regal, Mónica Díaz-Bao, Rocío Barreiro, Cristina Fente, and Alberto Cepeda present a rapid and selective method based on magnetic molecularly imprinted stir-bar (MMISB) extraction developed for the isolation of patulin, using 2-oxindole as a dummy template. Patulin is produced by a mold species that is normally related to vegetable-based products and fruit, mainly apple. Its ingestion may result in agitation, convulsions, edema, intestinal ulceration, inflammation, vomiting, and even immune, neurological, or gastrointestinal disorders. For this reason, the European Commission Regulation (EC) 1881/2006 established a maximum content for patulin of 10 ppb in infant fruit juice, 50 ppb for fruit juice for adults, and 25 ppb in fruit-derived products. The successful MMISB approach has been combined with high performance liquid chromatography coupled to tandem mass spectrometry (HPLC-MS/MS) to determine patulin [11].

Evangelos D. Trikas, Rigini M. Papi, Dimitrios A. Kyriakidis and George A. Zachariadis developed their research paper, "Sensitive LC-MS Method for Anthocyanins and Comparison of Byproducts and Equivalent Wine Content", for the detection and identification of these compounds in the solid wastes of the wine-making industry (red grape skins and pomace), using liquid–liquid extraction prior to the liquid chromatography–mass spectrometry technique (LC-MS). The complete process was investigated and optimized, starting from the extraction conditions (extraction solution selection, dried matter-to-solvent volume ratio, water bath extraction duration, and necessary consecutive extraction rounds), and continuing to the mobile phase selection [12].

Last but not least, Lingshuang Cai, Somchai Rice, Jacek A. Koziel and Murlidhar Dharmadhikari present "an Automated Method for Selected Aromas of Red Wines from Cold-Hardy Grapes Using Solid-Phase Microextraction and Gas Chromatography-Mass Spectrometry-Olfactometry". The effects of SPME coating selection, extraction time, extraction temperature, incubation time, sample volume, desorption time, and salt addition were studied. The developed method was used to determine the aroma profiles of seven selected red wines originating from four different cold-hardy grape cultivars. The presented method can be useful for grape growers and winemakers for the screening of aroma compounds in a wide variety of wines, and can be used to balance desired wine aroma characteristics. The aroma profile of red wine is complex, and research focusing on aroma compounds and their links to viticultural and enological practices is always of high importance [13].

As the Guest Editor of this Special Issue, I would like to thank all of the authors for their contributions, and reassure the readers that this field is expanding, so many other microextraction approaches are yet to evolve.

**Conflicts of Interest:** The author declares no conflict of interest.

## References

1. Kabir, A.; Locatelli, M.; Ulusoy, H.I. Recent Trends in Microextraction Techniques Employed in Analytical and Bioanalytical Sample Preparation. *Separations* **2017**, *4*, 36. [CrossRef]
2. Cárdenas, S.; Lucena, R. Recent Advances in Extraction and Stirring Integrated Techniques. *Separations* **2017**, *4*, 6. [CrossRef]
3. Chatzimitakos, T.; Stalikas, C. Carbon-Based Nanomaterials Functionalized with Ionic Liquids for Microextraction in Sample Preparation. *Separations* **2017**, *4*, 14. [CrossRef]
4. Kazantzi, V.; Anthemidis, A. Fabric Sol-gel Phase Sorptive Extraction Technique: A Review. *Separations* **2017**, *4*, 20. [CrossRef]
5. Kabir, A.; Mesa, R.; Jurmain, J.; Furton, K.G. Fabric Phase Sorptive Extraction Explained. *Separations* **2017**, *4*, 21. [CrossRef]
6. Saini, S.S.; Kabir, A.; Jagannadha Rao, A.L.; Malik, A.K.; Furton, K.G. A Novel Protocol to Monitor Trace Levels of Selected Polycyclic Aromatic Hydrocarbons in Environmental Water Using Fabric Phase Sorptive Extraction Followed by High Performance Liquid Chromatography-Fluorescence Detection. *Separations* **2017**, *4*, 22. [CrossRef]
7. Manousi, N.; Raber, G.; Papadoyannis, I. Recent Advances in Microextraction Techniques of Antipsychotics in Biological Fluids Prior to Liquid Chromatography Analysis. *Separations* **2017**, *4*, 18. [CrossRef]
8. Samanidou, V.; Bitas, D.; Charitonos, S.; Papadoyannis, I. On the Extraction of Antibiotics from Shrimps Prior to Chromatographic Analysis. *Separations* **2016**, *3*, 8. [CrossRef]
9. Kechagia, M.; Samanidou, V. Trends in Microextraction-Based Methods for the Determination of Sulfonamides in Milk. *Separations* **2017**, *4*, 23. [CrossRef]
10. Argente-García, A.I.; Moliner-Martínez, Y.; López-García, E.; Campíns-Falcó, P.; Herráez-Hernández, R. Application of Carbon Nanotubes Modified Coatings for the Determination of Amphetamines by In-Tube Solid-Phase Microextraction and Capillary Liquid Chromatography. *Separations* **2016**, *3*, 7. [CrossRef]
11. Regal, P.; Díaz-Bao, M.; Barreiro, R.; Fente, C.; Cepeda, A. Design of a Molecularly Imprinted Stir-Bar for Isolation of Patulin in Apple and LC-MS/MS Detection. *Separations* **2017**, *4*, 11. [CrossRef]
12. Trikas, E.D.; Papi, R.M.; Kyriakidis, D.A.; Zachariadis, G.A. A Sensitive LC-MS Method for Anthocyanins and Comparison of Byproducts and Equivalent Wine Content. *Separations* **2016**, *3*, 18. [CrossRef]
13. Cai, L.; Rice, S.; Koziel, J.A.; Dharmadhikari, M. Development of an Automated Method for Selected Aromas of Red Wines from Cold-Hardy Grapes Using Solid-Phase Microextraction and Gas Chromatography-Mass Spectrometry-Olfactometry. *Separations* **2017**, *4*, 24. [CrossRef]

*separations*

MDPI

*Review*

# Recent Trends in Microextraction Techniques Employed in Analytical and Bioanalytical Sample Preparation

**Abuzar Kabir [1], Marcello Locatelli [2,3,*] and Halil Ibrahim Ulusoy [4]**

[1] International Forensic Research Institute, Department of Chemistry and Biochemistry, Florida International University, 11200 SW 8th St, Miami, FL 33199, USA; akabir@fiu.edu
[2] Department of Pharmacy, University "G. d'Annunzio" of Chieti-Pescara, 66100 Chieti, Italy
[3] Interuniversity Consortium of Structural and Systems Biology INBB, 00136 Rome, Italy
[4] Department of Analytical Chemistry, Faculty of Pharmacy, Cumhuriyet University, 58140 Sivas, Turkey; hiulusoy@yahoo.com
* Correspondence: m.locatelli@unich.it; Tel.: +39-087-1355-4590

Received: 16 October 2017; Accepted: 27 November 2017; Published: 1 December 2017

**Abstract:** Sample preparation has been recognized as a major step in the chemical analysis workflow. As such, substantial efforts have been made in recent years to simplify the overall sample preparation process. Major focusses of these efforts have included miniaturization of the extraction device; minimizing/eliminating toxic and hazardous organic solvent consumption; eliminating sample pre-treatment and post-treatment steps; reducing the sample volume requirement; reducing extraction equilibrium time, maximizing extraction efficiency etc. All these improved attributes are congruent with the Green Analytical Chemistry (GAC) principles. Classical sample preparation techniques such as solid phase extraction (SPE) and liquid-liquid extraction (LLE) are being rapidly replaced with emerging miniaturized and environmentally friendly techniques such as Solid Phase Micro Extraction (SPME), Stir bar Sorptive Extraction (SBSE), Micro Extraction by Packed Sorbent (MEPS), Fabric Phase Sorptive Extraction (FPSE), and Dispersive Liquid-Liquid Micro Extraction (DLLME). In addition to the development of many new generic extraction sorbents in recent years, a large number of molecularly imprinted polymers (MIPs) created using different template molecules have also enriched the large cache of microextraction sorbents. Application of nanoparticles as high-performance extraction sorbents has undoubtedly elevated the extraction efficiency and method sensitivity of modern chromatographic analyses to a new level. Combining magnetic nanoparticles with many microextraction sorbents has opened up new possibilities to extract target analytes from sample matrices containing high volumes of matrix interferents. The aim of the current review is to critically audit the progress of microextraction techniques in recent years, which has indisputably transformed the analytical chemistry practices, from biological and therapeutic drug monitoring to the environmental field; from foods to phyto-pharmaceutical applications.

**Keywords:** MEPS; FPSE; DLLME; magnetic nanoparticles; MIP; extraction procedures; green analytical chemistry; quantitative analyses

---

## 1. Introduction

The development of analytical methods for quantitative analyses in environmental water, biological sample matrices, and in food or food supplements with a reduced amount of toxic solvents, and the replacing with non-toxic ones, without loss of efficacy in the extraction procedure, are important aims for contemporary researchers [1,2]. These aspects are deeply valued during method development and validation in all fields [3].

Often, these aspects lead to the development of new approaches for analyte extraction and clean-up involved in the development of better sorbent coating technology for solid phase microextraction and stir bar sorptive extraction. The use of novel devices, like packed sorbent (in microextraction by packed sorbent, MEPS), fabric phase sorptive extraction media (in fabric phase sorbent extraction, FPSE), and imprinted polymer (in molecularly imprinted polymer extraction), as well as the use of combined strategies with magnetic elements, can enhance the efficiency and the recovery of the target analyte.

Liquid phase microextraction methods have demonstrated important innovations for the extraction and pre-concentration of analytes from different matrices. Dispersive liquid-liquid microextraction (DLLME) and its modifications, such as ultrasound-assisted DLLME (UA-DLLME), ionic liquid-based dispersive liquid–liquid microextraction (IL-DLLME), deep eutectic solvent-based dispersive liquid-liquid microextraction (DES-DLLME), and sugaring-out assisted liquid-liquid extraction (SULLE), can offer unique benefits, such as a high pre-concentration factor for the target analytes, low cost, simplicity and combined use with almost every analytical measurement technique [4].

A large number of solvent microextraction techniques, including single-drop microextraction, DLLME, and liquid-phase microextraction (LPME), have been reported. Implementation of these techniques can vary widely, but common features remain the same, including the use of only a small amount of organic solvents and a high sample-to-acceptor volume ratio. The organic phase, which extracts and pre-concentrates the target analyte(s), can be used for quantification by means of different types of instrument configurations [5]. LPME is usually performed to analyze water samples or aqueous solutions. Analysis of solid samples is commonly done in two steps: the solid sample is converted to aqueous solution using a suitable pretreatment procedure, and then the LPME is applied. Direct analysis of solid samples is somewhat exceptional, rather than common. Several works have been reported for the determination of different analytes in complex matrices, such as phenolic compounds in plant materials [6] and food samples [7,8] by using DLLME in combination with High Performance Liquid Chromatography-UltraViolet/Visible detector (HPLC-UV/Vis) [6] and Gas Chromatography-Mass Spectrometry detector (GC-MS) [7] instrument configurations.

Replacing hazardous solvents with ionic liquids (IL) or natural deep eutectic solvents (NADES) is another important task available in DLLME, and was recently reviewed by Shishov and co-workers [9]. It is possible to modify the IL's properties depending on the analytical purpose, due to the cation's fine structure and the anion's identity [10], but high cost and toxicity remain as the main disadvantages [11]. Recently, NADESs have been rapidly developed as a new type of green solvents, as an alternative to ILs. NADESs are based on primary metabolites, such as organic acids, amino acids and sugars, but limited data are available for these solvents' properties.

The aim of this review is to report the recently applied protocols and devices used in the extraction (and clean-up) procedures for quantitative analyses in complex matrices, with the main goal being the reduction of time, sample manipulation, solvent consumption and use of non-toxic solvents, in accordance with Green Analytical Chemistry (GAC) concepts.

## 2. Sorbent-Based Sorptive Microextraction Techniques

Sorbent-based sorptive microextraction techniques utilize a solid/semi-solid organic polymer as the sorbent, immobilized on a substrate (such as fused silica fiber, silica particles, glass-coated bar magnet, cellulose/polyester/fiber glass fabric etc.), and include solid phase microextraction (SPME) and its different modifications and implementations, stir bar sorptive extraction (SBSE), microextraction by packed sorbent (MEPS), thin film microextraction (TFME), and fabric phase sorptive extraction (FPSE). Sampling and sample preparation using these techniques are often carried out either by (1) headspace extraction; or by (2) direct immersion extraction. Due to the glue-like, highly viscous polymeric sorbents are prone to irreversibly adsorb matrix interferents from the sample matrix, direct immersion extraction can only be done when the aqueous sample is free from particulates or macromolecules. As such, biological, environmental and food samples require rigorous sample pretreatment prior to analyte extraction such as filtration, centrifugation, protein precipitation, etc. Once the analytes are extracted into these devices, desorption can be carried out by applying thermal shock or by exposing to an organic solvent. Due to the special geometrical advantage (fiber retractable inside a syringe needle), SPME fiber can be introduced directly into GC inlet or into the HPLC system via a special interface. For SBSE or FPSE, a thermal desorption unit can be used. Alternatively, solvent mediated desorption can be used followed by injecting an aliquot into GC or HPLC for chromatographic separation and analysis.

### 2.1. Fiber-Based Solid-Phase Microextraction, Capillary Solid-Phase Microextraction, and Related Techniques

Solid-phase microextraction (SPME), invented by J. Pawliszyn in 1987, undoubtedly deserves the credit for beginning a new era in analytical sample preparation characterized by solvent-free extraction, miniaturization and automation. SPME integrates sampling, extraction and analyte preconcentration into a single step. Due to this ease of interfacing with other analytical systems, as well as many other advantages, SPME has been enjoying exponential growth in applications in many different areas since its inception.

The miniaturization of sample preparation techniques and the integration, particularly in *on-line* configuration, of chromatographic instruments that could allow a reduction in labor-intensive manual operation and help to enhance the overall analytical performance [12] still remain the major focus in academic and industrial research. In this scenario, even if MEPS is more easily automated than SPE, and more robust than solid-phase microextraction (SPME) [13], sorbent-based techniques and their different formats certainly represent a valid choice. These techniques provide simplicity in their operation, consume no solvent or minimize solvent usage, allow the separation and pre-concentration of the analytes using different commercial fibers, and the possibility of automating the entire process could be successfully applied in food, environmental, clinical, pharmaceutical and bioanalysis applications [14–16], as recently reviewed by Silva and co-workers [17].

The fibers, which are commercially available, may be different based on their type:

- non-bonded phases: stable with some water-miscible organic solvents, although some swelling may occur when used with non-polar solvents,
- bonded phases: stable with all organic solvents, except for some non-polar solvents,
- partially cross-linked phases: stable in most water-miscible organic solvents and some polar solvents,
- highly cross-linked phases: similar to the partially cross-linked phases, except that some bonding to the core may occur.

All of these phases have been strongly and deeply studied and implemented for analysis of volatile components and in different applications—e.g., food, food supplements, and bioanalysis—but all show the same limitation with regard to handling of large sample volumes as for the main SPE/SPME procedure. These procedures (also capillary-like) could easily be used in biological and food analyses due to the relatively low sample volume, but in environmental applications, where large sample volumes are required in order to obtain higher pre-concentration factors, their limitations are highlighted.

Even if these "limitations" are present, fiber-based solid-phase microextraction and capillary solid-phase microextraction represent valid alternatives to conventional approaches due to the wide range of phases commercially available, their better stability and reproducibility (both between lots and analyses), and their unique characteristic of being solvent-less.

The latter techniques, capillary solid-phase microextraction, consist of an inert liner with a packet of coated open capillary tubes inside. The main advantage in comparison to the other reported microextraction procedures is that the surface areas of the extraction phase are more than two orders and one order of magnitude higher, respectively, than that of fiber-SPME. Hence, an equal extraction quantity can be obtained in a lower time. Another advantage is represented by the large cross-section area, resulting in a lower flow resistance; consequently, water samples are able to flow through the cartridge independently, without the need for an auxiliary apparatus. Using capillary solid-phase microextraction, the extraction phase is protected from damage in the liner, so no heightened precautions are needed during application. Furthermore, the cartridge shows a small-bore diameter, which ensures the retention of trace and ultra-trace compounds using limited sample amounts, allowing high absolute recovery. The advances in SPME in terms of new coatings, formants and applications have been reviewed in a large number of articles; only a few are referenced here [18–21].

Although SPME offers numerous advantages over conventional sample preparation techniques, it suffers from significant shortcomings, including (1) relatively low operating temperature; (2) instability and swelling of the coating if exposed to organic solvents; (3) low sorbent loading results and poor extraction sensitivity; (4) high run-to-run and batch-to-batch variability; (5) the fact that the slow diffusion of the analyte(s) into viscous sorbents often leads to a long extraction equilibrium time; and (6) the fact that physically holding sorbent to the inert support results in a short life time for the SPME fiber. The majority of the shortcomings stem from the sorbent coating technology used in manufacturing the SPME fibers. However, the coating-related deficiency has been duly addressed by the sol-gel-based coating technology developed by Malik and his research groups [22]. This technology subsequently aided in the development of hundreds of sorbents possessing unique selectivity, as well as unprecedented thermal, solvent and chemical stability. A large number of review articles have critically evaluated these SPME coatings [23–27].

## 2.2. Stir Bar Sorptive Extraction (SBSE)

Stir bar sorptive extraction was developed by Pat Sandra and his research group [28] with the aim of increasing the extraction sensitivity of SPME by incorporating substantially higher sorbent loading compared to SPME. In the original invention, poly(dimethylsiloxane) (PDMS) was coated onto a glass-coated magnetic bar. The unique design of SBSE makes it an independent sample preparation device, capable of diffusing the sample matrix by itself on a magnetic stirrer without requiring any external magnet. Extraction and preconcentration of the analyte is carried out by introducing the SBSE device directly into the aqueous sample. The analytes are extracted and preconcentrated when the SBSE spins inside the solution. Following the analyte extraction (driven by equilibrium), the SBSE device is withdrawn from the sample, rinsed with deionized water to clean matrix interferents, and dried with a Kim wipe. Subsequently, the extracted analytes are desorbed using a thermal desorption unit coupled to gas chromatography, or can be subjected to solvent desorption by exposing it to a small volume of a suitable organic solvent. The eluent is typically dried under nitrogen, and the sample is reconstituted in smaller-volume solvent. The sample can be analyzed in a gas or liquid chromatographic system. SBSE devices are commercially available under the trade name Twister®. Among others, a major drawback of this technique is the availability of only two phases: PDMS and Poly(ethylene glycol) in PDMS [29]. The high viscosity of both of these phases slows down analyte diffusion during extraction, resulting in a long extraction equilibrium time. As such, the extraction sensitivity in SBSE has not been improved proportionately with the sorbent loading, compared to SPME. Several review articles have discussed recent developments in SBSE [29–33]. As in the case of SPME, the efficiency of SBSE has been substantially improved by adopting sol-gel coating technology [34–38].

## 2.3. Micro Extraction by Packed Sorbent Procedures (MEPS)

Recently, Abdel-Rehim and coworkers [39] reviewed the literature published on Micro Extraction by Packed Sorbent (MEPS) methods. This extraction procedure shows some very interesting potential benefits, such as low solvent consumption, small sample volume (10–250 μL), and the ability to be directly injected into the HPLC system without further treatments, with solvent volumes being compatible with several instrumental configurations and analyses. Figure 1 shows the device and the general procedure applied in MEPS extraction.

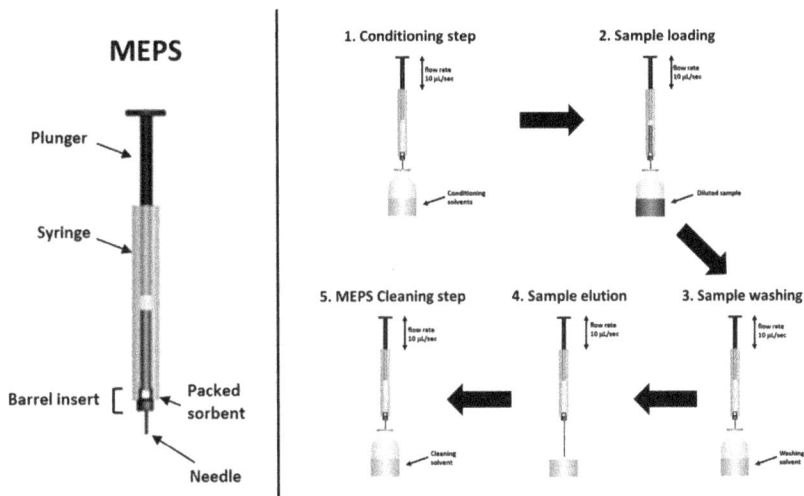

Figure 1. Device (**left**) and general procedure (**right**) applied in MEPS extraction.

This device is used in different fields, from biological applications to food and food supplement analyses. The main drawback is also related to its advantages. In fact, the possibility of using small sample volumes permits its application in analyses where only small volumes are available, for example plasma. In the case of higher volumes (such as in the environmental field), this device often shows its limitations.

Nowadays, several types of packing materials are available, including:

- Silica-based sorbents SIL (unmodified silica),
- C2(ethyl),
- C8 (octyl),
- C18 (octadecyl);
- Mixed-mode C8 and ion exchange (SCX),
- Mixed-mode M1 (80% C8 and 20% SCX with sulfonic acid bonded silica);
- Polystyrene-divinylbenzene (PS-DVB),
- Porous graphitic carbon,
- Molecular imprinted polymers (MIPs) based on different templates,
- Metal organic framework (MOF)-based MIPs [40,41],
- Monoclonal antibodies (mAbs) for immunoaffinity sorbents production.

Additionally, other commercial sorbents, such as new kinds of graphitic sorbent, polypyrrole/polyamide, polyaniline nanowires, CMK-3 nanoporous materials, functionalized silica monoliths, APS (amino-propyl silane), and cyanopropyl hybrid silica have been successfully applied in MEPS devices to extract different groups of analytes, as recently reviewed [39,42].

To aid and improve the reproducibility during the extraction process, MEPS devices are also coupled to syringes in semi-automated and/or fully automated configurations. In fact, the main critical point during extraction relates to the reproducibility of the flow rate (generally $\mu L\ s^{-1}$) used in the different steps (Figure 1). A MEPS syringe and a one-way check valve [43] was used for the realization of automated or semi-automated MEPS extraction, both for *on-line* and *off-line* instrument configurations.

Table 1 presents selected MEPS applications in different fields and the performances obtained when applying this device.

**Table 1.** Some recent MEPS applications (2012–2017, not previously reviewed) [39,43] in different fields, and the performances obtained when applying this device.

| Field | Analyte | MEPS | Matrix | Sample Volume | LOQ (LOD) | Reference |
|---|---|---|---|---|---|---|
| Biological | NSAIDs | C18 | Plasma Urine | 100 µL | 0.10 ng/mL (0.03 µg/mL) | [44] |
| | Fluoroquinolones | C18 | Sputum | 200 µL | 0.05 µg/mL (0.017 µg/mL) | [45] |
| | NSAIDs and Fluoroquinolones | C18 | Plasma Urine | 200 µL | 0.10 µg/mL (0.03 µg/mL) | [46] |
| | Imidazoles and Triazoles | C18 | Plasma Urine | 200 µL | 0.02 µg/mL (0.007 µg/mL) | [47] |
| | New psychoactive substances | mixed-mode C8/SCX | Oral fluid | 300 µL | 0.5 ng/mL (n.r.) | [48] |
| | Trans,trans-muconic acid | MIP-MEPS | Urine | 100 µL | 0.05 µg/mL (0.015 µg/mL) | [49] |
| | Statins | C18 | Plasma | 100 µL | 10–20 ng/mL (n.r.) | [50] |
| | Drugs of abuse | C8/SCX | Plasma | 300 µL | 0.01 µg/mL (0.005 µg/mL) | [51] |
| | Cocaine and metabolites | Mixed mode M1 | Urine | 200 µL | 25 ng/mL (n.r.) | [52] |
| Food and Food Supplements | Melatonin and other antioxidants | C8 | Foodstuffs | 100 µL | 0.05 ng/mL (0.02 ng/mL) | [53] |
| Environmental | Brominated diphenyl ethers | C18 | Sewage sludge | 15 mL reduced to 1 mL | n.r. (3 pg/mL) | [54] |
| | Chlorophenols | C18 | Soil samples | 1 mL | 0.353 µg/kg (0.118 µg/kg) | [55] |
| | Sulfonamides | C8 | Wastewater | n.r. | 5 ng/mL (n.r.) | [56] |
| | Phtalate esters | graphene and CNT/CNF-G nanostructures | Water | 10 mL reduced to dry | 0.02 ng/mL (0.004 ng/mL) | [57] |
| | Parabens | graphene supported on aminopropyl silica | Water | 1 mL | 0.2 µg/mL (n.r.) | [58] |

n.r. not reported.

As previously mentioned, applications in environmental fields are very limited due to the large volumes that are necessary for the trace analysis of pollutants, both organic and inorganic. Additionally, applications are relatively limited for foods and food supplements due to difficult application of MEPS, which requires a longer time in the pre-analytical steps.

### 2.4. Fabric Phase Sorptive Extraction Procedures (FPSE)

Fabric Phase Sorptive Extraction (FPSE) is a novel sample preparation procedure that mitigates the drawbacks of MEPS. In fact, it allows small and large volumes to be treated, and could be usefully applied in all fields where a very high pre-concentration factor is required, from environmental to biological, from toxicological to food and food supplement quality control. The common drawbacks encountered in conventional sample preparation techniques can be conveniently overcome by using FPSE—developed by Kabir and Furton [59], and recently reviewed by the inventors [60]—which does not require any matrix modifications or clean-up. FPSE successfully integrates the advantages of equilibrium-based extraction (SPME/SBSE) and exhaustive extraction (SPE) without the necessity of time-consuming sample pretreatment procedures such as protein precipitation. The sorbent is covalently bonded to the substrate surface, and therefore offers high chemical, physical, and thermal stability. In addition, the open geometry of the media facilitates fast analyte sorption and desorption. The substrate used in FPSE is not inert, and contributes synergistically to the overall polarity of the FPSE media. Fabric phase sorptive extraction substantially simplifies the sample preparation workflow in comparison to other available and recent techniques, as demonstrated in Figure 2.

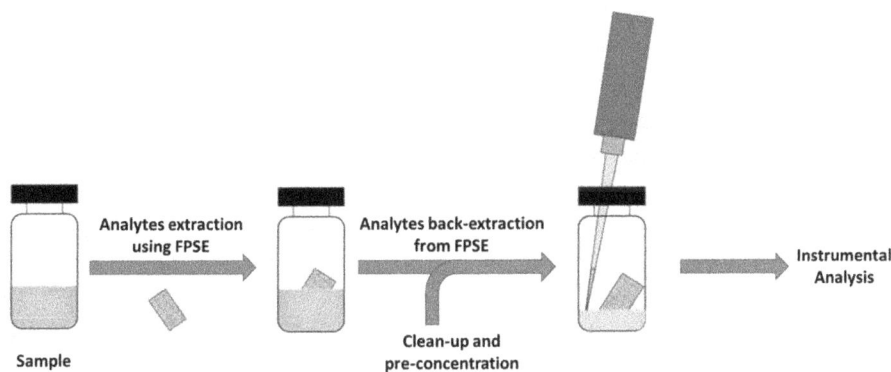

**Figure 2.** General procedure (**right**) applied in FPSE extraction.

Using FPSE devices, both large sample volumes (such as in the environmental field) and small sample volumes (generally applicable in the biological and pharmaceutical fields) can be easily handled, maintaining very good analytical performances in terms of LOQ, linearity and low pre-analytical steps, as reported in Table 2.

**Table 2.** Some recent FPSE applications in different fields, and the performances obtained when applying this device.

| Field | Analyte | FPSE | Matrix | Sample Volume | LOQ (LOD) | Reference |
|---|---|---|---|---|---|---|
| Biological | Imidazoles and Triazoles | sol-gel Carbowax® 20 M | Plasma Urine | 500 µL | 0.10 µg/mL (0.03 µg/mL) | [61] |
| | Ciprofloxacin Sulfasalazine Cortisone | sol-gel Carbowax® 20 M | Whole blood Plasma Urine | 100 µL.500 µL.500 µL | 0.05 µg/mL (0.015 µg/mL) 0.25 µg/mL (0.10 µg/mL) 0.10 µg/mL (0.03 µg/mL) | [62] |
| | Anastrozole Letrozole Exemestane | sol-gel PEG-PPG-PEG | Whole blood Plasma Urine | 200 µL.500 µL.1 mL | 0.1 µg/mL (0.03 µg/mL) 0.025 µg/mL (0.008 µg/mL) 0.025 µg/mL (0.008 µg/mL) | [63] |
| | Benzodiazepines | sol-gel PEG | Blood serum | 50 µL | 0.03 µg/mL (0.01 µg/mL) | [64] |
| | Selected estrogens | sol-gel PTHF | Urine | 10 mL | 0.066 ng/mL (0.020 ng/mL) | [65] |
| | Androgens and progestogens | sol-gel PTHF | Urine | 2 mL | 29.7 ng/L (8.9 ng/L) | [66] |
| Food and Food Supplements | Non-volatile plastic additives | sol-gel PDMS | Aqueous food simulants | 10 mL | 3 ng/g (1 ng/g) | [67] |
| | Amphenicols | sol-gel PEG | Milk | 0.5 g | 20 µg/kg | [68] |
| | Sulfonamides residues | sol-gel short-chain PEG | Milk | 1 g | 30 µg/kg (n.r.) | [69] |
| | Volatile compounds | sol-gel Carbowax® 20 M | Orange | 75 mL | n.r. | [70] |
| | Penicillin antibiotics | sol-gel PEG | Milk | 0.5 g | 10 µg/kg (3 µg/kg) | [71] |
| | Bisphenol A and residual dental restorative material | sol-gel graphene | Cow and human breast milk | 0.5 g | 50 µg/kg (16.7 µg/kg) | [72] |
| Environmental | Pharmaceuticals and personal care products | sol-gel Carbowax® 20 M | Water | 50 mL | 20 ng/mL (2 ng/mL) | [73] |
| | Selected estrogens | sol-gel PTHF | Water | n.r. | 0.066 ng/mL (0.020 ng/mL) | [65] |
| | Alkyl phenols | sol-gel PTHF | Water Soil | n.r. 1 g | n.r. (0.161 ng/mL) n.r (1 ng/g) | [74] |
| | NSAIDs | sol-gel PTHF | Water | 30 mL | 3 ng/L (0.8 ng/L) | [75] |
| | Triazine herbicides | sol-gel PTHF | Water | 100 mL | 0.26 µg/L | [76] |
| | Benzotriazole UV stabilizers | sol-gel PDMDPS | Sewage | 10 mL | 24.5 ng/L (7.34 ng/L) | [77,78] |
| | Pharmaceuticals and personal care products | sol-gel Carbowax® 20 M | Water | 10 mL | 0.1 µg/L (0.01 µg/L) | [79] |
| | Cadmium | sol-gel PDMDPS | Water | 13.5 mL | 1.2 µg/L (0.4 µg/L) | [80] |
| | Androgens and progestogens | sol-gel PTHF | Waters | 2 L | 5.7 ng/L (1.7 ng/L) | [66] |
| | Co(II), Ni(II) and Pd(II) | sol-gel PTHF | Water | 10 mL | 1 ng/mL (n.r.) | [81] |
| | Pheromones | sol-gel PDMDPS | Air | - | 2.6 µg (0.8 µg) | [82] |

n.r. not reported; PEG polyethylenglycol; PDMS poly(dimethylsiloxane); PTHF polytetrahydrofuran; PDMDPS polydimethyldiphenylsiloxane.

*2.5. Magnetic Nanoparticle Extraction*

Recently these applications were reviewed [9], and in particular, a review focused on using magnetic nanoparticles for the selective extraction of trace species from a complex matrix was reported [83].

The main advantage of this last configuration is the possibility of retaining the analytes adsorbed on magnetic stationary phase directly in the tube, cleaning the sample from the matrix and the interference compounds, and analyzing the extract directly for trace species.

To date, no other innovative applications—except for those reported in very recent review papers—have been reported in the literature for food analysis [84], for drugs in biological matrices [85], or in other research fields [86,87].

All of these papers clearly report the great advantages in using magnetic devices to allow the total recovery of the extracted analytes by using a strong magnet during the cleaning process. In this way, it could be possible to retain the analytes without any loss related to the wash step.

Additionally, it is also possible to dry the extracted samples and re-suspend them in a mobile phase more suitable for the instrumental analysis, thus also obtaining a great pre-concentration factor for trace analyses.

## 3. Solvent-Based Microextraction Techniques

Due to its high toxicity, expensive disposal requirements, and contribution to further environmental pollution, liquid-liquid extraction and its various modifications have undergone critical evaluation during the last decade, leading to the introduction of liquid phase microextraction (LPME). In a very short period, a number of techniques evolved, with the common goal of minimizing solvent consumption in the sample preparation process.

*3.1. Liquid-Liquid Micro Extraction (LPME)*

Considering the principles of Green Analytical Chemistry [88], the development of analytical methods that reduce the amount of toxic solvents, or replace them with non-toxic alternatives without sacrificing the efficacy of the extraction procedure, is a major aim for researchers.

Liquid phase microextraction methods (LPME), in comparison to solid phase (micro) extraction, have shown important innovations for trace analytes in different matrices. LPME is utilized for organic compounds and inorganic trace elements in several application fields, such as the environmental, biological, and food fields. LPME can be divided into three different procedure modes: headspace LPME (HS-LPME), direct-immersed LPME (DI-LPME), and hollow fiber LPME (HF-LPME). In HS-LPME, a drop of extraction solvent—which can be either an organic solvent or a water solution—is suspended at the tip of a micro-syringe needle and exposed to the headspace of the sample; this is very suitable for analyses of volatile compounds. DI-LPME is very similar, except that the extraction solvent must be immiscible with aqueous solutions, and is directly immersed into a stirred sample solution. HF-LPME uses a hollow fiber in order to stabilize and protect the extraction solvent, while the small fiber pore size avoids the interference of large molecules and particles, which could result in a more extensive clean-up of the sample during the extraction process. Sharifi and co-workers [89] recently reviewed the principal applications in which these LPME techniques are applied.

*3.2. Dispersive Liquid-Liquid Microextraction (DLLME)*

Dispersive liquid-liquid microextraction (DLLME) and its modifications—such as ultrasound-assisted DLLME (UA-DLLME), ionic liquid-based dispersive liquid-liquid microextraction (IL-DLLME), deep eutectic solvent-based dispersive liquid-liquid microextraction (DES-DLLME), and sugaring-out assisted liquid-liquid extraction (SULLE)—offer unique benefits, such as a high pre-concentration factor for the target analytes, low cost, simplicity, and the possibility of combined use with almost every analytical measurement technique [4,90,91]. A large number of solvent microextraction techniques, including single drop microextraction, DLLME, liquid phase microextraction (LPME), have been reported. This extraction procedure allows a better analytical performance in comparison to HF-LPME, as reported by Xiong and Hu [92]. The organic phase, which contains the target analyte(s), can be used for quantification by means of different types of instrument configurations [5]. LPME is usually performed to analyze water samples or aqueous solutions. Analysis of solid samples is commonly done in two steps; the solid sample is converted to aqueous solution using a suitable pretreatment procedure, and then LPME is applied.

## 4. Conclusions

As is clearly highlighted in this review paper, the extraction (and clean-up) procedures applied to complex matrices are the real rate-limiting step in sample preparation, particularly related to the overall analytical performance of the developed (and validated) method. Several procedures that have recently been applied, their main aim being the reduction of time, sample manipulation, solvent consumption, and use of toxic solvents, in accordance with the Green Analytical Chemistry (GAC) concepts. A good idea of the advantages/disadvantages of the different procedures treated herein is reported in Table 3.

**Table 3.** A comparison of some characteristics of sample preparation techniques [17,93,94].

| Feature | MEPS | FPSE | DLLME | SPE | SPME |
|---|---|---|---|---|---|
| Phase amount | 0.5–4 mg | n.a. | n.a. | 50–10,000 mg | 150 mm thickness |
| Principle-separation | no emulsion | no emulsion | emulsion | no emulsion | no emulsion |
| Procedure time | 1–2 min | 5–30 min | 5–15 min | 10–15 min | 10–40 min |
| Re-use | 40–100 times | 30–50 times | Single use | Single use | 50–100 times |
| Recovery | + | + | + | + | − |
| Carryover | − | − | n.a. | + | + |
| Solvent consumption | − | +/− | + | + | solventless |
| Sensitivity | − | + | + | + | − |
| Easy-to-use | − | + | − | + | − |
| Sample quantity | − | +/− | +/− | + | + |
| Easily adaptable to | GC or HPLC | GC or HPLC | GC or HPLC | GC or HPLC | GC |
| Automatable | + | − | − | + | + |
| Target analytes | polar and charged analytes may be extracted | polar and charged analytes may be extracted | polar analytes difficult to extract | polar and charged analytes may be extracted | polar and charged analytes may be extracted |
| Cost | − | n.a. | + | + | + |
| Commercially available | + | − | + | + | + |

n.a. not applicable; + high; − low; +/− high or low depend to the application field.

These innovative procedures also allow analytical performance to be improved by using well-known instrument configurations—such as HPLC-UV/Vis—while avoiding the use of more complex and expensive ones (HPLC-MS, UPLC-MS, etc.). Additionally, these instrumentations can also be used by non-expert operators in routine analyses, both in clinical and in quality-control procedures.

**Acknowledgments:** This work was supported by University "G. d'Annunzio" of Chieti-Pescara, Chieti, Italy.

**Author Contributions:** All Authors contributed equally to the present work.

**Conflicts of Interest:** The authors declare no conflict of interest.

## References

1. Kabir, A.; Furton, K.J. Sample preparation in Food Analysis: Practices, Problems and Future Outlook. In *Analytical Chemistry: Developments, Applications and Challenges in Food Analysis*; Locatelli, M., Celia, C., Eds.; Nova Science Publishers, Inc.: Hauppauge, NY, USA, 2017; pp. 23–54, ISBN 978-1-53612-267-1.
2. Locatelli, M.; Cifelli, R.; Vitalei, S.; Santini, P.; De Luca, E.; Bellagamba, G.; Celia, C.; Carradori, S.; Di Marzio, L.; Mollica, A. Method validation and hyphenated techniques: Recent trends and future perspectives. In *Analytical Chemistry: Developments, Applications and Challenges in Food Analysis*; Locatelli, M., Celia, C., Eds.; Nova Science Publishers, Inc.: Hauppauge, NY, USA, 2017; pp. 1–22, ISBN 978-1-53612-267-1.

3.      Locatelli, M.; Sciascia, F.; Cifelli, R.; Malatesta, L.; Bruni, P.; Croce, F. Analytical methods for the endocrine disruptor compounds determination in environmental water samples. *J. Chromatogr. A* **2016**, *1434*, 1–18. [CrossRef] [PubMed]

4.      Campillo, N.; Viñas, P.; Šandrejová, J.; Andruch, V. Ten years of dispersive liquid-liquid microextraction and derived techniques. *Appl. Spectrosc. Rev.* **2017**, *52*, 267–415. [CrossRef]

5.      Yan, H.; Wang, H. Recent development and applications of dispersive liquid-liquid microextraction. *J. Chromatogr. A* **2013**, *1295*, 1–15. [CrossRef] [PubMed]

6.      Hao, Y.; Chen, X.; Hu, S.; Bai, X.; Gu, D. Utilization of dispersive liquid-liquid microextraction coupled with HPC-UV as a sensitive and efficient method for the extraction and determination of oleanolic acid and ursolic acid in Chinese medicinal herbs. *Am. J. Anal. Chem.* **2012**, *3*, 675–682. [CrossRef]

7.      Yang, P.; Li, H.; Wang, H.; Han, F.; Jing, S.; Yuan, C.; Guo, A.; Zhang, Y.; Xu, Z. Dispersive liquid-liquid microextraction method for HPLC determination of phenolic compounds in wine. *Food Anal. Methods* **2017**, *10*, 1–15. [CrossRef]

8.      Fariña, L.; Boido, E.; Carrau, F.; Dellacassa, E. Determination of volatile phenols in red wines by dispersive liquid-liquid microextraction and gas chromatography-mass spectrometry detection. *J. Chromatogr. A* **2007**, *1157*, 46–50. [CrossRef] [PubMed]

9.      Shishov, A.; Bulatov, A.; Locatelli, M.; Carradori, S.; Andruch, V. Application of deep eutectic solvents in analytical chemistry. A review. *Microchem. J.* **2017**, *135*, 33–38. [CrossRef]

10.     Tang, B.; Bi, W.; Tian, M.; Row, K.H. Application of ionic liquid for extraction and separation of bioactive compounds from plants. *J. Chromatogr. B* **2012**, *904*, 1–21. [CrossRef] [PubMed]

11.     Khezeli, T.; Daneshfar, A.; Sahraei, R. A green ultrasonic-assisted liquid-liquid microextraction based on deep eutectic solvent for the HPLC-UV determination of ferulic, caffeic and cinnamic acid from olive, almond, sesame and cinnamon oil. *Talanta* **2016**, *150*, 577–585. [CrossRef] [PubMed]

12.     Alves, G.; Rodrigues, M.; Fortuna, A.; Falcão, A.; Queiroz, J. A critical review of microextraction by packed sorbent as a sample preparation approach in drug bioanalysis. *Bioanalysis* **2013**, *5*, 1409–1442. [CrossRef] [PubMed]

13.     Zhang, X.; Wang, C.; Yang, L.; Zhang, W.; Lin, J.; Li, C. Determination of eight quinolones in milk using immunoaffinity microextraction in a packed syringe and liquid chromatography with fluorescence detection. *J. Chromatogr. B* **2017**, *1064*, 68–74. [CrossRef] [PubMed]

14.     Oliveira e Silva, H.; de Pinho, P.G.; Machado, B.P.; Hogg, T.; Marques, J.; Câmara, J.S.; Albuquerque, F.; Silva Ferreira, A.C. Impact of forced-aging process on madeira wine flavor. *J. Agric. Food Chem.* **2008**, *56*, 11989–11996. [CrossRef] [PubMed]

15.     Perestrelo, R.; Caldeira, M.; Rodrigues, F.; Camara, J.S. Volatile flavour constituent patterns of terras madeirenses red wines extracted by dynamic headspace solid-phase microextraction. *J. Sep. Sci.* **2008**, *31*, 1841–1850. [CrossRef] [PubMed]

16.     Câmara, J.; Marques, J.; Alves, A.; Ferreira, A.S. Heterocyclic acetals in madeira wines. *Anal. Bioanal. Chem.* **2003**, *375*, 1221–1224. [CrossRef] [PubMed]

17.     Silva, C.; Cavaco, C.; Perestrelo, R.; Pereira, J.; Câmara, J.S. Microextraction by Packed Sorbent (MEPS) and Solid-Phase Microextraction (SPME) as Sample Preparation Procedures for the Metabolomic Profiling of Urine. *Metabolites* **2014**, *4*, 71–97. [CrossRef] [PubMed]

18.     Abdulra'uf, L.B.; Hammed, W.A.; Tan, G.H. SPME Fibers for the Analysis of Pesticide Residues in Fruits and Vegetables: A Review. *Crit. Rev. Anal. Chem.* **2012**, *42*, 152–161. [CrossRef]

19.     Hou, X.D.; Wang, L.C.; Guo, Y. Recent Developments in Solid-phase Microextraction Coatings for Environmental and Biological Analysis. *Chem. Lett.* **2017**, *46*, 1444–1455. [CrossRef]

20.     Piri-Moghadam, H.; Alam, M.N.; Pawliszyn, J. Review of geometries and coating materials in solid phase microextraction: Opportunities, limitations, and future perspectives. *Anal. Chim. Acta* **2017**, *984*, 42–65. [CrossRef] [PubMed]

21.     Zhang, Q.H.; Zhou, L.D.; Chen, H.; Wang, C.Z.; Xia, Z.N.; Yuan, C.S. Solid-phase microextraction technology for in vitro and in vivo metabolite analysis. *TrAC* **2016**, *80*, 57–65. [CrossRef] [PubMed]

22.     Wang, D.X.; Chong, S.L.; Malik, A. Sol-gel column technology for single-step deactivation, coating, and stationary-phase immobilization in high-resolution capillary gas chromatography. *Anal. Chem.* **1997**, *69*, 4566–4576. [CrossRef]

23.     Amiri, A. Solid-phase microextraction-based sol–gel technique. *TrAC* **2016**, *75*, 57–74. [CrossRef]

24. Bagheri, H.; Piri-Moghadam, H.; Naderi, M. Towards greater mechanical, thermal and chemical stability in solid-phase microextraction. *TrAC* **2012**, *34*, 126–139. [CrossRef]

25. Dietz, C.; Sanz, J.; Camara, C. Recent developments in solid-phase microextraction coatings and related techniques. *J. Chromatogr. A* **2006**, *1103*, 183–192. [CrossRef] [PubMed]

26. Kabir, A.; Furton, K.G.; Malik, A. Innovations in sol-gel microextraction phases for solvent-free sample preparation in analytical chemistry. *TrAC* **2013**, *45*, 197–218. [CrossRef]

27. Malik, A. Advances in sol-gel based columns for capillary electrochromatography: Sol-gel open-tubular columns. *Electrophoresis* **2002**, *23*, 3973–3992. [CrossRef] [PubMed]

28. Baltussen, E.; Sandra, P.; David, F.; Cramers, C. Stir bar sorptive extraction (SBSE), a novel extraction technique for aqueous samples: Theory and principles. *J. Microcolumn Sep.* **1999**, *11*, 737–747. [CrossRef]

29. Gilart, N.; Marce, R.M.; Borrull, F.; Fontanals, N. New coatings for stir-bar sorptive extraction of polar emerging organic contaminants. *TrAC* **2014**, *54*, 11–23. [CrossRef]

30. Cardenas, S.; Lucena, R. Recent Advances in Extraction and Stirring Integrated Techniques. *Separations* **2017**, *4*, 6. [CrossRef]

31. Nazyropoulou, C.; Samanidou, V. Stir bar sorptive extraction applied to the analysis of biological fluids. *Bioanalysis* **2014**, *7*, 2241–2250. [CrossRef] [PubMed]

32. He, M.; Chen, B.B.; Hu, B. Recent developments in stir bar sorptive extraction. *Anal. Bioanal. Chem.* **2014**, *406*, 2001–2026. [CrossRef] [PubMed]

33. Prieto, A.; Basauri, O.; Rodil, R.; Usobiaga, A.; Fernandez, L.A.; Etxebarria, N.; Zuloaga, O. Stir-bar sorptive extraction: A view on method optimisation, novel applications, limitations and potential solutions. *J. Chromatogr. A* **2010**, *1217*, 2642–2666. [CrossRef] [PubMed]

34. Amlashi, N.E.; Hadjmohammadi, M.R. Sol-gel coating of poly(ethylene glycol)-grafted multiwalled carbon nanotubes for stir bar sorptive extraction and its application to the analysis of polycyclic aromatic hydrocarbons in water. *J. Sep. Sci.* **2016**, *39*, 3445–3456. [CrossRef] [PubMed]

35. Fan, W.; Mao, X.; He, M.; Chen, B.; Hu, B. Development of novel sol-gel coatings by chemically bonded ionic liquids for stir bar sorptive extraction-application for the determination of NSAIDS in real samples. *Anal. Bioanal. Chem.* **2014**, *406*, 7261–7273. [CrossRef] [PubMed]

36. Fan, W.; Mao, X.J.; He, M.; Chen, B.B.; Hu, B. Stir bar sorptive extraction combined with high performance liquid chromatography-ultraviolet/inductively coupled plasma mass spectrometry for analysis of thyroxine in urine samples. *J. Chromatogr. A* **2013**, *1318*, 49–57. [CrossRef] [PubMed]

37. Duy, S.V.; Fayad, P.B.; Barbeau, B.; Prevost, M.; Sauve, S. Using a novel sol-gel stir bar sorptive extraction method for the analysis of steroid hormones in water by laser diode thermal desorption/atmospheric chemical ionization tandem mass spectrometry. *Talanta* **2012**, *101*, 337–345. [CrossRef] [PubMed]

38. Liu, W.M.; Wang, H.M.; Guan, Y.F. Preparation of stir bars for sorptive extraction using sol-gel technology. *J. Chromatogr. A* **2004**, *1045*, 15–22. [CrossRef] [PubMed]

39. Yang, L.; Said, R.; Abdel-Rehim, M. Sorbent, device, matrix and application in microextraction by packed sorbent (MEPS): A review. *J. Chromatogr. B* **2017**, *1043*, 33–43. [CrossRef] [PubMed]

40. Iskierko, Z.; Sharma, P.S.; Prochowicz, D.; Fronc, K.; D'Souza, F.; Toczydłowska, D.; Stefaniak, F.; Noworyta, K. Molecularly Imprinted Polymer (MIP) Film with Improved Surface Area Developed by Using Metal-Organic Framework (MOF) for Sensitive Lipocalin (NGAL) Determination. *ACS Appl. Mater. Interfaces* **2016**, *8*, 19860–19865. [CrossRef] [PubMed]

41. Qian, K.; Fang, G.; Wang, S. A novel core-shell molecularly imprinted polymer based on metal-organic frameworks as a matrix. *Chem. Commun.* **2011**, *47*, 10118–10120. [CrossRef] [PubMed]

42. Ribeiro, C.; Ribeiro, A.R.; Maia, A.S.; Gonçalves, V.M.; Tiritan, M.E. New trends in sample preparation techniques for environmental analysis. *Crit. Rev. Anal. Chem.* **2014**, *44*, 142–185. [CrossRef] [PubMed]

43. Elmongy, H.; Ahmed, H.; Wahbi, A.-A.; Amini, A.; Colmsjö, A.; Abdel-Rehim, M. Determination of metoprolol enantiomers in human plasma and saliva samples utilizing microextraction by packed sorbent and liquid chromatography-tandem mass spectrometry. *Biomed. Chromatogr.* **2016**, *30*, 1309–1317. [CrossRef] [PubMed]

44. Locatelli, M.; Ferrone, V.; Cifelli, R.; Barbacane, R.C.; Carlucci, G. MicroExtraction by Packed Sorbent and HPLC determination of seven non-steroidal anti-inflammatory drugs in human plasma and urine. *J. Chromatogr. A* **2014**, *1367*, 1–8. [CrossRef] [PubMed]

45. Locatelli, M.; Ciavarella, M.T.; Paolino, D.; Celia, C.; Fiscarelli, E.; Ricciotti, G.; Pompilio, A.; Di Bonaventura, G.; Grande, R.; Zengin, G.; et al. Determination of Ciprofloxacin and Levofloxacin in Human Sputum Collected from Cystic Fibrosis Patients using Microextraction by Packed Sorbent-High Performance Liquid Chromatography PhotoDiode Array Detector. *J. Chromatogr. A* **2015**, *1419*, 58–66. [CrossRef] [PubMed]

46. D'Angelo, V.; Tessari, F.; Bellagamba, G.; De Luca, E.; Cifelli, R.; Celia, C.; Primavera, R.; Di Francesco, M.; Paolino, D.; Di Marzio, L.; et al. MicroExtraction by Packed Sorbent and HPLC-PDA quantification of multiple anti-inflammatory drugs and fluoroquinolones in human plasma and urine. *J. Enzyme Inhib. Med. Chem.* **2016**, *31*, 110–116. [CrossRef] [PubMed]

47. Campestre, C.; Locatelli, M.; Guglielmi, P.; De Luca, E.; Bellagamba, G.; Menta, S.; Zengin, G.; Celia, C.; Di Marzio, L.; Carradori, S. Analysis of imidazoles and triazoles in biological samples after MicroExtraction by Packed Sorbent. *J. Enzyme Inhib. Med. Chem.* **2017**, *32*, 1053–1063. [CrossRef] [PubMed]

48. Ares, A.M.; Fernández, P.; Regenjo, M.; Fernández, A.M.; Carro, A.M.; Lorenzo, R.A. A fast bioanalytical method based on microextraction by packed sorbent and UPLC-MS/MS for determining new psychoactive substances in oral fluid. *Talanta* **2017**, *174*, 454–461. [CrossRef] [PubMed]

49. Soleimani, E.; Bahrami, A.; Afkhami, A.; Shahna, F.G. Determination of urinary trans,trans-muconic acid using molecularly imprinted polymer in microextraction by packed sorbent followed by liquid chromatography with ultraviolet detection. *J. Chromatogr. B* **2017**, *1061–1062*, 65–71. [CrossRef] [PubMed]

50. Ortega, S.N.; Santos-Neto, A.J.; Lancas, F.M. Development and optimization of a fast method for the determination of statins in human plasma using microextraction by packed sorbent (MEPS) followed by ultra high-performance liquid chromatography-tandem mass spectrometry (UHPLC-MS/MS). *Anal. Methods* **2017**, *9*, 3039–3048. [CrossRef]

51. Fernández, P.; González, M.; Regenjo, M.; Ares, A.M.; Fernández, A.M.; Lorenzo, R.A.; Carro, A.M. Analysis of drugs of abuse in human plasma using microextraction by packed sorbents and ultra-high-performance liquid chromatography. *J. Chromatogr. A* **2017**, *1485*, 8–19. [CrossRef] [PubMed]

52. Rosado, T.; Gonçalves, A.; Margalho, C.; Barroso, M.; Gallardo, E. Rapid analysis of cocaine and metabolites in urine using microextraction in packed sorbent and GC/MS. *Anal. Bioanal. Chem.* **2017**, *409*, 2051–2063. [CrossRef] [PubMed]

53. Mercolini, L.; Mandrioli, R.; Raggi, M.A. Content of melatonin and other antioxidants in grape-related foodstuffs: Measurementusing a MEPS-HPLC-F method. *J. Pineal Res.* **2012**, *53*, 21–28. [CrossRef] [PubMed]

54. Martínez-Moral, M.P.; Tena, M.T. Use of microextraction bypacked sorbents following selective pressurised liquid extraction for thedetermination of brominated diphenyl ethers in sewage sludge by gaschromatography-mass spectrometry. *J. Chromatogr. A* **2014**, *1364*, 28–35. [CrossRef] [PubMed]

55. Paredes, R.M.G.; Pinto, C.G.; Pavón, J.L.P.; Cordero, B.M. In situ derivatization combined to automatedmicroextraction by packed sorbents for the determination of chlorophenolsin soil samples by gas chromatography mass spectrometry. *J. Chromatogr. A* **2014**, *1359*, 52–59. [CrossRef] [PubMed]

56. Salami, F.H.; Queiroz, M.E.C. Microextraction inpacked sorbent for analysis of sulfonamides in poultry litter wastewatersamples by liquid chromatography and spectrophotometric detection. *J. Liq. Chromatogr. Relat. Technol.* **2014**, *37*, 2377–2388. [CrossRef]

57. Amiri, A.; Ghaemi, F. Microextraction in packed syringe by using a three-dimensional carbon nanotube/carbon nanofiber-graphene nanostructure coupled to dispersive liquid-liquid microextraction for the determination of phthalate esters in water samples. *Microchim. Acta* **2017**, *184*, 3851–3858. [CrossRef]

58. Fumes, B.H.; Lanças, F.M. Use of graphene supported on aminopropyl silica for microextraction of parabens from water samples. *J. Chromatogr. A* **2017**, *1487*, 64–71. [CrossRef] [PubMed]

59. Kabir, A.; Furton, K.G. Fabric Phase Sorptive Extractor (FPSE). U.S. Patent US 20140274660 A1, 18 September 2014.

60. Kabir, A.; Mesa, R.; Jurmain, J.; Furton, K.G. Fabric Phase Sorptive Extraction Explained. *Separations* **2017**, *4*, 21. [CrossRef]

61. Locatelli, M.; Kabir, A.; Innosa, D.; Lopatriello, T.; Furton, K.G. A Fabric Phase Sorptive Extraction-High Performance Liquid Chromatography-Photo Diode Array Detection Method for the Determination of Twelve Azole Antimicrobial Drug Residues in Human Plasma and Urine. *J. Chromatogr. B* **2017**, *1040*, 192–198. [CrossRef] [PubMed]

62. Kabir, A.; Furton, K.G.; D'Ovidio, C.; Grossi, R.; Innosa, D.; Macerola, D.; Tartaglia, A.; Di Donato, V.; Locatelli, M. Fabric phase sorptive extraction-high performance liquid chromatography-photo diode array detection method for triple therapy in the treatment of inflammatory bowel diseases. *Anal. Chim. Acta* **2017**, submitted.

63. Locatelli, M.; Kabir, A.; Tinari, N.; Macerola, D.; Tartaglia, A.; Furton, K.G. A Fabric Phase Sorptive Extraction-High Performance Liquid Chromatography-Photo Diode Array Detection Method for the Determination of three antitumoral Drugs. *Sci. Rep.* **2017**, submitted.

64. Samanidou, V.; Kaltzi, I.; Kabir, A.; Furton, K.G. Simplifying sample preparation using fabric phase sorptive extraction technique for the determination of benzodiazepines in blood serum by high-performance liquid chromatography. *Biomed. Chromatogr.* **2016**, *30*, 829–836. [CrossRef] [PubMed]

65. Kumar, R.; Gaurav; Heena; Malik, A.K.; Kabir, A.; Furton, K.G. Efficient analysis of selected estrogens using fabric phase sorptive extraction and high performance liquid chromatography-fluorescence detection. *J. Chromatogr. A* **2014**, *1359*, 16–25. [CrossRef] [PubMed]

66. Guedes-Alonso, R.; Ciofi, L.; Sosa-Ferrera, Z.; Santana-Rodríguez, J.J.; Del Bubba, M.; Kabir, A.; Furton, K.G. Determination of androgens and progestogens in environmental and biological samples using fabric phase sorptive extraction coupled to ultra-high performance liquid chromatography tandem mass spectrometry. *J. Chromatogr. A* **2016**, *1437*, 116–126. [CrossRef] [PubMed]

67. Aznar, M.; Alfaro, P.; Nerin, C.; Kabir, A.; Furton, K.G. Fabric phase sorptive extraction: An innovative sample preparation approach applied to the analysis of specific migration from food packaging. *Anal. Chim. Acta* **2016**, *936*, 97–107. [CrossRef] [PubMed]

68. Samanidou, V.; Galanopoulos, L.-D.; Kabir, A.; Furton, K.G. Fast extraction of amphenicols residues from raw milk using novel fabric phase sorptive extraction followed by high-performance liquid chromatography-diode array detection. *Anal. Chim. Acta* **2015**, *855*, 41–50. [CrossRef] [PubMed]

69. Karageorgou, E.; Manousi, N.; Samanidou, V.; Kabir, A.; Furton, K.G. Fabric phase sorptive extraction for the fast isolation of sulfonamides residues from raw milk followed by high performance liquid chromatography with ultraviolet detection. *Food Chem.* **2016**, *196*, 428–436. [CrossRef] [PubMed]

70. Aznar, M.; Úbeda, S.; Nerin, C.; Kabir, A.; Furton, K.G. Fabric phase sorptive extraction as a reliable tool for rapid screening and detection of freshness markers in oranges. *J. Chromatogr. A* **2017**, *1500*, 32–42. [CrossRef] [PubMed]

71. Samanidou, V.; Michaelidou, K.; Kabir, A.; Furton, K.G. Fabric phase sorptive extraction of selected penicillin antibiotic residues from intact milk followed by high performance liquid chromatography with diode array detection. *Food Chem.* **2017**, *224*, 131–138. [CrossRef] [PubMed]

72. Samanidou, V.; Filippou, O.; Marinou, E.; Kabir, A.; Furton, K.G. Sol-gel-graphene-based fabric-phase sorptive extraction for cow and human breast milk sample cleanup for screening bisphenol A and residual dental restorative material before analysis by HPLC with diode array detection. *J. Sep. Sci.* **2017**, *40*, 2612–2619. [CrossRef] [PubMed]

73. Lakade, S.S.; Borrull, F.; Furton, K.G.; Kabir, A.; Marcé, R.M.; Fontanals, N. Dynamic fabric phase sorptive extraction for a group of pharmaceuticals and personal care products from environmental waters. *J. Chromatogr. A* **2016**, *1456*, 19–26. [CrossRef] [PubMed]

74. Kumar, R.; Gaurav; Kabir, A.; Furton, K.G.; Malik, A.K. Development of a fabric phase sorptive extraction with high-performance liquid chromatography and ultraviolet detection method for the analysis of alkyl phenols in environmental samples. *J. Sep. Sci.* **2015**, *38*, 3228–3238. [CrossRef] [PubMed]

75. Racamonde, I.; Rodil, R.; Quintana, J.B.; Sieira, B.J.; Kabir, A.; Furton, K.G.; Cela, R. Fabric phase sorptive extraction: A new sorptive microextraction technique for the determination of non-steroidal anti-inflammatory drugs from environmental water samples. *Anal. Chim. Acta* **2015**, *865*, 22–30. [CrossRef] [PubMed]

76. Roldán-Pijuán, M.; Lucena, R.; Cárdenas, S.; Valcárcel, M.; Kabir, A.; Furton, K.G. Stir fabric phase sorptive extraction for the determination of triazine herbicides in environmental waters by liquid chromatography. *J. Chromatogr. A* **2015**, *1376*, 35–45. [CrossRef] [PubMed]

77. Montesdeoca-Esponda, S.; Sosa-Ferrera, Z.; Kabir, A.; Furton, K.G.; Santana-Rodríguez, J.J. Fabric phase sorptive extraction followed by UHPLC-MS/MS for the analysis of benzotriazole UV stabilizers in sewage samples. *Anal. Bioanal. Chem.* **2015**, *407*, 8137–8150. [CrossRef] [PubMed]

78. García-Guerra, R.B.; Montesdeoca-Esponda, S.; Sosa-Ferrera, Z.; Kabir, A.; Furton, K.G.; Santana-Rodríguez, J.J. Rapid monitoring of residual UV-stabilizers in seawater samples from beaches using fabric phase sorptive extraction and UHPLC-MS/MS. *Chemosphere* **2016**, *164*, 201–207. [CrossRef] [PubMed]

79. Lakade, S.S.; Borrull, F.; Furton, K.G.; Kabir, A.; Fontanals, N.; Marcé, R.M. Comparative study of different fabric phase sorptive extraction sorbents to determine emerging contaminants from environmental water using liquid chromatography-tandem mass spectrometry. *Talanta* **2015**, *144*, 1342–1351. [CrossRef] [PubMed]

80. Anthemidis, A.; Kazantzi, V.; Samanidou, V.; Kabir, A.; Furton, K.G. An automated flow injection system for metal determination by flame atomic absorption spectrometry involving on-line fabric disk sorptive extraction technique. *Talanta* **2016**, *156–157*, 64–70. [CrossRef] [PubMed]

81. Heena; Kaur, R.; Rani, S.; Malik, A.K.; Kabir, A.; Furton, K.G. Determination of cobalt(II), nickel(II) and palladium(II) Ions via fabric phase sorptive extraction in combination with high-performance liquid chromatography-UV detection. *Sep. Sci. Technol.* **2017**, *52*, 81–90. [CrossRef]

82. Alcudia-León, M.C.; Lucena, R.; Cárdenas, S.; Valcárcel, M.; Kabir, A.; Furton, K.G. Integrated sampling and analysis unit for the determination of sexual pheromones in environmental air using fabric phase sorptive extraction and headspace-gas chromatography–mass spectrometry. *J. Chromatogr. A* **2017**, *1488*, 17–25. [CrossRef] [PubMed]

83. Ulusoy, H.I. Applications of magnetic nanoparticles for the selective extraction of trace species from a complex matrix. In *Analytical Chemistry: Developments, Applications and Challenges in Food Analysis*; Locatelli, M., Celia, C., Eds.; Nova Science Publishers, Inc.: Hauppauge, NY, USA, 2017; pp. 55–76, ISBN 978-1-53612-267-1.

84. Hernández-Hernández, A.A.; Álvarez-Romero, G.A.; Contreras-López, E.; Aguilar-Arteaga, K.; Castañeda-Ovando, A. Food Analysis by Microextraction Methods Based on the Use of Magnetic Nanoparticles as Supports: Recent Advances. *Food Anal. Methods* **2017**, *10*, 2974–2993. [CrossRef]

85. Vasconcelos, I.; Fernandes, C. Magnetic solid phase extraction for determination of drugs in biological matrices. *TrAC* **2017**, *89*, 41–52. [CrossRef]

86. Rocío-Bautista, P.; González-Hernández, P.; Pino, V.; Pasán, J.; Afonso, A.M. Metal-organic frameworks as novel sorbents in dispersive-based microextraction approaches. *TrAC* **2017**, *90*, 114–134. [CrossRef]

87. Ríos, Á.; Zougagh, M. Recent advances in magnetic nanomaterials for improving analytical processes. *TrAC* **2017**, *84*, 72–83. [CrossRef]

88. Płotka-Wasylka, J.; Owczarek, K.; Namieśnik, J. Modern solutions in the field of microextraction using liquid as a medium of extraction. *TrAC* **2016**, *85*, 46–64. [CrossRef]

89. Sharifi, V.; Abbasi, A.; Nosrati, A. Application of hollow fiber liquid phase microextraction and dispersive liquid-liquid microextraction techniques in analytical toxicology. *J. Food Drug Anal.* **2016**, *24*, 264–276. [CrossRef] [PubMed]

90. Yilmaz, E.; Soylak, M. Latest trends, green aspects, and innovations in liquid-phase-based microextraction techniques: A review. *Turkish J. Chem.* **2016**, *40*, 868–893. [CrossRef]

91. Diuzheva, A.; Carradori, S.; Andruch, V.; Locatelli, M.; De Luca, E.; Tiecco, M.; Germani, R.; Menghini, L.; Nocentini, A.; Gratteri, P.; et al. Use of innovative (micro)extraction techniques to characterize harpagophytum procumbens root and its commercial food supplements. *Phytochem. Anal.* **2017**, in press. [CrossRef] [PubMed]

92. Xiong, J.; Hu, B. Comparison of hollow fiber liquid phase microextraction and dispersive liquid-liquid microextraction for the determination of organosulfur pesticides in environmental and beverage samples by gas chromatography with flame photometric detection. *J. Chromatogr. A* **2008**, *1193*, 7–18. [CrossRef] [PubMed]

93. Abdel-Rehim, M. Recent advances in microextraction by packed sorbent for bioanalysis. *J. Chromatogr. A* **2010**, *1217*, 2569–2580. [CrossRef] [PubMed]

94. Abdel-Rehim, M. New trend in sample preparation: On-line microextraction in packed syringe for liquid and gas chromatography applications: I. Determination of local anaesthetics in human plasma samples using gas chromatography-mass spectrometry. *J. Chromatogr. B* **2004**, *801*, 317–321. [CrossRef]

*separations*

MDPI

*Review*

# Recent Advances in Extraction and Stirring Integrated Techniques

**Soledad Cárdenas and Rafael Lucena \***

Department of Analytical Chemistry, Institute of Fine Chemistry and Nanochemistry, Marie Curie Building, Campus of Rabanales, University of Córdoba, 14071 Córdoba, Spain; qa1caarm@uco.es
\* Correspondence: q62luror@uco.es; Tel.: +34-957-211-066

Academic Editor: Victoria F. Samanidou
Received: 31 January 2017; Accepted: 7 March 2017; Published: 15 March 2017

**Abstract:** The extraction yield of a microextraction technique depends on thermodynamic and kinetics factors. Both of these factors have been the focus of intensive research in the last few years. The extraction yield can be increased by synthesizing and using novel materials with favorable distribution constants (one of the thermodynamic factors) for target analytes. The extraction yield can also be increased by improving kinetic factors, for example, by developing new extraction modes. Microextraction techniques are usually non-exhaustive processes that work under the kinetic range. In such conditions, the improvement of the extraction kinetics necessarily improves the performance. Since the extraction yield and efficiency is related to how fast the analytes diffuse in samples, it is crucial to stir the sample during extraction. The stirring can be done with an external element or can be integrated with the extraction element in the same device. This article reviews the main recent advances in the so-called extraction/stirring integrated techniques with emphasis on their potential and promising approaches rather than in their applications.

**Keywords:** extraction/stirring integrated techniques; stir bar sorptive extraction; stir membrane extraction; stir cake sorptive extraction; rotating disk sorptive extraction

## 1. Introduction

Microextraction techniques are physicochemical processes based on the mass transference between, at least, two different phases. Mass transference is driven by thermodynamic and kinetics factors, both of which have a dramatic impact on the extraction yield [1]. Thermodynamics defines the maximum amount of analytes that can be extracted by a technique while kinetics defines the rate at which this transference occurs. The microextraction thermodynamics is mainly described by the distribution constant and many factors, such as the use of secondary reactions or the proper selection of the working pH, may affect this partitioning equilibrium. However, microextraction techniques usually work under diffusion controlled conditions. This situation, a direct consequence of the size difference between the sample and extractant phases (the diffusion paths of the analytes become larger), is of paramount practical importance. In short, a thermodynamically favored but slow technique is not suitable for analytical purposes, as it would provide a very low sample throughput. Therefore, kinetic factors must be deeply considered in any microextraction development. Among these factors, the extraction surface and the Nernst boundary layer can be highlighted. On one hand, the extraction kinetics are dependent on the area of the interface between the sample and the extractant. This phenomenon is behind the development of dispersive techniques [2] where the liquid or solid extractant is dispersed in the form of fine droplets or small-sized particles into the sample. On the other hand, the thickness of the boundary layer between the bulk sample and extractant also affects the extraction rate. In this case, the higher the thickness, the lower the rate. The boundary layer thickness can be effectively reduced if the sample is agitated, enhancing the mass transference during the extraction. In fact, agitation is common to all of the microextraction techniques.

Agitation can be done in two different ways. In most cases, the sample is agitated using an internal mechanical element (e.g., inert stir bar) or an external energy source (e.g., ultrasound) independent of the extraction element. However, some techniques are based on the integration of the agitation and extraction elements in the same device. This integration simplifies the extraction to a large extent, avoids analyte losses due to retention on external devices, and enhances the extraction yields. These techniques, that can be named as extraction/stirring integrated techniques, are the topic of this review article. This contribution reviews the recent developments of these techniques rather than focusing on the applications reported in the field.

## 2. Stir Bar Sorptive Extraction

Stir bar sorptive extraction (SBSE) was the first technique based on the integration of the extraction and stirring elements in the same device. It was proposed in 1999 as an alternative to solid phase microextraction (SPME) with which it shares some extraction principles [3]. The basic SBSE device consists of a stir bar coated with a polymeric phase that acts as the extractant. The stir bar is introduced and stirred into the sample and the analytes are extracted in an enhanced diffusion medium. The similarities between SPME and SBSE are based on the use of almost the same coatings which makes the application field similar. However, both techniques operate with different workflows and also differ based on the sorptive phase thickness. The last fact is of paramount importance and has thermodynamic and kinetics connotations. On the one hand, the extraction recoveries in SBSE are higher than in SPME, as the volume of the sorptive phase is 50–250 times higher in the first technique. However, the extraction kinetics are somewhat affected by the thickness since the diffusion of the analytes in the polymeric coating is hindered [4].

Polydimethylsiloxane (PDMS), a classic SPME coating, is extensively reported in SBSE applications, which are focused, as a direct consequence of the hydrophobic nature of the coating, on the extraction of non-polar compounds [5]. The in-lab synthesis of new coatings to extend the applicability of SBSE has been the main research topic during the last decade. The most salient coatings are presented in the following sections. However, it is necessary to highlight that some of the new coatings are not commercially available since they are not as robust as the classical PDMS one.

### 2.1. Selective Coatings in SBSE

Conventional PDMS lacks extraction selectivity as the analytes are isolated by hydrophobic interactions. However, selectivity is an important analytical feature in extraction methods that becomes critical in the microextraction context where the sorptive capacity is limited. This situation is even more complicated when complex samples comprising hundreds of interferents are processed [6]. Under these circumstances, a selective coating allows the use of the limited sorption capacity to the target analytes, avoiding sorbent saturation by the matrix components. The enhancement of sorption selectivity in SBSE has been accomplished following several strategies, the most important among them being the use of molecularly imprinted polymers (MIPs) or selective molecules.

Molecularly imprinted polymers can be defined as polymeric networks containing selective chemical cavities for the target analyte and structurally related compounds. These cavities are ad-hoc prepared by incubation, prior to the polymerization, of a special monomer with the target compound. In these conditions, the polymer grows around the template creating a cavity sterically and chemically compatible with the template. This cavity is released by the washing of the polymer, a critical step.

The use of MIPs as SBSE coatings was first proposed by Zhu et al. in 2006 [7]. This approach exploited the switchable solubility of nylon in order to create a multi-cavity network towards monocrotophos. A precursor solution containing the template and the polymer was prepared in formic acid where nylon was completely soluble. The stir bar was subsequently immersed into the solution and finally in pure water. The solvent change over induces the gelation of the polymer, which is not soluble in water, around the template molecules creating the selective cavities. The resulting MIP showed a high porosity which was beneficial for the extraction kinetics.

Although the previous approach is simple and useful, MIPs-bars have been mainly synthesized by co-polymerization of an appropriate monomer and a crosslinker in the presence of the template molecule. The bar must be previously treated in order to achieve a strong retention of the MIP over its surface which is crucial during the extraction (mechanical stability) and elution (chemical stability) steps. This strategy was followed by Xu et al. in order to fabricate MIP-bar towards ractopamine [8]. In the last few years, this workflow has been proposed for the extraction of triazines in rice [9], thiabendazole in citrus samples [10], and melamine in powdered milk [11]. The selectivity of MIPs restricts their applicability to compounds with similar chemical structures. In some cases, the analyst may be interested in the extraction of two or more different families of compounds while maintaining a high selectivity level. Dual template MIPs try to address this situation by creating a polymeric network with two types of cavities, each one selective for a family of compounds. The process is quite similar to that previously described although in this case two templates, instead of one, are added to the synthesis medium. This approach has been proposed to create stir bars that are selective towards different estrogenic compounds in water and plastic samples [12]. Seng et al. have also proposed an alternative synthetic method that consists of the preparation of silica particles coated with the MIP which are finally loaded into the stir bar. This approach was applied for the extraction of bisphenol A from water with excellent results [13].

Other chemical structures can be used to enhance the extraction selectivity. In this sense, a polymeric composite containing $\alpha$-cyclodextrin as a selective element has been reported as an SBSE coating for the extraction of polychlorinated biphenyls (PCBs) from water samples [14]. $\alpha$-Cyclodextrin is a cyclic ($R$-1,4)-linked oligosacharide that consist of six glucopyranose subunits that presents a cage-like structure. The hydrophobic cavity can host analytes with a mechanism where the molecular size plays a key-role.

Aptamers, which are artificial nucleic acid (DNA or RNA) ligands towards specific analytes, have been used as biorecognition elements in SBSE for the extraction of selected PCBs from fish samples [15].

## 2.2. The Potential of Nanoparticles in SBSE

Nanoparticles (NPs) can be defined as those particles that present one or more dimensions in the nanometric range (using 100 nm as an arbitrary reference), which provide them with special properties not observed in the bulk material [16]. The use of NPs in sample treatment has been the focus of extensive research during the last decade [17,18], since their properties are different than the classical materials. In SBSE, NPs may play two different and, in some cases, complementary roles. On the one hand, they can act as the active sorptive phase introducing new interaction chemistries that boost the extraction of the target analytes. On the other hand, the inclusion of NPs in the polymeric SBSE coating avoids the normal stacking of the polymer network, creating a more porous structure which makes the diffusion of the analytes easier with evident kinetics benefits.

Nanoparticles can be classified according to different criteria. Considering their chemical composition, they can be divided into inorganic or carbon based NPs. In this sense, inorganic NPs have been proposed as components of SBSE coatings. Li et al. reported the use of zirconia NPs as the sorptive phase for the extraction of polar organophosphorous compounds in water samples [19]. The location of $ZrO_2$ NPs, with sizes in the range of 10–20 nm, over a PDMS phase generates a rough coating with a superficial area of 103 $m^2$/g and a pore size of 5 nm. Figure 1 shows an SEM picture of the dumbbell-shaped stir bar and a closer view of the coating. $ZrO_2$ NPs prevail over PDMS in the interaction with the polar analytes, which can be extracted by three mechanisms (cation/anion/ligand exchange) depending on the working pH. The enrichment factors (742–1583) were excellent.

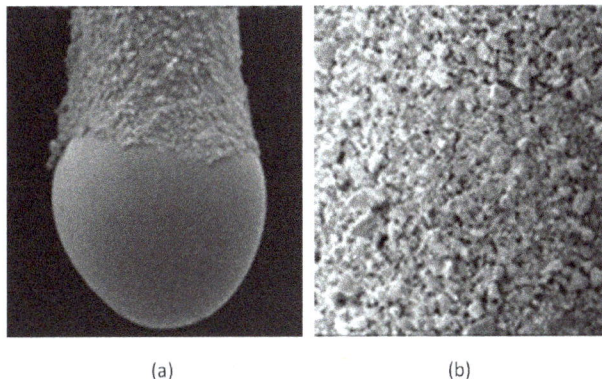

**Figure 1.** Scanning electron microscopy (SEM) picture of the (**a**) dumbbell-shaped stir bar (magnification 40×) and (**b**) the surface of the coating (magnification 150×). Reproduced with permission from [19], copyright Elsevier, 2012.

Magnetic nanoparticles have also found applications in SBSE in an elegant approach proposed by Benedé et al. [20], which combines the benefits of classical dispersive phase microextraction and SBSE. The extraction workflow, which is schematically shown in Figure 2, consists of several subsequent steps. Initially, the oleic acid coated $CoFe_2O_4$ NPs are attached to the stir bar by magnetic forces. Once the stir bar is introduced into the sample, the bar is stirred at high velocities inducing the detachment of the magnetic nanoparticles (MNPs) which are dispersed into the sample, extracting the target analytes. After that, the stirring rate is reduced inducing again the attachment of the MNPs on the bar for the final elution. This approach reduces the treatment time as the isolation takes place under a perfect dispersion of the sorbent. In addition, it is versatile since virtually any MNP can be applied. In fact, the same authors have evaluated the potential of magnetic nylon 6 composites under this format [21].

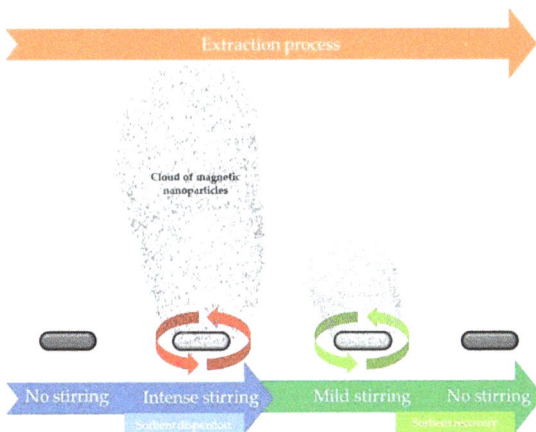

**Figure 2.** Scheme of stir bar sorptive-dispersive microextraction mediated by magnetic nanoparticles. Reproduced with permission of the Microextraction Tech blog.

Carbon based nanomaterials can develop special interaction chemistries that can be used to enhance the extraction selectivity. The capacity of carbon nanotubes (CNTs) to interact by π–π bonds with analytes containing aromatic domains has been exploited in SBSE [22–24]. The introduction of

CNTs in the sorptive phase can be done in several ways. Hu et al. [22] introduced amino-modified CNTs in a PDMS matrix, where they are entrapped, in order to make the extraction of polar phenols feasible. The presence of CNTs in the polymer creates a rougher surface with a higher extraction capacity. Farhadi and co-workers preferred a different approach based on the electro-polymerization of aniline over a steel pin in the presence of CNTs, which are also introduced in the polymeric network [23]. Although the inclusion of CNTs enhances the extraction capacity, the amount of CNTs that can be loaded is somewhat limited by mechanical stability issues. In fact, the resulting composite becomes unstable when the amount of CNTs is high. This stability can be improved if covalent bonding is selected to anchor the CNTs into the polymeric network [24]. This synthetic path requires a modification of the CNTs which are previously oxidized to include carboxylic groups on their surface. Carboxyl groups are finally transformed into the acyl chloride form, allowing their inclusion into a polyethylene glycol (PEG) phase.

Graphene oxide (GO) has been also proposed as a SBSE coating element. Fan et al. evaluated a polyethylene glycol (PEG)/GO composite for the extraction of fluoroquinolones from chicken muscle and liver [25]. The composite, which is synthesized by co-blending, is finally immobilized over the stir bar by the sol-gel reaction. It provides the best extraction of the target analytes compared to commercial and lab-made PDMS coatings thanks to its superior hydrophobic/hydrophilic balance. The presence of PEG also stabilizes GO avoiding losses of the nanomaterial during stirring. The introduction of GO into the coating can be also done with a different strategy. Zhang et al. proposed the use of polydopamine coated over a stainless steel bar as a support for the covalent GO immobilization [26], providing a better distribution of the nanomaterial in the polymeric network. The resulting composites provided better efficiencies (signal enhanced by factors higher than 10) than the naked polymeric coating. Although GO has an active role on the extraction in the above reported applications, it may also play other secondary but positive roles in the extraction. In this sense, it has been also proposed as a dopant for enhancing the extraction capabilities of an MIP stir bar coating [27] through improvement of the coating porosity.

*2.3. ICE Concentration Linked with Extractive Stirrer*

The improvement of the extraction yield can be also achieved by the proper modification of the experimental conditions. Recently, Logue et al. have proposed the so-called ice concentration linked with an extractive stirrer (ICECLES) which combines the advantages of freeze concentration and SBSE [28].

Freeze concentration can be considered as an innovation in analytical sample preparation although it is frequently used in the food industry to concentrate liquid products. It is based on a simple and well known principle, freezing-point depression. This principle states that solutions have lower freezing points than pure water. As a consequence, when a sample is cooled down, ice starts to form, excluding the solutes which are concentrated in the remaining liquid.

ICECLES operates in a simple and well established workflow (see Figure 3). A sorptive bar is introduced into a liquid sample and is stirred continuously. During the extraction, the sample is cooled down, inducing the freeze concentration. As a result, the solution is enriched with the analytes establishing a higher concentration gradient with the stir bar, thus increasing the extraction yield. In addition, the lower temperature increases the sorption of the analytes as this process is exothermic.

ICECLES provides signal enhancement factors of 450 over conventional SBSE for the extraction of (semi)volatile compounds, which makes it a very promising technique in the coming years.

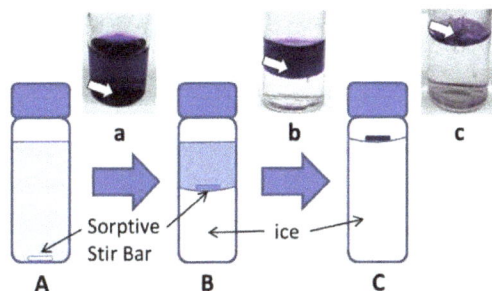

**Figure 3.** Scheme of ice concentration linked with the extractive stirrer technique. The stirrer is located in the bottom of the vial (**A**) which is cooled down. As a consequence, ice is created (**B** and **C**) leaving a more concentrated solution than the original sample. The pictures show the extraction of methyl violet as the model analyte: (**a**) initial state, (**b**) extraction and (**c**) final state. Reproduced with permission from [28], copyright Elsevier, 2016.

## 3. Stir Membrane Extraction

Stir membrane extraction (SME) was proposed by our research group in 2009 as a new stirring technique based on the use of polymeric membranes as sorptive phases [29]. Membranes present a high surface-to-volume ratio and permeability, especially if they are compared with classical SBSE coatings, which are advantageous characteristics from the kinetics point of view. To be applied, SME requires a special device that integrates stirring (a protected iron wire) and extraction (membrane) elements. The first unit, schematically shown in Figure 4A, was designed in polypropylene using commercial products as precursors. In its first application, aimed at determining selected polycyclic aromatic hydrocarbons (PAHs) in water samples, SME provided better results than stir bars fabricated by the same membrane polymer (PTFE). Although SME can be applied in the normal extraction/elution workflow, it can be directly combined with spectroscopic techniques that allow the in-membrane detection of the targets without a previous elution. This strategy was reported for the determination of the hydrocarbon index in water by infrared spectroscopy [30]. For this purpose, the SME unit was built in stainless steel (Figure 4B) to promote the retention of the analytes into the membrane which presents a low thickness (40 µm), to permit the transmission of the infrared beam. The versatility of SME is based on the great availability of commercial and lab-made membranes. In fact, the use of fabric phases, a porous hybrid inorganic-organic sorbent material chemically bonded to the flexible and permeable substrate matrix, [31] has been also proposed as a sorptive phase in SME [32]. This combination allowed the extraction of several triazine herbicides from water samples with enrichment factors in the range from 444 to 1410.

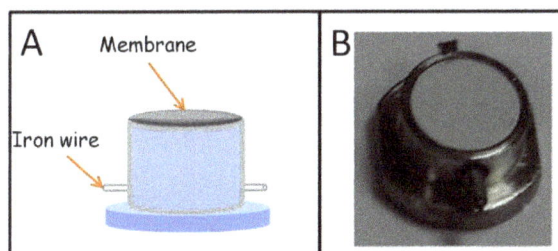

**Figure 4.** Stir membrane extraction device. (**A**) Diagram of the unit built in plastic; (**B**) picture of the unit built in stainless steel (reproduced with permission from [30], copyright Springer, 2010).

## 3.1. Stir Membrane Extraction in the Liquid Phase Microextraction Context

The original SME device can be adapted to the liquid phase microextraction (LPME) context with slight modifications. The simple introduction of a cap in the lower part of the unit creates a small chamber where an extractant phase can be located, as can be observed in Figure 5. The membrane is used for achieving the confinement of the liquid extractant phase in the chamber while the unit is stirred in the sample. The stir membrane unit in LPME can operate under the two [33] and three [34,35] phase mode. In the two phase mode, an organic solvent is located in the chamber, wetting the membrane, and the analytes are extracted by the different solubilities they present in the aqueous (sample) and organic (extractant) phases. This mode is specially indicated for the extraction of non-polar compounds and it is fully compatible with gas chromatography. In the three phase mode, the organic solvent only wets the membrane (forming the so-called supported liquid membrane, SLM) while an aqueous phase is located in the internal chamber. This mode, which is based on the transference of the analytes through the SLM thanks to an existing pH gradient between the sample and the extractant (both of aqueous nature), is useful for the extraction of ionisable non-polar compounds.

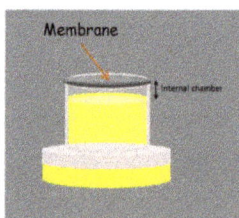

**Figure 5.** Stir membrane extraction device working in the liquid phase microextraction context.

The conventional stir membrane unit, both in the solid and liquid phase microextraction context, is stirred in the sample. Sample volumes larger than 20 mL (the exact volume depends on the extraction vessel) are required which restricts the application to some biological samples with limited volume. For this reason, the stir membrane extraction has been also adapted to this scenario. This was accomplished by slightly modifying the device, as can be observed in Figure 6, by just increasing the height of the plastic adapter. This modification allows the creation of an upper chamber with a small volume where the sample is located [36]. Although this approach, which was evaluated for the extraction of paracetamol from saliva samples, cannot be strictly considered as a stirring device (the unit is agitated in a vortex), it clearly shows the versatility of the membrane-based units.

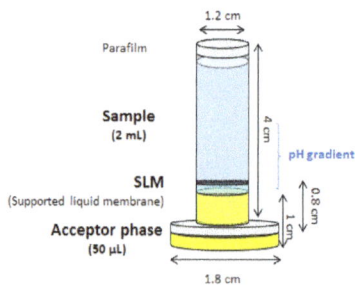

**Figure 6.** Modification of the stir membrane liquid-liquid-liquid extraction device to process a low volume of sample. SLM, supported liquid membrane.

In the same way, the original device has been adapted for solid-liquid-liquid extraction [37]. Although the main idea remains the same, the unit was built using a different construction block.

In this case, a conventional plastic eppendorf tube is used as extraction device (Figure 7) where the solid sample is dispersed in an organic solvents mixture. An aqueous extractant phase is situated in the eppendorf cap and protected by a polymeric membrane. Once deployed, the unit is rotated and agitated. During the extraction, the analytes are transferred from the solid sample to the organic solvent and then from the organic solvent to the aqueous phase, passing through the membrane. This approach was initially used for the determination of parabens in lyophilized human breast milk with very good sensitivity.

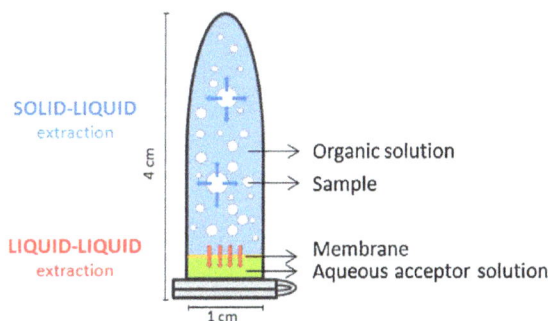

**Figure 7.** Device for stir membrane solid-liquid-liquid extraction. Reproduced with permission from [37], copyright Elsevier, 2014.

### 3.2. Adaptations of Stir Membrane Units

The stir membrane unit has been adapted to other formats. For example, the polymeric membrane can be substituted by a borosilicate disk to develop liquid phase microextraction in a solvent film. In this case, the disk is derivatized with octadecyl groups that make the physical immobilization of an organic solvent, the active extractant, on the disk surface easier. This approach provides enrichment factors towards selective triazine herbicides in the range from 79 to 839 [38].

In addition, the use of small magnets within the unit (Figure 8) has allowed the development of the so called magnetically confined extraction [39,40]. The magnet allows the stirring of the device and the confinement of the hydrophobic magnetic nanoparticles (extractant) over the unit. This approach exploits the extraction capacity of these nanoparticles which are hard to disperse into an aqueous media due to hydrophobic attractions.

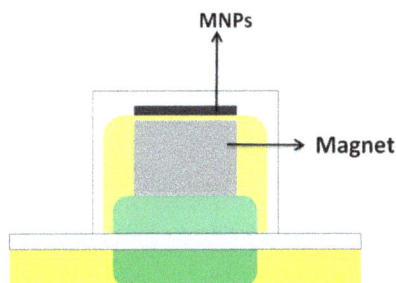

**Figure 8.** Device for microextraction of magnetically confined hydrophobic nanoparticles. MNPs, magnetic nanoparticles.

## 4. Stir Cake Sorptive Extraction

Stir cake sorptive extraction (SCSE) was first proposed in 2011 by Huang et al. [41]. It is very similar to stir bar sorptive extraction but, in this case, the sorbent is a monolithic cake placed in a homemade holder with a protected iron wire (Figure 9). This configuration avoids the direct contact of the sorptive phase with the sample vessel. As a result, higher stirring rates can be used during the extraction. In addition, the life span of the monolith is increased allowing its reuse up to 300 times, while the typical reusability of a stir bar is ca. 60.

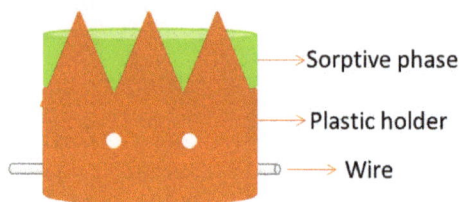

**Figure 9.** Device for stir cake sorptive extraction. Reproduced with permission of the Microextraction Tech blog.

Monoliths are continuous porous structures which are easily obtained by the polymerization of a monomer mixture with a porogen solvent into a given holder (capillary, spin column, pipette tip, stir cake). Their chemistry can be tailored by the proper selection of the monomers and also the porosity can be controlled during the synthetic process. Monoliths can be classified in three groups: organic, silica, and hybrid monoliths, depending on the nature of their ingredients. Their potential in separation techniques has been widely studied [42].

The preparation of the monolith-based SCSE unit is very simple [41]. The monolithic cake is synthesized and located in a cake-shaped plastic holder (see Figure 9). The holder is finally pierced by an iron wire that allows for magnetic stirring of the system. In addition, small holes can be drilled through the holder to promote the flow of the sample through the monolithic cake. Before its first use, the monolith must be conditioned with the appropriate solvents. Once conditioned, the sorptive cake is introduced in the sample for the isolation of the analytes, which are finally eluted for instrumental analysis.

Different monoliths have been used in this SCSE format, with polymeric ionic liquids and organic-based polymers being the most reported in the literature.

Ionic liquids (ILs) can be considered as a group of non-molecular solvents which present a melting point below 100 °C [43]. Ionic liquids have been extensively used in microextraction techniques, both in the solid and liquid phase formats. In recent years polymeric ionic liquids (PILs), obtained by the polymerization of IL cations, have been also proposed in this context. In comparison with the ILs, PILs exhibit higher viscosity, thermal stability, and mechanical strength which are important characteristics for sorptive phases in miniaturized solid-phase extraction techniques [44]. PILs were first proposed as an active phase in solid phase microextraction (SPME) in 2010 [45]. In 2012, Wang et al. reported the first application of PILs-monolith (PILM) in SCSE for the extraction of inorganic anions from water [46]. The PILM was in situ obtained by the copolymerization of 1-allyl-3-methylimidazolium chloride (AMIC) and ethylene dimethacrylate (EDMA) providing a final structure that exhibited a strong anionic exchange capacity towards the analytes ($F^-$, $Cl^-$, $Br^-$ $NO_2^-$, $NO_3^-$, $SO_4^-$, $PO_4^{3-}$). The presence of imidazolium cations in the structure was behind this capacity. Following a similar procedure, traces of antimony can be extracted from environmental water using a PILM containing 3-(1-ethyl imidazolium-3-y)propyl-methacryamido bromide (EPB) and EDMA as precursors [47]. In this specific example, the amino groups of the PILM coordinate to Sb, favoring its extraction. PILM-SCSE has also demonstrated its applicability for the pre-concentration of organic compounds.

The selection of the monomers is crucial to obtain an adequate sensitivity and selectivity. The monolith described in [46] was also efficient for the extraction of preservatives (sorbic, benzoic, and cinnamic acids) in fruit juices and soft drinks [48]. In this case, the hydrophobic and anion exchange interactions are the driving forces of the extraction procedure. Similar analytes can also be extracted using 1-ally-3-vinylimidazolium chloride (AV) polymerized in-situ with divinylbenzene (DVB) to form the PILM-SCSE [49]. Other relevant contributions in the field are the determination of benzimidazole anthelmintics in water, honey, and milk samples [50], and estrogens in environmental water [51].

PILM-SCSE can be ad-hoc synthesized depending on the target analytes. A multi interaction cake can even be fabricated if analytes of different nature are intended to be extracted. [52]. In this sense, if 1-vinylbenzyl-3-methylimidazolium chloride (VBMI) and divinylbenzene (DB) are used as precursors, the resulting PILM may interact with the target compounds via $\pi$–$\pi$ hydrophobic, hydrogen bonding, dipole-dipole, and anion exchange interactions.

Conventional monolithic phases have also been synthesized for SCSE. Polar phenols were extracted from environmental water by means of an allylthiourea and DB polymer using DMF as a porogen solvent and AIBN as an initiator [53]. B-agonists have been extracted in milk and swine urine samples by SCSE using a poly(4-vinylbenzoic acid-divinylbenzene) monolith [54], while a novel boron-rich monolith has been synthesized and characterized for the determination of fluoroquinolones in environmental water and milk samples [55].

## 5. Stir Disk Extractions

Disks offer a higher superficial area than bars and, in some cases, they present a higher porosity that enhances the diffusion of the analytes through the sorptive phase. Rotating disk sorptive extraction (RDSE) was the first approach in this context [56]. The RDSE unit (Figure 10A) consists of a thin film of PDMS deposited over a PTFE disk containing an integrated magnetic bar. Similarly to SCSE, this configuration protects the sorptive phase from direct contact with the vial, allowing for the application of faster stirring rates than conventional SBSE. RDSE operates in a similar fashion than SBSE since the unit is stirred into the sample for a defined period of time for the isolation of the target compounds. After the extraction the unit is recovered for the final instrumental analysis. When chromatographic techniques are employed, the analytes must be eluted with a proper solvent [57,58]. However, the configuration of the unit allows for the detachment of the film after the extraction for its direct spectroscopic analysis, avoiding the dilution inherent to the elution step [59,60]. RDSE has evolved in the last few years following two different trends, namely: the development of new extraction phases and the potential automation of the technique [61].

PDMS has a clear potential as a sorptive phase but presents some disadvantages that have been previously described. The development of new phases in RSDE will increase the versatility of the technique, opening the door to the extraction of analytes of intermediate or high polarity. The first approach is this context involved the use of conventional C18 solid phase extraction (SPE) disks for the extraction of hexachlorobenzene from water [62]. These SPE disks, which are membranes with embedded particles, are linked to the device using silicone as a binder. The results were promising but these disks have non-polar compounds as targets. Some polymeric SPE sorbents have been specially designed for the extraction of polar compounds. Richter et al. proposed an evolution of the classic RDSE in order to exploit this characteristic. In this case, a cavity is dug into the unit and finally loaded with polymeric sorbent (HLB) particles (Figure 10B) [63]. The cavity is covered with a filter to confine the particles, avoiding their losses during the extraction. The great variety of commercially available SPE sorbents and their easy loading on the unit makes this approach highly versatile. In addition, lab-synthesized sorbents can be also applied. In this sense, researchers have proposed the use of MIPs [64] and ionic liquids intercalated in montmorillonite [65] as a sorptive phase in this device.

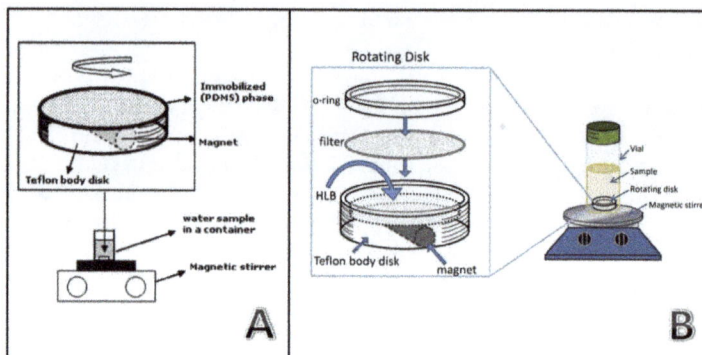

**Figure 10.** Device for rotating disk sorptive extraction using thin films (Reproduced with permission from [56], copyright Elsevier, 2009) (**A**) and sorbent particles (Reproduced with permission from [63], copyright Springer, 2014) (**B**).

Polyethylene disks, conventionally used as frits in solid phase extraction cartridges, can be also used as stirred units by simply piercing them with a metallic wire. This material is hydrophobic and it is indicated for the extraction of hydrophobic compounds [66]. The material exhibits sufficient thermal stability to allow for its mild thermal desorption in a conventional headspace vial.

Borosilicate disks can also be excellent supports for RDSE due to their mechanical strength. However, as polar materials, they are not able to extract organic compounds from aqueous samples. Borosilicate disks have been modified with carbon nanohorns on the surface for the extraction of benzophenone-3 from swimming pool water [67]. These modified disks can be easily adapted to a portable drill (Figure 11) that stirs them into the sample for the on-site extraction of the target compound. The enrichment factor and extraction recovery, which are 1379 and 68.9%, respectively, reveal the great potential of the technique.

**Figure 11.** Borosilicate disk modified with carbon nanohorns extraction device. Reproduced with permission of from [67], copyright Elsevier, 2014.

## 6. Conclusions

The approaches described in this review article clearly demonstrate the advantages of extraction/stirring integrated techniques. Since its proposal in 1999, new microextraction formats and new sorptive materials have been developed. No doubt, the combination of both aspects results in an analytical measurement processes with enhanced basic (sensitivity, selectivity) and productivity-related (rapidity, environmental friendship) analytical properties. Monolithic solids, polymers, nanoparticles, and ionic liquids can be cited among the most novel and efficient sorptive phases. SCSE and RDSE are, on the other hand, the most competitive approaches in solid phase approaches as they minimizes the friction of the coating with the extraction vessel, allowing for the use of higher agitation speeds. SME has proven to be a versatile configuration as it is compatible with solid and liquid phase microextraction approaches. Moreover, it permits the processing of low sample volumes by a simple re-design of the unit.

Future trends will be focused on the application of new materials and new extraction strategies. In this context, ICECLES is especially useful.

Figure 12 shows a brief summary of the techniques presented in this review article. The discussion has been focused on the main advantages provided by the novel techniques compared with classical SBSE. Despite these advantages, it is necessary to point out that SBSE is a consolidated technique due to its commercial availability and robustness.

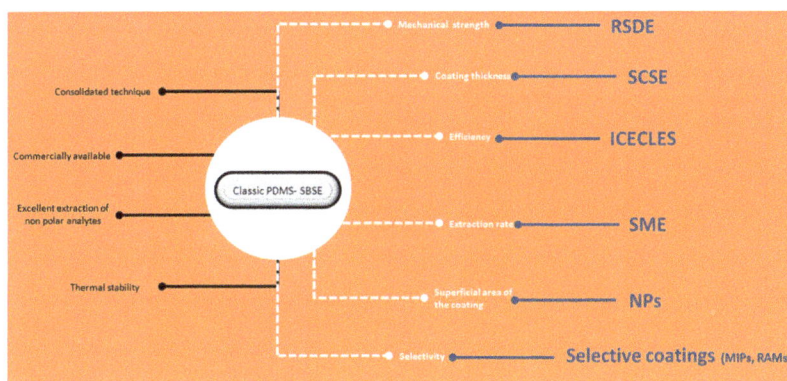

**Figure 12.** Brief summary of the techniques described in the article. The advantages of classic polydimethylsiloxane (PDMS) based stir bar sorptive extraction (SBSE) are shown in black letters, the characteristics that may be compromised in several applications are indicated in white characters, and the techniques/material that can overcome these limitations are presented in blue. For further details, read the text.

**Acknowledgments:** Financial support from the Spanish Ministry of Economy and Competitiveness (CTQ2014-52939R) is gratefully acknowledged.

**Author Contributions:** Both authors contributed equally to the article.

**Conflicts of Interest:** The authors declare no conflict of interest.

## References

1. Lucena, R. Extraction and stirring integrated techniques: Examples and recent advances. *Anal. Bioanal. Chem.* **2012**, *403*, 2213–2223. [CrossRef] [PubMed]
2. Cruz-Vera, M.; Lucena, R.; Cárdenas, S.; Valcárcel, M. Sample treatments based on dispersive (micro)extraction. *Anal. Methods* **2011**, *3*, 1719–1728. [CrossRef]

3. Baltussen, E.; Sandra, P.; David, F.; Cramers, C. Stir bar sorptive extraction (SBSE), a novel extraction technique for aqueous samples: Theory and principles. *J. Microcol. Sep.* **1999**, *11*, 737–747. [CrossRef]

4. Wells, M.J.M. *Sample Preparation Techniques in Analytical Chemistry*; Mitra, S., Ed.; John Wiley & Sons: Hobojen, NY, USA, 2003.

5. Nogueira, J.M.F. Stir bar sorptive extraction and related techniques. In *Analytical Microextraction Techniques*; Valcárcel, M., Cárdenas, S., Lucena, R., Eds.; Bentham Science: Sharjah, UAE, 2017.

6. Lucena, R. Making biosamples compatible with instrumental analysis. *J. Appl. Bioanal.* **2015**, *1*, 72–75. [CrossRef]

7. Zhu, X.; Cai, J.; Yang, J.; Su, Q.; Gao, Y. Films coated with molecular imprinted polymers for the selective stir bar sorption extraction of monocrotophos. *J. Chromatogr. A* **2006**, *1131*, 37–44. [CrossRef] [PubMed]

8. Xu, Z.; Hu, Y.; Hu, Y.; Li, G. Investigation of ractopamine molecularly imprinted stir bar sorptive extraction and its application for trace analysis of β2-agonists in complex samples. *J. Chromatogr. A* **2010**, *1217*, 3612–3618. [CrossRef] [PubMed]

9. Zhong, Q.; Hu, Y.; Hu, Y.; Li, G. Online desorption of molecularly imprinted stir bar sorptive extraction coupled to high performance liquid chromatography for the trace analysis of triazines in rice. *J. Sep. Sci.* **2012**, *35*, 3396–3402. [CrossRef] [PubMed]

10. Turiel, E.; Martín-Esteban, A. Molecularly imprinted stir bars for selective extraction of thiabendazole in citrus samples. *J. Sep. Sci.* **2012**, *35*, 2962–2969. [CrossRef] [PubMed]

11. Zhu, L.; Xu, G.; Wei, F.; Yang, J.; Hu, Q. Determination of melamine in powdered milk by molecularly imprinted stir bar sorptive extraction coupled with HPLC. *J. Colloid Interface Sci.* **2015**, *454*, 8–13. [CrossRef] [PubMed]

12. Xu, Z.; Yang, Z.; Liu, Z. Development of dual-templates molecularly imprinted stir bar sorptive extraction and its application for the analysis of environmental estrogens in water and plastic samples. *J. Chromatogr. A* **2014**, *1358*, 52–59. [CrossRef] [PubMed]

13. Sheng, N.; Wei, F.; Zhan, W.; Cai, Z.; Du, S.; Zhou, X.; Li, F.; Hu, Q. Dummy molecularly imprinted polymers as the coating of stir bar for sorptive extraction of bisphenol A in tap water. *J. Sep. Sci.* **2012**, *35*, 707–712. [CrossRef] [PubMed]

14. Lei, Y.; He, M.; Chen, B.; Hu, B. Polyaniline/cyclodextrin composite coated stir bar sorptive extraction combined with high performance liquid chromatography-ultraviolet detection for the analysis of trace polychlorinated biphenyls in environmental waters. *Talanta* **2016**, *150*, 310–318. [CrossRef] [PubMed]

15. Lin, S.; Gan, N.; Zhang, J.; Qiao, L.; Chen, Y.; Cao, Y. Aptamer-Functionalized stir bar sorptive extraction coupled with gas chromatography-mass spectrometry for selective enrichment and determination of polychlorinated biphenyls in fish samples. *Talanta* **2016**, *149*, 266–274. [CrossRef] [PubMed]

16. Auffan, M.; Rose, J.; Bottero, J.; Lowry, G.V.; Jolivet, J.; Wiesner, M.R. Towards a definition of inorganic nanoparticles from an environmental, health and safety perspective. *Nat. Nanotechnol.* **2009**, *4*, 634–641. [CrossRef] [PubMed]

17. Garcia-Valverde, M.T.; Lucena, R.; Cardenas, S.; Valcarcel, M. Titanium-Dioxide nanotubes as sorbents in (micro)extraction techniques. *Trends Anal. Chem.* **2014**, *62*, 37–45. [CrossRef]

18. Lasarte Aragonés, G.; Lucena, R.; Cardenas, S.; Valcarcel, M. Nanoparticle-based microextraction techniques in bioanalysis. *Bioanalysis* **2011**, *3*, 2533–2548. [CrossRef] [PubMed]

19. Li, P.; Hu, B.; Li, X. Zirconia coated stir bar sorptive extraction combined with large volume sample stacking capillary electrophoresis-indirect ultraviolet detection for the determination of chemical warfare agent degradation products in water samples. *J. Chromatogr. A* **2012**, *1247*, 49–56. [CrossRef] [PubMed]

20. Benedé, J.L.; Chisvert, A.; Giokas, D.L.; Salvador, A. Development of stir bar sorptive-dispersive microextraction mediated by magnetic nanoparticles and its analytical application to the determination of hydrophobic organic compounds in aqueous media. *J. Chromatogr. A* **2014**, *1362*, 25–35. [CrossRef] [PubMed]

21. Benedé, J.L.; Chisvert, A.; Giokas, D.L.; Salvador, A. Stir bar sorptive-dispersive microextraction mediated by magnetic nanoparticles–nylon 6 composite for the extraction of hydrophilic organic compounds in aqueous media. *Anal. Chim. Acta* **2016**, *926*, 63–71. [CrossRef] [PubMed]

22. Hu, C.; Chen, B.; He, M.; Hu, B. Amino modified multi-walled carbon nanotubes/polydimethylsiloxane coated stir bar sorptive extraction coupled to high performance liquid chromatography-ultraviolet detection for the determination of phenols in environmental samples. *J. Chromatogr. A* **2013**, *1300*, 165–172. [CrossRef] [PubMed]

23. Farhadi, K.; Firuzi, M.; Hatami, M. Stir bar sorptive extraction of propranolol from plasma samples using a steel pin coated with a polyaniline and multiwall carbon nanotube composite. *Microchim. Acta* **2015**, *182*, 323–330. [CrossRef]

24. Ekbatani Amlashi, N.; Hadjmohammadi, M.R. Sol-Gel coating of poly(ethylene glycol)-grafted multiwalled carbon nanotubes for stir bar sorptive extraction and its application to the analysis of polycyclic aromatic hydrocarbons in water. *J. Sep. Sci.* **2016**, *39*, 3445–3456. [CrossRef] [PubMed]

25. Fan, W.; He, M.; Wu, X.; Chen, B.; Hu, B. Graphene oxide/polyethyleneglycol composite coated stir bar for sorptive extraction of fluoroquinolones from chicken muscle and liver. *J. Chromatogr. A* **2015**, *1418*, 36–44. [CrossRef] [PubMed]

26. Zhang, W.; Zhang, Z.; Zhang, J.; Meng, J.; Bao, T.; Chen, Z. Covalent immobilization of graphene onto stainless steel wire for jacket-free stir bar sorptive extraction. *J. Chromatogr. A* **2014**, *1351*, 12–20. [CrossRef] [PubMed]

27. Fan, W.; He, M.; You, L.; Zhu, X.; Chen, B.; Hu, B. Water-Compatible graphene oxide/molecularly imprinted polymer coated stir bar sorptive extraction of propranolol from urine samples followed by high performance liquid chromatography-ultraviolet detection. *J. Chromatogr. A* **2016**, *1443*, 1–9. [CrossRef] [PubMed]

28. Maslamani, N.; Manandhar, E.; Geremia, D.K.; Logue, B.A. ICE concentration linked with extractive stirrer (ICECLES). *Anal. Chim. Acta* **2016**, *941*, 41–48. [CrossRef] [PubMed]

29. Alcudia-León, M.C.; Lucena, R.; Cárdenas, S.; Valcárcel, M. Stir membrane extraction: A useful approach for liquid sample pretreatment. *Anal. Chem.* **2009**, *81*, 8957–8961. [CrossRef] [PubMed]

30. Alcudia-León, M.C.; Lendl, B.; Lucena, R.; Cárdenas, S. Valcárcel, Sensitive in-surface infrared monitoring coupled to stir membrane extraction for the selective determination of total hydrocarbon index in waters. *Anal. Bioanal. Chem.* **2010**, *398*, 1427–1433. [CrossRef] [PubMed]

31. Kabir, A.; Furton, K.G. Fabric Phase Sorptive Extractors (FPSE). US Patent Application No. 14.216,121, 2014.

32. Roldán-Pijuán, M.; Lucena, R.; Cárdenas, S.; Valcárcel, M.; Kabir, A.; Furton, K.G. Stir fabric phase sorptive extraction for the determination of triazine herbicides in environmental waters by liquid chromatography. *J. Chromatogr. A* **2015**, *1376*, 35–45. [CrossRef] [PubMed]

33. Alcudia-León, M.C.; Lucena, R.; Cárdenas, S.; Valcárcel, M. Stir membrane liquid-liquid microextraction. *J. Chromatogr. A* **2011**, *1218*, 869–874. [CrossRef] [PubMed]

34. Alcudia-León, M.C.; Lucena, R.; Cárdenas, S.; Valcárcel, M. Determination of phenols in waters by stir membrane liquid–liquid–liquid microextraction coupled to liquid chromatography with ultraviolet detection. *J. Chromatogr. A* **2011**, *1218*, 2176–2181. [CrossRef] [PubMed]

35. Riaño, S.; Alcudia-León, M.C.; Lucena, R.; Cárdenas, S.; Valcárcel, M. Determination of non-steroidal anti-inflammatory drugs in urine by the combination of stir membrane liquid–liquid–liquid microextraction and liquid chromatography. *Anal. Bioanal. Chem.* **2012**, *403*, 2583–2589. [CrossRef] [PubMed]

36. Roldán-Pijuán, M.; Alcudia-León, M.C.; Lucena, R.; Cárdenas, S. Valcárcel, Stir-membrane liquid microextraction for the determination of paracetamol in human saliva samples. *Bioanalysis* **2013**, *5*, 307–315. [CrossRef] [PubMed]

37. Rodríguez-Gómez, R.; Roldán-Pijuán, M.; Lucena, R.; Cárdenas, S.; Zafra-Gómez, A.; Ballesteros, O.; Navalón, A.; Valcárcel, M. Stir-Membrane solid–liquid–liquid microextraction for the determination of parabens in human breast milk samples by ultrahigh performance liquid chromatography-tandem mass spectrometry. *J. Chromatogr. A* **2014**, *1354*, 26–33. [CrossRef] [PubMed]

38. Roldán-Pijuán, M.; Lucena, R.; Alcudia-León, M.C.; Cárdenas, S.; Valcárcel, M. Stir octadecyl-modified borosilicate disk for the liquid phase microextraction of triazine herbicides from environmental waters. *J. Chromatogr. A* **2013**, *1307*, 58–65. [CrossRef] [PubMed]

39. Alcudia-León, M.C.; Lucena, R.; Cárdenas, S. Valcárcel, M. Magnetically confined hydrophobic nanoparticles for the microextraction of endocrine-disrupting phenols from environmental waters. *Anal. Bioanal. Chem.* **2013**, *405*, 2729–2734. [CrossRef] [PubMed]

40. Alcudia-León, M.C.; Lucena, R.; Cárdenas, S.; Valcárcel, M. Determination of parabens in waters by magnetically confined hydrophobic nanoparticle microextraction coupled to gas chromatography/mass spectrometry. *Microchem. J.* **2013**, *110*, 643–648. [CrossRef]

41. Huang, X.; Chen, L.; Lin, F.; Yuan, D. Novel extraction approach for liquid samples: Stir cake sorptive extraction using monolith. *J. Sep. Sci.* **2011**, *34*, 2145–2151. [CrossRef] [PubMed]

42. Dario Arruda, R.; Causon, T.J.; Hilder, E.F. Recent developments and future possibilities for polymer monoliths separation science. *Analyst* **2012**, *137*, 5179–5189.

43. Welton, T. Room-Temperature ionic liquids. Solvents for synthesis and catalysis. *Chem. Rev.* **1999**, *99*, 2071–2084. [CrossRef] [PubMed]

44. Trujillo-Rodríguez, M.J.; Pino, V.; Ayala, J.H.; Afonso, A.M. Ionic liquids in the microextraction context. In *Analytical Microextraction Techniques*; Valcárcel, M., Cárdenas, S., Lucena, R., Eds.; Bentham Science: Sharjah, UAE, 2017; pp. 70–134.

45. López-Darias, J.; Pino, V.; Anderson, J.L.; Graham, C.M.; Afonso, A.M. Determination of water pollutants by direct-immersion solid-phase microextraction using polymeric ionic liquid coatings. *J. Chromatogr. A* **2010**, *1217*, 1236–1243. [CrossRef] [PubMed]

46. Huang, X.; Chen, L.; Yuan, X.; Si, S. Preparation of a new polymeric ionic liquid-based monolith for stir cake sorptive extraction and its application in the extraction of inorganic anions. *J. Chromatogr. A* **2012**, *1248*, 67–73. [CrossRef] [PubMed]

47. Zhang, Y.; Mei, M.; Ouyang, T.; Huang, X. Preparation of a new polymeric ionic-liquid-based sorbent for stir cake sorptive extraction of trace antimony in environmental water samples. *Talanta* **2016**, *161*, 377–383. [CrossRef] [PubMed]

48. Lin, F.; Nong, S.; Huang, X.; Yuan, D. Sensitive determination of organic acid preservatives in juices and soft drinks treated by monolith-based stir cake sorptive extraction and liquid chromatography analysis. *Anal. Bioanal. Chem.* **2013**, *405*, 2077–2081. [CrossRef] [PubMed]

49. Chen, L.; Huang, X. Preparation of a polymeric ionic liquid-based adsorbent for stir cake sorptive extraction of preservatives in orange juices and tea drinks. *Anal. Chim. Acta* **2016**, *916*, 33–41. [CrossRef] [PubMed]

50. Wang, Y.; Zhang, J.; Huang, X.; Yuan, D. Preparation of stir cake sorptive extraction based on polymeric ionic liquid for the enrichment of benzimidazole anthelmintics in water, honey and milk samples. *Anal. Chim. Acta* **2014**, *840*, 33–41. [CrossRef] [PubMed]

51. Chen, L.; Mei, M.; Huang, X.; Yuan, D. Sensitive determination of estrogens in environmental waters treated with polymeric ionic liquid-based stir cake sorptive extraction and liquid chromatographic analysis. *Talanta* **2016**, *152*, 98–104. [CrossRef] [PubMed]

52. Huang, X.; Wang, Y.; Hong, Q.; Liu, Y.; Yuan, D. Preparation a new sorbent based on polymeric ionic liquid for stir cake sorptive extraction of organic compounds and inorganic anions. *J. Chromatogr. A* **2013**, *1314*, 7–14. [CrossRef] [PubMed]

53. Huang, X.; Wang, Y.; Yuan, D.; Li, X.; Nong, S. New monolithic stir-cake-sorptive extraction for the determination of polar phenols by HPLC. *Anal. Bioanal. Chem.* **2013**, *405*, 2185–2193. [CrossRef] [PubMed]

54. Huang, X.; Chen, L.; Yuan, D. Preparation of stir cake sorptive extraction based on poly(4-vinylbenzoic acid-divinylbenzene) monolith and its application in sensitive determination of β-agonists in milk and swine urine samples. *J. Hazard. Mater.* **2013**, *262*, 121–129. [CrossRef] [PubMed]

55. Mei, M.; Huang, X. Determination of fluoroquinolones in environmental water and milk samples treated with stir cake sorptive extraction based on a boron-rich monolith. *J. Sep. Sci.* **2016**, *39*, 1908–1918. [CrossRef] [PubMed]

56. Richter, P.; Leiva, C.; Choque, C.; Giordano, A.; Sepúlveda, B. Rotating-Disk sorptive extraction of nonylphenol from water samples. *J. Chromatogr. A* **2009**, *1216*, 8598–8602. [CrossRef] [PubMed]

57. Giordano, A.; Richter, P.; Ahumada, I. Determination of pesticides in river water using rotating disk sorptive extraction and gas chromatography-mass spectrometry. *Talanta* **2011**, *85*, 2425–2429. [CrossRef] [PubMed]

58. Jachero, L.; Sepúlveda, B.; Ahumada, I.; Fuentes, E.; Richter, P. Rotating disk sorptive extraction of triclosan and methyl-triclosan from water samples. *Anal. Bioanal. Chem.* **2013**, *405*, 7711–7716. [CrossRef] [PubMed]

59. Manzo, V.; Navarro, O.; Honda, L.; Sánchez, K.; Toral, M.I.; Richter, P. Determination of crystal violet in water by direct solid phase spectrophotometry after rotating disk sorptive extraction. *Talanta* **2013**, *106*, 305–308. [CrossRef] [PubMed]

60. Muñoz, C.; Toral, M.I.; Ahumada, I.; Richter, P. Rotating disk sorptive extraction of Cu-bisdiethyldithiocarbamate complex from water and its application to solid phase spectrophotometric quantification. *Anal Sci.* **2014**, *30*, 613–617. [PubMed]

61. Manzo, V.; Miró, M.; Richter, P. Programmable flow-based dynamic sorptive microextraction exploiting an octadecyl chemically modified rotating disk extraction system for the determination of acidic drugs in urine. *J. Chromatogr. A* **2014**, *1368*, 64–69. [CrossRef] [PubMed]

62. Cañas, A.; Richter, P. Solid-Phase microextraction using octadecyl-bonded silica immobilized on the surface of a rotating disk: Determination of hexachlorobenzene in water. *Anal. Chim. Acta* **2012**, *743*, 75–79. [CrossRef] [PubMed]

63. Cañas, A.; Valdebenito, S.; Richter, P. A new rotating-disk sorptive extraction mode, with a copolymer of divinylbenzene and N-vinylpyrrolidone trapped in the cavity of the disk, used for determination of florfenicol residues in porcine plasma. *Anal. Bioanal. Chem.* **2014**, *406*, 2205–2210. [CrossRef] [PubMed]

64. Manzo, V.; Ulisse, K.; Rodríguez, I.; Pereira, E.; Richter, P. A molecularly imprinted polymer as the sorptive phase immobilized in a rotating disk extraction device for the determination of diclofenac and mefenamic acid in wastewater. *Anal. Chim. Acta* **2015**, *889*, 130–137. [CrossRef] [PubMed]

65. Fiscal-Ladino, J.A.; Obando-Ceballos, M.; Rosero-Moreano, M.; Montaño, D.G.; Cardona, W.; Giraldo, L.F.; Richter, P. Ionic liquids intercalated in montmorillonite as the sorptive phase for the extraction of low-polarity organic compounds from water by rotating-disk sorptive extraction. *Anal. Chim. Acta* **2017**, *953*, 23–31. [CrossRef] [PubMed]

66. Roldán-Pijuán, M.; Alcudia-León, M.C.; Lucena, R.; Cárdenas, S.; Valcárcel, M. Stir frit microextraction: An approach for the determination of volatile compounds in water by headspace-gas chromatography/mass spectrometry. *J. Chromatogr. A* **2012**, *1251*, 10–15. [CrossRef] [PubMed]

67. Roldán-Pijuán, M.; Lucena, R.; Cárdenas, S.; Valcárcel, M. Micro-Solid phase extraction based on oxidized single-walled carbon nanohorns immobilized on a stir borosilicate disk: Application to the preconcentration of the endocrine disruptor benzophenone-3. *Microchem. J.* **2014**, *115*, 87–94. [CrossRef]

*separations*

MDPI

*Review*

# Carbon-Based Nanomaterials Functionalized with Ionic Liquids for Microextraction in Sample Preparation

**Theodoros Chatzimitakos and Constantine Stalikas ***

Laboratory of Analytical Chemistry, Department of Chemistry, University of Ioannina, Ioannina 45110, Greece; chatzimitakos@outlook.com
* Correspondence: cstalika@cc.uoi.gr; Tel.: +30-26510-08414; Fax: +30-26510-08796

Academic Editors: Jared L. Anderson and Victoria F. Samanidou
Received: 9 February 2017; Accepted: 4 April 2017; Published: 10 April 2017

**Abstract:** A large number of carbon-based nanomaterials has been investigated as sorbents in sample preparation, including fullerenes, carbon nanotubes, nanofibers, nanohorns and graphene, as well as their functionalized forms. Taking into account their properties, carbon-based nanomaterials have found a wide range of applications in different sample preparation techniques. Ionic liquids, as an alternative to environmentally-harmful ordinary organic solvents, have attracted extensive attention and gained popularity in analytical chemistry covering different fields like chromatography, electrochemistry and (micro)extraction. Some of the properties of ionic liquids, including polarity, hydrophobicity and viscosity, can be tuned by the proper selection of the building cations and anions. Their tunable nature allows the synthesis of tailor-made solvents for different applications. This review provides a snapshot of the most important features and applications of different carbon-based nanomaterials functionalized with ionic liquids for sample preparation. Emphasis is placed on the description of the different works that have provided interesting results for the use of graphene and carbon nanotubes, in this analytical field.

**Keywords:** graphene; carbon nanotubes; ionic liquid; magnetic; microextraction; sample preparation

---

## 1. Introduction

The selection and optimization of the appropriate sample preparation procedure qualifies as essential for the development of a successful method. The modern trend for 'greener' chemical analyses has led to the development of microscale extraction approaches towards minimizing the organic solvent consumption, while maximizing sample throughput and extraction efficiency of the analytes. Microextraction is an extraction technique where the volume of the extracting phase is very small in relation to the volume of the sample [1]. It is not necessarily an exhaustive extraction procedure, in contrast to the classical liquid-liquid extraction and solid-phase extraction (SPE), signifying that possibly only a fraction of the analyte may finally be subjected to analysis. Hence, it is not surprising that microextraction, in the modes of solid-phase microextraction (SPME), dispersive solid-phase microextraction (DSPE), magnetic solid-phase extraction (MSPE), pipette-tip solid-phase extraction (PT-SPE), etc., has come to the forefront of analytical chemistry in the past few years.

Carbon-based nanomaterials have been extensively used in analytical applications [2]. A large number of them have been investigated as sorbents in sample preparation, including fullerenes, carbon nanotubes, nanofibers, nanohorns and graphene, as well as their chemically-modified analogues. The characteristic structures of carbon-based nanomaterials allow them to interact with molecules via non-covalent forces, such as hydrogen bonding, $\pi-\pi$ stacking, electrostatic forces, van der Waals forces and hydrophobic interactions. Taking into account the aforementioned possibilities, carbon-based nanomaterials have found a wide range of applications in different sample preparation techniques. Although the reasons for the selection of a particular allotrope over another are still being discussed, a wide variety of carbon-based materials is available and applicable to analytical procedures.

Since the first report in 1991 [3], carbon nanotubes (CNTs) have shown great possibilities for a wide variety of processes and applications. The combination of structures, dimensions and topologies has provided physical and chemical properties that are unparalleled by most known materials. Their applications have also reached the analytical chemistry field, in which CNTs are being used as matrices in matrix-assisted laser desorption ionization, as stationary phases in separation techniques (gas chromatography, high performance liquid chromatography and capillary electrochromatography and capillary electrophoresis), as well as new SPE materials [4,5]. As regards this last application, the number of works has considerably increased in the last ten years.

Graphene, discovered in 2004, is another kind of novel and particularly fascinating carbon material, which has sparked a tremendous amount of research from both the experimental and theoretical scientific communities in recent years [6]. Because of its extraordinary electrical properties and very high specific surface area, graphene (G) and graphene oxide (GO) have been widely applied as electrode materials, adsorbents in solid-phase (micro)extraction and magnetic solid-phase extraction.

Ionic liquids (ILs) have attracted extensive attention and gained popularity in analytical chemistry, as an alternative to environmentally-harmful ordinary organic solvents, covering different application fields like chromatography, electrochemistry and (micro)extraction [7]. The potential of ILs in chemistry is related to their unique properties as non-molecular solvents, negligible vapor pressure associated with high thermal stability, tunable viscosity, miscibility with water and organic solvents and good extractability for various organic compounds and metal ions. Their polarity, hydrophobicity, viscosity and other chemical and physical properties can be tailored through the choice of the cationic and anionic constituents [8]. Functional materials, which consist of ILs and magnetic composites, presented excellent properties for extraction [9]. However, at elevated temperatures, the viscosity of the IL is reduced, resulting in a flowing state in which the IL can be lost. To overcome it, polymeric ILs (PILs) were synthesized and studied extensively. The PILs have higher thermal stability compared to monomeric ILs and, therefore, more resistance to flow. Additionally, PILs are tunable through functionalization of the IL monomers, therefore altering their physicochemical properties along with their extractive capabilities [10].

Combining the two above-mentioned major fields, i.e., carbon-based nanomaterials and ILs, it is possible to design and develop new extracting phases with outstanding properties for microextraction purposes. The carbon-based nanomaterials modified with ILs are expected to possess the advantages of both components of the functionalized material, resulting in new, advanced adsorbents with tunable microextraction capabilities. This review provides an updated and critical snapshot of the applications of the carbon-based nanomaterials in microextraction, as an integral step of sample preparation. Attention is paid to the discussion of cases dealing with IL-functionalized graphene and multi-walled carbon nanotubes, which are examined according to the microextraction mode they support (Figure 1) and illustrate novel concepts and promising applications.

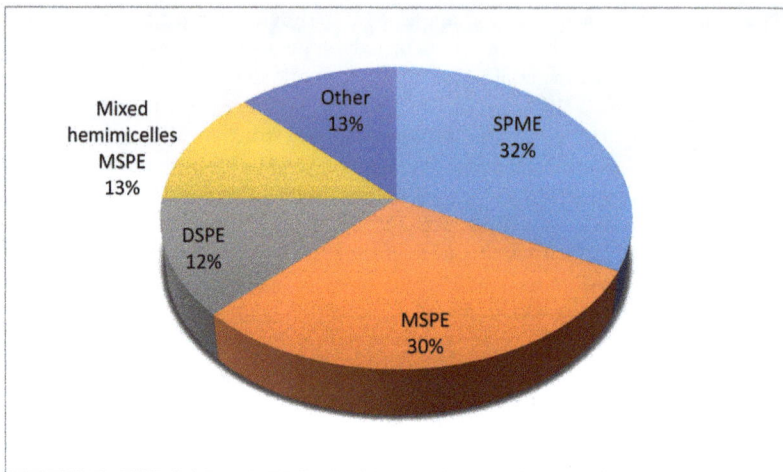

**Figure 1.** Main microextraction techniques based on carbon nanomaterials functionalized with ionic liquids. Pie chart showing the percentage of published articles referring to them during the period of 2013–2016.

## 2. Ionic Liquid-Coated Graphene-Based Nanomaterials

### 2.1. Solid-Phase Microextraction

In SPME, the fiber coating, which is the heart of the technique to extract the compounds from samples, consists of a liquid (polymer), a solid (sorbent) or a combination of both. The type of coating plays a crucial role in the extraction and desorption process, since its efficiency depends on the distribution constant of the analytes between the sample and the stationary phase. In addition, it provides some selectivity to the SPME process towards the analytes versus other matrix compounds. Fibers containing the IL-G/CNTs composites are reported to be stable, under headspace and direct injection mode. So far, various procedures have been proposed to prepare stainless steel fibers for SPME coated with GO, functionalized either with monomeric or polymeric ILs. Hou et al. coated a stainless steel fiber with GO and then tried four different coatings based on ILs: two monomeric coatings with 1-aminoethyl-3-methylimidazolium bromide or bis(trifluoromethanesulfonyl)imide ([NTf$_2$]) and two coatings based on poly(1-vinyl-3-hexylimidazolium) bromide or [NTf$_2$] [10]. Then, they evaluated their capabilities for the headspace SPME ((HS)SPME) of polycyclic aromatic hydrocarbons (PAHs) and phthalate esters. It was proven that an in situ anion exchange process to alter the anion of the IL from bromide to [NTf$_2$] has merits in terms of the simplicity of synthesis and extraction capability towards the target analytes. Obviously, altering the anion of the ILs (bromide) to a more hydrophobic analogue ([NTf$_2$]) directly affects the extraction capability of the fiber towards the selected analytes, enhances the overall extraction efficiency and lowers the detection limits. The superiority of the PIL-coated GO-stainless steel fibers over the monomeric analogues was underlined, as regards the SPME extraction of PAHs and phthalates.

To improve the extraction efficiency of the SPME procedure for PAHs, 1-methyl-3-[3-(trimethoxysilyl) propyl] imidazolium chloride was employed as a cross-linking agent, between a silanol functionalized silver-coated stainless steel fiber and GO. In this way, layers of GO grew on the fiber [11]. The IL greatly enhanced the bonding strength between the fiber and the GO layers, resulting in a stable coating. The coating on the fiber was achieved via a layer-by-layer method, by repeating a cycle of immersing the silanol functionalized fiber in an IL solution, drying and then immersing it in a GO solution, five times. The layer-by-layer method provides convenience to control the number of GO layers and, hence, the extraction efficiency of the SPME procedure. The developed SPME fiber was used to extract effectively certain PAHs from rain and river water, in a direct-immersion SPME ((DI)SPME) procedure.

Another interesting case is the immobilization of a metal-organic framework-IL functionalized graphene on an etched fiber [12]. The metal-organic frameworks have a high surface area, porosity and enable the in-pore and outer-surface functionalization, which in turn, allows their use as sorbents in sample preparation. The low mechanical and chemical stability of the metal-organic frameworks is improved by introducing mechanically-durable materials, such as graphene. In the present case, the morphology of the resulting fiber showed that the presence of the IL-G did not interfere with the growth of the metal-organic framework, in terms of structure and crystal shape. On the contrary, the presence of the IL-G enhanced the uniformity of the microcrystals' structure, along with their mechanical stability. The combination of a metal-organic framework and an IL-G resulted in a fiber that combined the advantageous properties of all three components. To take advantage of the aforementioned benefits, an analytical method for the (HS)SPME extraction and detection of antibiotics in food and biological matrixes was developed. Because of the presence of IL, the results were improved a great deal, as the enrichment factors (EFs) revealed after conducting relevant experiments. In addition, the presence of G enhanced the EFs by 4–7-fold compared to a fiber coated only with the metal-organic framework.

An alternative approach of functionalized SPME fiber was employed by Sun et al. [13], who prepared a PIL monolith doped with GO attached to a fiber, prior to the (DI)SPME, instead of directly coating the stainless steel fiber. This novel type of coating was used to extract phenols from environmental aqueous samples with success. Finally, Wu et al. developed a poly(3,4-ethylenedioxythiophene) (PEDOT) coating doped with a GO-IL composite (1-hydroxyethyl-3- methyl imidazolium [NTf$_2$] was the IL used), for SPME purposes [14]. The GO-IL-PEDOT coating was electrodeposited on the surface of the fiber using cyclic voltammetry, and the resulting coating, which had many pores and a wrinkled structure, was improved compared to similar materials, in terms of surface area. The high porosity of the coating, alongside with the rich π-electron stacking of the coating, favored the applicability of the new fiber to SPME procedures. To prove this, the authors developed an analytical method aiming to determine benzene derivatives in industrial wastewater and environmental water samples, after their (HS)SPME. The fiber was superior to commercially available fibers (2–4-times better for the extraction of benzene derivatives compared to a polydimethylsiloxane fiber) and had a long lifetime, greatly reducing in this way the cost of the proposed method.

Summarizing, only a few analytical methods have been developed that utilize IL-G composite nanomaterials in SPME, either in headspace or direct-immersion mode (a complete list can be seen in Table 1). The developed methods have been applied to the determination of PAHs, phthalate esters, phenols, benzene derivatives and antibiotics in water (mainly), food and biological samples. The facile approaches have the merits of low limits of detection (LODs) (lower than others from previously reported methods), short extraction time and a wide linear range. Additionally, the analytical performances of the new methods are better than those of commercially available fibers, in terms of LODs and EFs [11,14], while the functionalized fibers can be reused with an insignificant decrease in the extraction efficiency [12]. From the studies conducted, it was apparent that the presence of the IL-G composites on the surfaces of the fibers is advantageous and facilitates the improvement of the SPME efficiency, since the resulting fibers combine the advantages of both ILs and G. The comparison of the analytical performances of the experiments, conducted in the absence of the IL or the G, proved that the results are less satisfactory than those in the presence of IL-G, underlining the role of IL.

**Table 1.** Microextraction techniques employing IL-G/IL-GO composite nanomaterials.

| Ionic Liquid | Microextraction Technique | Matrix | Target Analytes | LOD (μg/L) | Recoveries (%) | Instrumental Analytical System | Reference |
|---|---|---|---|---|---|---|---|
| poly(1-vinyl-3-hexylimidazolium-[NTf$_2$]) | (HS)SPME | food-wrap, potato | PAHs and phthalate esters | 0.015–0.025 | 78.3–101.7 | GC-FID | [10] |
| 1-methyl-3-[3-(trimethoxysilyl)propyl] imidazolium chloride | (D)SPME | rain and river water | PAHs | 0.05–0.10 | 92.3–120 | GC-FID | [11] |
| 1-(3-aminopropyl)-3-methylimidazolium bromide | (HS)SPME | milk, honey, urine and serum | antibiotics | 0.014–0.019 | 82.3–103.2 | GC-FID | [12] |
| 1-(3-aminopropyl)-3-(4-vinylbenzyl) imidazolium 4-styrenesulfonate | (D)SPME | groundwater of industrial park and river water | phenols | 0.2–0.5 | 75.5–113 | HPLC-DAD | [13] |
| 1-hydroxyethyl-3-methyl imidazolium-[NTf$_2$] | (HS)SPME | petrochemical, printing, dyeing wastewater and lake water | benzene derivatives | 0.010–0.019 | 82.3–108.3 | GC-FID | [14] |
| 1-butyl-3-aminopropyl imidazolium chloride | DSPE | effluent municipal wastewater treatment plant water, river and lake water | steroids, β-blockers | 0.007–0.023 | 87–98 | HPLC-DAD | [15] |

*2.2. Other Microextraction Technique*

Aside from immobilizing the IL-G on the surface of a fiber, GO-IL composite nanomaterials can also be used directly as sorbent materials. In this context, a microextraction method has been published, utilizing an IL-GO composite material in DSPE, which objectively follows the principles of the microextraction procedure. Specifically, Serrano et al. functionalized graphene with 1-butyl-3-aminopropyl imidazolium chloride in a one-pot synthesis, where the IL, graphene oxide and *N,N'*-dicyclohexylcarbodiimide were stirred for 24 h [15]. The characterization of the synthesized material verified that the structure of the final product was similar to that of graphene, hinting at a high specific surface area of the final product and making it suitable for microextraction purposes. The monomeric IL-GO composite was used as an adsorbent, for the first time, in the DSPE procedure, and the authors developed a method for the simultaneous detection of steroids and β-blockers. Under the same procedure, G was unable to extract any of the target analytes, while GO could extract only three out of the ten analytes. Therefore, the presence of the IL was responsible for the successful extraction of the target analytes, and exceedingly high EFs were achieved for all analytes. The LODs achieved rival those of the classical SPE and SPME methods.

## 3. Ionic Liquid-Coated Carbon Nanotubes-Based Materials

*3.1. Solid-Phase Microextraction*

In addition to graphene, which has already been commented on, adsorbents for microextraction purposes have been developed based on multi-walled CNTs (Table 2). The functionalization of CNTs with PILs follows two main, simple procedures, which are worth discussion: the covalent and the non-covalent functionalization [16]. A representative example of the covalent PIL functionalization of CNTs is the work presented by Cordero-Vaca et al. who functionalized nitinol wires (Ni/Ti alloy) with a crosslinked PIL-CNT coating [17]. The crosslinked PIL consisted of the monocationic IL 1-vinyl-3-butylimidazolium [NTf$_2$] as the monomer and the dicationic 1,12-di(3-vinylimidazolium) dodecane [NTf$_2$] as the IL crosslinker. Automation for the (DI)SPME of phenols and PAHs was possible once the wires were attached to a holder.

Bucky gels are gelatinous composite materials consisting of CNTs and ILs. In a different way than before, Zhang et al. coated a fused silica SPME fiber with a cross-linked PIL bucky gel containing CNTs [18]. To do so, the monomer 1-vinyl-3-butylimidazolium [NTf$_2$] and the cross-linker 1,12-di(3-vinylimidazolium) dodecane [NTf$_2$], along with CNTs (between 3% and 8% (*w/w*)) were ground to achieve complete mixing of the components, followed by the addition of an initiator and dip-coating of the fiber. The manufactured coated fibers were tested for the (HS)SPME of eight PAHs. A PIL-based coating without the addition of CNTs was found to be inferior compared to the commercially available polydimethylsiloxane fiber for the extraction of PAHs. The addition of the CNTs in the coating was found to be highly advantageous, since the PIL bucky gel-coated fiber achieved the same extraction efficiency as the polydimethylsiloxane coating for four analytes. Much higher extraction efficiency was achieved for the rest of the analytes. This was attributed to π−π interactions between the CNTs and the PAHs. Furthermore, the higher the quantity of the CNTs in the PIL bucky gel coating, the better the extraction efficiency. To elucidate the extraction mechanism (partition/adsorption) of the novel coating, extractions of various concentrations of 1-octanol, in the presence of naphthalene (an interfering compound), were conducted. The results indicated no competition between the two compounds, and therefore, a partition extraction mechanism (more preferable for complex matrices) was put forward.

**Table 2.** Microextraction techniques employing CNT-IL composite nanomaterials.

| Ionic Liquid | Microextraction Technique | Matrix | Target Analytes | LOD (µg/L) | Recoveries (%) | Instrumental Analytical System | Reference |
|---|---|---|---|---|---|---|---|
| poly(1-vinyl-3-hexyl-imidazolium [NTf$_2$]) | (HS)SPME | Petrochemical waste water, lake and tap water | benzene derivatives | 0.0177–0.0326 | 84.0–106.9 | GC-FID | [16] |
| monomer: (1-vinyl-3-butylimidazolium [NTf$_2$]), crosslinker: (1, 12-di(3-vinylimidazolium)dodecane [NTf$_2$]) | (DI)SPME | - | phenols and PAHs | 0.75–121.0 | - | GC-FID | [17] |
| poly(1-vinyl-3-octylimidazolium bromide) | (HS)SPME | - | alcohols | 0.015–0.05 | - | GC-FID | [19] |
| | | - | n-alkanes | 0.02–1 | - | | |
| poly(1-vinyl-3-octylimidazolium [NTf$_2$]) | | - | n-alkanes | 0.01–0.2 | - | | |
| | | - | phthalate esters | 0.005–0.02 | - | | |
| poly(1-vinyl-3-octylimidazolium 2-naphthalene-sulfonate) | | industrial park groundwater | halogenated aromatic hydrocarbons | 0.05–2 | 75–113 | | |
| monomer: (1-vinyl-3-ethylimidazolium hexafluorophosphate), crosslinker: 1,1'-(1,6-hexanediyl)bis(1-vinylimidazolium) bis(hexafluoro-phosphate) | (HS)SPME | citrus fruits | 2-naphthol | - | 81.9–110 | GC-FID | [20] |
| 1-methyl 3-[(3-octylamino)propyl] imidazolium [NTf$_2$] | (HS)SPME | urine | methamphetamine, ephedrine | 0.07–0.1 | 94.0–104.0 | GC-MS | [21] |
| poly(1-vinyl-3-ethylimidazole bromide) | (DI)SPME | apple, lettuce | carbamate pesticides | 0.0152–0.0272 | 87.5–106.5 | GC-FID | [22] |
| 1-(3-aminopropyl)-3-methylimidazolium bromide | (HS)SPME | perfume | benzoic acid esters | 0.0015–0.0061 | 87.8–110.8 | GC-FID | [23] |
| monomer: (1-vinyl-3-butylimidazolium [NTf$_2$], crosslinker: 1,12-di(3-vinylimidazolium) dodecane [NTf$_2$] | (HS)SPME | river and tap water | PAHs | 0.001–0.0025 | 60.0–122.3 | GC-MS | [18] |
| 1-butyl-3-methylimidazolium hexafluorophosphate | ultrasound-assisted DSPE | wine, grape juice, blueberry juice and chili oil | rhodamine B | 0.28 | 85.1–96.0 | HPLC-DAD | [24] |
| trihexyl (tetradecyl)phosphonium chloride | DSPE | garlic | As(V) | 0.0071 | 98.0–106.0 | Electrothermal atomic absorption spectrometer | [25] |
| tri-iso-octylammonium chloride | DSPE | acidic aqueous solutions | Cd(II) | - | - | Atomic absorption spectrometer | [26] |
| 1-butyl-3-methylimidazolium hexafluorophosphate | DSPE | Nitric acid solutions | lanthanide ions | - | - | Inductively-coupled plasma -mass spectrometer | [27] |
| 1-hexyl-3-methylimidazolium hexafluorophosphate | fabric-sorptive phase extraction | river water | PAHs | - | 87.0–105.0 | Spectrofluorometer | [28] |
| 1-hexyl-3-methylimidazolium hexafluorophosphate | in-line micro-SPE | river water | nitrophenols | 0.22–0.28 | 90.0–112.0 | Capillary electrophoresis | [29] |
| 1-(3-aminopropyl)-3-methylimidazolium chloride | PT-SPE | urine | (Z)-3-(chloromethylene)-6-fluorothiochroman-4-one | 0.009 | 73.9–93.9 | HPLC-UV | [30] |

Despite the advantages that accompany the covalent PIL functionalization of CNTs, it has been stated that this functionalization has drawbacks, since it is time consuming and the chemical treatment employed can disrupt the π conjugation of carbon nanotubes, causing destabilization of the structure, alteration of the electrical and mechanical properties, surface defects and shortening of the carbon nanotubes [16,31]. The non-covalent functionalization, based on van der Waals and electrostatic interactions, along with π−π stacking has the merits of simplicity and versatility. Such an example is the case of the non-covalent coating of an SPME fiber with CNTs, functionalized with poly(1-vinyl-3-octylimidazolium bromide) [19]. The fabrication of the CNTs-PILs fiber is illustrated in Figure 2. The prepared fiber was initially used for the extraction of alcohols from aqueous matrixes. As the nature of the anion of the PIL alters the extraction properties of the fiber, an in situ, on-fiber anion exchange protocol was followed, instead of altering the anion of the PIL prior to the functionalization of the fiber. When bromide was substituted for [NTf$_2$], the coating was rendered more hydrophobic and, therefore, more suitable for *n*-alkanes. This was evidenced by relevant experiments assessing the performance of the two fibers (with bromide and [NTf$_2$]) towards the extraction of *n*-alkanes. When the bromide anion was exchanged for 2-naphthalene-sulfonate, the interactions with aromatic compounds were favored, due to the π−π, n−π and hydrophobic interactions that may occur. This was supported experimentally by the (HS)SPME of phthalate esters and halogenated aromatic hydrocarbons. From the three cases already mentioned, it is apparent that the composition of the IL is crucial, since the selection of the proper anion can tune the selectivity towards various chemical groups and assist the overall performance of the method, in terms of the achieved LODs and EFs.

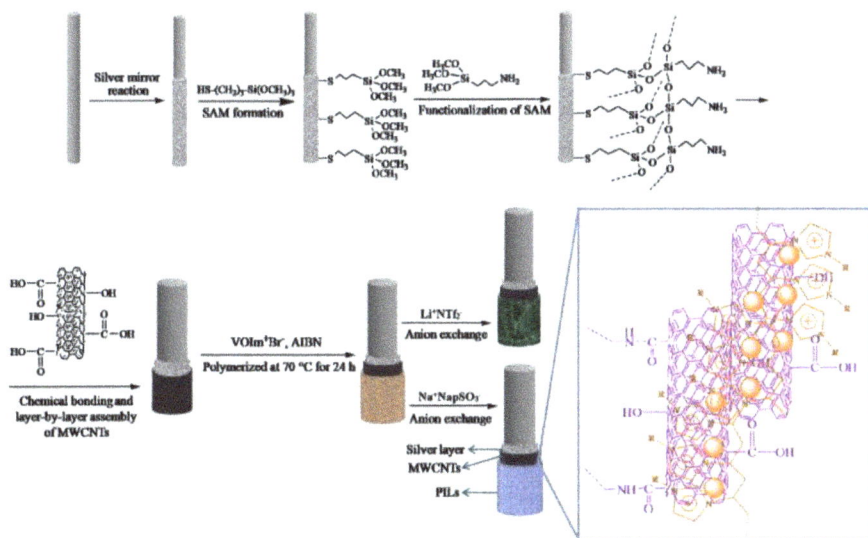

**Figure 2.** Preparation schema of the CNT-PILs SPME fiber. Reproduced with permission from [19]. Elsevier. Copyright Elsevier, 2015.

Another case of non-covalent coating of CNTs with IL is that presented by Feng et al. [20], who coated CNTs with 1-vinyl-3-ethylimidazolium hexafluorophosphate, after an in situ anion exchange. The coating of the CNTs with this IL facilitated the dispersibility of the CNTs in the pre-polymerization solution, which was used to fabricate a modified SPME fiber. Namely, the polymerization solution consisted of 1-vinyl-3-octylimidazolium bromide (IL monomer), pre-functionalized CNT-PIL, 1,1'-(1,6-hexanediyl)bis(1-vinylimidazolium) bis(hexafluoro-phosphate) (cross-linking agent) and azobisisobutyronitrile (initiator). The prepared fiber was used for the multiple (HS)SPME of 2-napthol from fruit samples, and the resulting method was found to be independent of matrix effects, although matrix variations were found to greatly influence the single (HS)SPME.

Taking advantage of the properties of other absorbing phases, synthesized in order to enhance the properties of the final material. Under this concept, two studies exploited the advantageous properties of nafion to prepare SPME fibers coated with nafion-CNT-PIL [21,22]. Narimani et al. firstly functionalized CNTs with amino groups to enhance the overall performance of the (HS)SPME fiber towards methamphetamine and ephenephrine [21]. Afterwards, the amino functionalized CNTs were coated with 1-methyl 3-[(3-octylamino)propyl] imidazolium [NTf$_2$]. However, since the direct coating of the stainless steel fiber with the CNT-IL complex was not a successful choice due to the absence of binding between them, nafion was used as the first coating layer of the fiber, benefiting from its cation exchange properties. This way, the quantity of CNT-IL was increased, due to the electrostatic interactions between nafion and the IL. The proposed fiber was compared with fibers coated with the single components of the proposed composite material. The nafion-coated fiber, on its own, has low extraction ability despite being mechanically strong. The nafion-IL-coated fiber exhibited slightly better extraction ability compared to bare nafion coating, but bleeding of the IL during desorption was a serious defect. The nafion-CNT-coated fiber showed extraction ability between the two previous fibers. However, the authors found that prior functionalization of the CNTs could further enhance the extraction capability of the coating. The nafion-CNT-IL coating was nearly twice as efficient as the previous coatings. In another case presented by Wu et al., the role of nafion was to enhance the stability and durability of the coating [22]. The CNTs coated with poly(1-vinyl-3-ethylimidazole bromide) were mixed with 3,4-ethylenedioxythiophene (EDOT), and the mixture was electrodeposited on a stainless steel fiber leading to a PEDOT-PIL-CNT coating. Then, the coated fiber was immersed in a nafion solution to form an outer layer, and the modified fiber was used in a (DI)SPME procedure to extract carbamate pesticides from apple and lettuce. As reported, the fiber could be reused for more than 150 extractions without obvious loss of extraction efficiency. This was attributed to the nafion coating, since a fiber lacking the outer nafion coating could not be used more than 60 times.

Polyaniline (PANI) is another polymer with a porous structure and large specific area, but with low thermal stability, which is used for preparing composite coatings [23]. It was found that CNTs could enhance the physical-chemical properties of the polymer. Ai et al. proposed the coating of an SPME fiber with CNTs@IL (1-(3-aminopropyl)-3-methylimidazolium bromide), doped with PANI [23]. In order to achieve this, the fiber was immersed in a mixture of CNTs@IL and aniline, containing hexadecyl trimethyl ammonium bromide, and the CNTs@IL-polyaniline-doped coating was electrodeposited. PANI is a good electron donor, whereas the IL selected by the authors acts as an electron acceptor, facilitating the adsorption and immobilization of the CNTs@IL composite during the electropolymerization of aniline. The prepared fiber was tested for its extraction selectivity towards various groups of compounds, including: benzoic acid esters, amines, benzene compounds, alcohols, PAHs and phenolic compounds after (HS)SPME. In a similar way, Wu et al. synthesized a PANI-PIL-CNT composite, consisting of poly(1-vinyl-3-hexyl-imidazolium [NTf$_2$]) [16], where the PIL was used to enhance the dispersibility of the CNTs in water and organic solvents, as well as to prevent aggregation. The CNT-PIL composite was coated on a stainless steel fiber, which was firstly coated with PANI, via electrodeposition. The applicability of the prepared fiber to the (HS)SPME of compounds with varying polarity was tested. The fiber showed high affinity towards benzene derivatives, due to the simultaneous $\pi-\pi$ interactions, hydrogen bonding and the hydrophobic effect occurring between the extractant and the adsorbent. The proposed coated fiber was able to be used for 200 repeatable extractions without loss of the extraction performance.

## 3.2. Dispersive Solid-Phase Microextraction

So far, only a few cases of the utilization of CNT-IL composite materials in DSPE procedures have been reported in the literature, focusing mainly on the determination of metal species [24–27]. In three out of the four cases, the CNTs were coated with monomeric ILs following a straightforward synthetic

*Separations* **2017**, *4*, 14

route. Usually, an organic solvent is added to a mixture of CNT-IL to ensure the good dispersion of IL and the reproducibility of the synthesis [24]. Under this principle, functionalized CNTs with tri-*iso*-octylammonium bromide were used as an adsorbent for cadmium ions [26]. Unmodified CNTs were unable to extract cadmium ions, highlighting the necessity of the IL. Extraction of lanthanide ions, which is a rare case of target analytes, was the aim of the study of Bazhenov et al. [27]. The authors functionalized CNTs with 1-butyl-3-methylimidazolium hexafluorophosphate, following the aforementioned procedure. The images from scanning electron microscopy revealed that the IL did not only cover the CNTs from the outside, but also filled partially their inner cavities. The extraction of the lanthanide ions was performed with 0.1 M $HNO_3$, in the presence of tetraphenyl methylene diphosphine dioxide. The bare CNTs were unable to extract any amount, while the addition of the IL greatly increased the extraction efficiency.

The last reported case focused on the determination of As(V) and As(III) in garlic samples [25]. The authors extracted arsenic from garlic with an acidic solution, and then, they added ammonium molybdate to form an arsenomolybdate complex. In this case, instead of directly using a CNT-IL composite material, the IL tetradecyl(trihexyl)phosphonium chloride was added to the extract to form a low-polarity complex with the arsenomolybdate, via the ion pair, followed by the addition of the CNTs. The whole composite material was directly inserted in a graphite furnace of an electrothermal atomic absorption spectrometer. This was attainable as CNTs are pyrolyzed prior to the atomization step, and therefore, no back-extraction/elution or digestion was needed. A drawback of the method was the potential interferences caused by phosphate anions during the step of arsenomolybdate complex formation. This hindrance was overcome by removing the phosphate anions with a mixture of organic solvents, prior to the addition of ammonium molybdate. The advantages of this approach were the low sample consumption (5 mL of the extract) and the low preconcentration/extraction time needed (20 min) compared to previous methods.

*3.3. Other Microextraction Techniques*

CNT-IL composite materials were also employed for other microextraction procedures. The first of the reported cases focused on the fabric-sorptive phase extraction of PAHs from river water [28]. A cotton fiber with fixed dimensions was immersed in a bucky gel, prepared by mortar agitation, which consisted of CNTs and 1-hexyl-3-methylimidazolium hexafluorophosphate. The critical gel concentration of CNTs was found to be 0.5%–1%. Excessive quantities of CNTs led to separation of the IL and gel phase, when centrifuged, demonstrating that a limited amount of IL could be trapped in the bucky gel. The next case of microextraction employing a CNT-IL composite is the study described by Polo-Luque et al. who aimed at the in-line preconcentration (in-tube SPME) of nitrophenols prior to capillary electrophoresis [29]. The preconcentration took place in the autosampler of the capillary electrophoresis, where a "spin-column" moved inside the sample vial. In the center of the moving part, there was a channel (1 mm in diameter and 4 mm in length), through which the sample was circulated. The adsorbent material was laid on the channel walls, and by applying mechanical pressure, the liquid sample was forced to pass through the channel, leading to the analyte extraction. The velocity towards the bottom of the vial and the rotating speed directly affected the preconcentration rate of the analytes, as well as the reproducibility. With the proposed method, 400–600 µL of sample could pass through the extraction unit, in 5 min. After the completion of extraction, the sample was rejected and an elution solvent passed through the channel in order to elute the adsorbed compounds, prior to the capillary electrophoresis. Good reproducibility, good analytical performance, low cost and automation are the main advantages of this microextraction technique.

Chen et al. developed and used an IL-G-CNT composite as an adsorbent in a PT-SPE procedure. The combination of IL, G and CNTs yielded satisfactory extraction of thiochromanones, which was much less effective in the absence of any of the above components. The results are undoubtedly better than those with G-CNTs, since the IL enables the ion exchange and electrostatic interactions of the analytes with the adsorbent, instead of only the $\pi-\pi$ interactions occurring in the case of G-CNTs [30].

## 4. Ionic Liquid-Coated Magnetic Graphene-Based Nanomaterials

### 4.1. Magnetic Dispersive Solid-Phase Extraction

Despite the progress in developing G/GO and CNTs materials modified with various ILs, it is obvious from the above discussion and Figure 1 that novel extraction procedures based on the principle of SPME hold the largest share among the microextraction procedures. This can be partially justified by the limited potential that composite materials containing GO have, as regards their applicability to DSPE methods, due to their high dispersibility, as a result of the high hydrophilicity of GO [32]. On the contrary, the low dispersibility of G in water minimizes the adsorption potential of the material. To overcome these hindrances, magnetic nanoparticles were embodied in the composite materials to convert them into being magnetic, thus enabling easier isolation from the matrix or increasing the wettability of graphene-based materials [33]. The magnetic materials can be employed to develop MSPE methods, under the principles of microextraction, which possess advantages over magnetic G/GO-based MSPE methods. The main advantage of the MSPE methods over the conventional DSPE is the shorter time needed for the sample preparation, since time-consuming steps, like filtration/centrifugation, are omitted.

Hitherto, miscellaneous IL-coated, magnetic graphene-based materials have been synthesized (Table 3). Almost exclusively, magnetite ($Fe_3O_4$) was used to endow nanomaterials with magnetic properties, although, virtually, other iron oxides are present. In most cases, magnetite nanoparticles were coated with silica, to form core-shell magnetic nanoparticles, which are less prone to oxidation compared to magnetite. The next step consisted of the reaction with an amino-terminated silane, so that free amino groups are formed on the surface of the silica-coated magnetite nanospheres. Through the free amino groups, GO was chemically bonded to form a $Fe_3O_4@SiO_2-NH_2@GO$ composite. The next step was the reduction of GO to G, often with hydrazine. However, this step was optional and hinges on two factors: (i) the ease of synthetic route to modify the G or the GO with the desirable IL; depending on the functional groups exploited and the method employed, the best option is selected; (ii) since the material is going to be used as an adsorbent, the affinity of the analytes with the G or the GO, shall be considered. According to this common procedure, $Fe_3O_4@SiO_2-NH_2@GO@IL$ composite nanomaterials were obtained, with enhanced properties. Up to now, two studies based on this procedure synthesized the described composite materials with varying IL coatings: (i) Cai et al. prepared the $Fe_3O_4@SiO_2.NH_2@GO$ composite and coated it with 1-carboxymethyl-3-methylimidazolium chloride [34] for the determination of chlorophenols; and (ii) Huang et al. synthesized the above-mentioned composite material, omitting the formation of silica on the surface or magnetite nanoparticles [35]. Following functionalization with a betaine-based IL, the nanomaterial was successfully applied to the MPSE of bovine serum albumin from bovine calf whole blood.

**Table 3.** Microextraction techniques employing magnetic G-IL composite nanomaterials.

| Ionic Liquid | Microextraction Technique | Matrix | Target Analytes | LOD (µg/L) | Recoveries (%) | Instrumental Analytical System | Reference |
|---|---|---|---|---|---|---|---|
| 1-carboxymethyl-3-methylimidazolium chloride | MSPE | tap, river and well water | chlorophenols | 0.0002–0.0026 | 85.3–99.3 | LC–MS/MS | [34] |
| N,N,N-trimethylglycine butanoate | MSPE | bovine calf whole blood | bovine serum albumin | - | - | UV-Vis spectrophotometer | [35] |
| di(6-hydroxyhexyl) tetramethyl guanidinium chloride | MSPE | - | trypsin, lysozyme, ovalbumin and bovine serum albumin | - | - | UV-Vis spectrophotometer | [36] |
| tetraoctylammonium bromide | MSPE | - | Pb | - | - | Atomic absorption spectrometer | [37] |
| N,N'-bis(2-aminoethyl)-N,N,N',N'-tetramethyl-chloride | MSPE | porcine and bovine blood | hemoglobin | 11,870 | - | UV-Vis spectrophotometer | [38] |
| 1,4-diazabicyclo[2.2.2]octane | Ultrasound assisted-MSPE | medicine capsules | Pb(II), Cd(II), Ni(II), Cu(II) and Cr(II) | 0.2–1.811 | 95.4–102.4 | Flame atomic absorption spectrometer | [39] |
| 1-dodecyl-3-methylimidazolium hexafluorophosphate | mixed hemimicelles MSPE | urine | cephalosporins | 0.0006–0.0019 | 84.3–101.7 | HPLC-UV | [40] |
| 1-hexadecyl-3-methylimidazolium bromide | mixed hemimicelles MSPE | human urine, environmental water and pharmaceutical formulation | fluoxetine | 0.21 | 95.3–100.6 | Angled mode-mismatched thermal lens spectrometer | [41] |
| 1-heptyl-3-methylimidazolium hexafluorophosphate | ultrasound-assisted mixed hemimicelles MSPE | river, lake and rain water | nitrobenzenes | 1.35–4.57 | 80.35–102.77 | HPLC-UV | [42] |
| 1-hexadecyl-3-methylimidazolium chloride | MSPE | sea and river water, tea, spinach, cacao powder and cigarette | nickel | 0.16 | 96.8–99.2 | Flame atomic absorption spectrometer | [43] |
| 1,3-didecyl-2-methylimidazolium chloride | MSPE | blood serum | hemin | 3.0 | 97.3–105.6 | Flame atomic absorption spectrometer | [44] |
| 1-vinyl-3-octylimidazolium bromide | QuEChERS | purple cabbage, bitter gourd, sponge gourd, tomatoes and cabbage | preservatives residues | 0.82–6.64 µg/kg | 81.7–118.3 | GC-MS | [45] |
| 1-ethyl-3-methylimidazolium tetrafluoroborate | ferrofluid-based DSPE | river and sea water, carrot, lettuce and tobacco | Cd | 0.12 | 98.2–101.5 | Flame atomic absorption spectrometer | [46] |

Chitosan, a polysaccharide polymer bearing hydroxyl and amino groups, may be a judicious alternative to the silica coating of the magnetite nanoparticles. It can easily coat magnetite nanoparticles, as well as form bonds with GO via amine-carboxyl groups bonding. Two research groups have synthesized $Fe_3O_4$@chitosan@GO@IL composites with different ILs, for the MSPE of different analytes. In the first case, the IL used was di(6-hydroxyhexyl) tetramethyl guanidinium-chloride, and the composite material was suitable for the extraction of trypsin, lysozyme, ovalbumin and bovine serum albumin [36]. In the second case, the magnetic GO was functionalized with tetraoctylammonium bromide, and the material was found suitable for the MSPE of Pb [37].

Finally, in an effort to synthesize functional adsorbent materials, there are some reported cases that diverge from the above procedures. Wen et al. chose to use nano-iron, covered with polyethylene glycol-4000, instead of magnetite [38]. The vibrating sample magnetometer curves demonstrated that the produced Fe@GO@IL complex demonstrates satisfactory saturation magnetization values, enabling its use in MSPE procedures. Wen et al. used an amino functional dicationic IL ($N,N'$-bis(2-aminoethyl)-$N,N,N',N'$-tetramethyl-chloride) as a coating to the magnetic GO, and the whole adsorbent was used to develop a method for the determination of hemoglobin [38]. When compared to other IL coatings with fewer amino-functional groups or a shorter alkyl chain, the extraction capability of the material was reduced. Therefore, the presence of the amino functional groups promotes the hydrogen bonding interactions between the adsorbent and hemoglobin. The last case revolves around the use of a double-charged IL immobilized on a magnetic GO complex [39]. A magnetite-grafted GO was modified with (3-mercaptopropyl)trimethoxysilane, and finally, diazoniabicyclo[2.2.2]octane chloride was attached covalently for the ultrasound-assisted MSPE of heavy metals. Due to the double-charged IL used, the adsorbent shows high adsorption capacity. The amount of sorbent used for the extraction of five heavy metals is nearly five-times lower than that of other methods.

*4.2. Other Microextraction Techniques*

Another mode of microextraction, which makes use of the magnetic properties of the adsorbent nanoparticles, is the mixed hemimicelles-MSPE. In this technique, mixed hemimicelles (hemimicelles and admicelles) of the IL are formed on the surface of the magnetic G/GO, using a proper IL-based surfactant. Hitherto, a few studies have been published relying on the above-mentioned principle. The first study focused on the extraction of cephalosporins from urine, using 1-dodecyl-3-methylimidazolium hexafluorophosphate to form the mixed hemimicelles [40]. The new method shows high extractability, a relatively wide linear range and satisfactory analytical figures of merit. The method developed by Kazemi et al. for the extraction of fluoxetine from environmental, biological and pharmaceutical samples proved highly time saving, since the overall procedure is completed in less than 15 min [41]. A small amount of magnetic graphene modified with 1-hexadecyl-3-methylimidazolium bromide was used, due to the high capacity of the material, rendering the method cost-effective. Similar merits were noticed in the method developed by Cao et al. for the determination of nitrobenzenes in environmental samples, based on $Fe_3O_4$@graphene coated with 1-heptyl-3-methylimidazolium hexafluorophosphate [42]. The developed method uses an infinitesimal amount of extraction solvent and small sample volumes. The study of Aliyari et al. revolved around the in situ synthesis of magnetite, based on the chemical co-precipitation of $Fe^{3+}$, directly on the surface of GO, prior to the modification with mixed IL hemimicelles [43]. Based on an electrostatic self-assembly technique, 1-hexadecyl-3-methylimidazolium chloride adhered on the GO, forming hemimicelles/admicelles through electrostatic attraction of the negative surface charge of the GO and the cationic IL. The long alkyl chain of the IL facilitated the assembly due to the hydrophobic interactions. Taking advantage of its property to solubilize organic molecules, a hybrid nanocomposite material with dimethylglyoxime was produced and used as an adsorbent for traces of nickel, for a wide variety of matrices. Under the same synthetic principle, Farzin et al. formed mixed hemimicelles, by way of 1,3-didecyl-2-methylimidazolium chloride, on the surface of magnetic GO, facilitating the fast MSPE of hemin from blood serum [44]. The developed technique was compared with a commercially available hemin assay kit. The results obtained were not statistically different, suggesting that the proposed method is a reliable and promising alternative to the determination of hemin.

The applicability of the magnetic G-IL composites in microextraction techniques is broad, as revealed by the discussion above. However, there are two studies that make use of composite magnetic nanomaterials and that deserve consideration, because of their potential for alternative microextraction purposes. In the first case, Chen et al. synthesized the $Fe_3O_4@SiO_2$-$NH_2@G$ composite [45]. Then, they added 1-vinyltriethoxysilane on the surface of the composite, via hydrogen bonds, and finally, they added 1-vinyl-3-octylimidazolium bromide to obtain a PIL coating. The magnetic adsorbent was used in the clean-up step of a Quick, Easy, Cheap, Effective, Rugged and Safe (QuEChERS) method to purify the extract obtained from vegetables, in order to determine their content in preservative residues. The developed method can simultaneously extract twenty preservatives, twice more than the number of analytes of previously developed methods, which also make use of magnetic nanoparticles. In another study, instead of using a solid adsorbent, the authors developed a method based on the principle of DSPE, using a ferrofluid, which was synthesized using a $Fe_3O_4@GO@IL$ complex and 1-ethyl-3-methylimidazolium tetrafluoroborate as IL [46]. As evidenced by the authors, the type of the IL is important for the ferrofluid in order to prevent aggregation and sedimentation. The resulting ferrofluid is stable, with high water dispersibility, thus facilitating its use for microextraction purposes. The IL, in the last case, is exclusively used as the carrier fluid without any profound role in the extraction process, in contrast with the IL mixed hemimicelles formed on a solid surface, which are predominantly used for extraction purposes.

## 5. Ionic Liquid Coated Magnetic CNTs-Based Materials

### 5.1. Magnetic Solid-Phase Extraction MSPE

Till now, a limited number of studies have been conducted, with respect to the synthesis and employment of magnetic CNTs-IL composites in analytical procedures (Table 4) [47–52]. In the studies examined, CNTs were magnetized before or after their modification with IL, either synthesizing magnetite nanoparticles directly on the surface of acid-treated CNTs or spontaneously bonding them with the CNT-IL composite. In one of them, unlike all previous synthetic routes, the authors synthesized stepwise the IL on the CNTs, instead of covalently or non-covalently functionalizing them with the IL. A necessary step was the functionalization of the CNTs with acid chloride functional groups, through the reaction with thionyl chloride. Afterwards, 1-(3-aminopropyl)imidazole was chemically bonded on the functionalized CNTs, and 1-chlorobutane [47] or 1-bromobutane [48] was added to complete the synthesis. The composites were used in MSPE procedures of triazole fungicides [47] and aryloxyphenoxy-propionate herbicides (and their metabolites) [48] from environmental water samples, after they had been converted into being magnetic. Both extraction procedures were fast (5 and 8 min) and used small amounts of sorbent materials (2 and 12.5 mg). As a result of the presence of the IL, the two methods achieved LODs ranging between 0.05 and 0.2 $ng \cdot mL^{-1}$ for triazole fungicides and 0.002–3.4 $mg \cdot L^{-1}$ for aryloxyphenoxy-propionate herbicides. The recoveries achieved were different, despite the similarity of the adsorbent material (84.7%–105.3% for triazole fungicides and 66.1%–89.6% for aryloxyphenoxy-propionate herbicides).

**Table 4.** Microextraction techniques employing magnetic CNT-IL composite nanomaterials.

| Ionic Liquid | Microextraction Technique | Matrix | Target Analytes | LOD (μg/L) | Recoveries (%) | Instrumental Analytical System | Reference |
|---|---|---|---|---|---|---|---|
| 1-butyl-3-aminopropyl imidazolium chloride | MSPE | canal water | triazole fungicides | 0.05–0.22 | 84.7–105.3 | GC-MS | [47] |
| 1-butyl-3-aminopropyl imidazolium bromide | MSPE | ground and reservoir water | aryloxyphenoxy-propionate herbicides | 2.8–14.3 / 0.002–3.4 | 66.1–89.6 | HPLC-DAD / HPLC-MS/MS | [48] |
| 1-(4-vinylbenzyl)-(3-aminopropyl)-imidazolium chloride | MSPE | porcine whole blood | Cu, Zn-superoxide | - | 82.7–102.3 | UV-Vis spectrophotometer | [49] |
| 6-hydroxy-N-(2-hydroxyethyl)-N,N-dimethylhexan-1-aminium chloride | MSPE | - | lysozyme | - | - | UV-Vis Spectrophotometer | [50] |
| 1-hexadecyl-3-methyl-imidazolium bromide | mixed hemimicelles MSPE | human urine | flavonoids | 0.20–0.75 | 90.1–97.6 | HPLC-UV | [51] |
| 1-butyl-3-methylimidazolium hexafluorophosphate | USA-IL-LDMME | tap and well water, cow milk, fish liver | Cd, As | 0.003–0.005 | 94.6–98.0 | Atomic absorption spectrometer | [52] |

In the study of Wen et al., a magnetic CNTs-PIL composite was used to extract Cu, Zn-superoxide dismutase from porcine whole blood [49]. The composite material was proven to be superior compared to magnetic CNT-IL composite and magnetic CNTs, in terms of extraction capacity for Cu, Zn-superoxide dismutase, which was three-times higher than that of the two latter materials. Additionally, the adsorbent showed lower affinity towards lysozyme and bovine hemoglobin, the two major components of the studied matrix. The proposed mechanism of interaction between adsorbent and analyte included hydrogen bonding, $\pi-\pi$ and electrostatic interactions. It is noteworthy that instead of organic solvents, a solution of sodium chloride was used for the elution of Cu, Zn-superoxide dismutase from the adsorbent. The efficient desorption was attributed to the weakening of the electrostatic interactions between the analyte and the adsorbent. In the last example, CNTs were rendered magnetic by synthesizing magnetite nanoparticles on the surface, then coating them with silicon dioxide. A dual hydroxyl IL (6-hydroxy-*N*-(2-hydroxyethyl)-*N*,*N*-dimethylhexan-1-aminium chloride) was synthesized on the surface, resulting in an IL-CNTs@$Fe_3O_4$@$SiO_2$ composite material [50]. The synthesized material was tested for the extraction capability of four different proteins: bovine serum albumin, lysozyme, trypsin and ovalbumin. Lysozyme was extracted more easily. The proposed adsorbent was superior to magnetic CNTs-$SiO_2$ because of the electrostatic and hydrophobic interactions between lysozyme and the IL. Additionally, the contribution of the two hydroxyl groups on the surface of the IL was examined by comparing with the respective adsorbent, without hydroxyl groups. The dual hydroxyl-functionalized IL was preferable, since it enhanced the extraction capability by forming hydrogen bond interactions with the aliphatic carbon residue of the lysozyme. Overall, the DSPE procedures carried out with the CNT-IL composite were endowed with excellent repeatability, high recoveries and specificity.

*5.2. Other Microextraction Techniques*

Xiao et al. reported on the extraction of PAHs from urine samples using magnetic CNTs and 1-hexadecyl-3-methylmidazolium bromide, in a mixed hemimicelles MSPE procedure [51]. Figure 3 illustrates the whole procedure of the preparation of surfactant-coated CNTs and the application to simultaneous microextraction and preconcentration of analytes. In the beginning, the IL was mixed with the magnetic CNTs via ultrasonication, so that mixed hemimicelles were formed on the surface of the magnetic CNTs. Subsequently, the urine sample was added and the mixture was further sonicated to disperse the adsorbent material and extract the analytes. The adsorbent was compared with IL-coated $Fe_3O_4$@$SiO_2$ nanoparticles; it was found that at low concentrations of analytes (<40 ng/mL), the recoveries of the two materials were the same, while for higher concentrations, the adsorbent containing the CNTs was more efficient. This was attributed to the great number of hemimicelles formed on the surface of each material. Therefore, the CNTs enhance the extraction mainly indirectly, by increasing the amount of hemimicelles available for the extraction. The analytical characteristics of the proposed method were quite acceptable, and the proposed method is highly time saving, low cost and environmentally friendly, since small amounts of organic solvent are used for the elution, while the adsorbent could be used for more than one extraction-desorption cycle. The last reported case deals with the extraction of Cd and As from water, milk and fish liver samples [52]. In this case, sodium diethyldithiocarbamate trihydrate was added to the prepared sample, functioning as a chelating agent, and then, 1-butyl-3-methylimidazolium hexafluorophosphate was added. The mixture was ultrasonicated, and then, the magnetic CNTs were added, followed by another short ultrasonication period. The magnetic particles were harvested with the aid of a magnet, and then, the metal-diethyldithiocarbamate complex was eluted from the adsorbent. In this case (termed ultrasound-assisted, ionic liquid-linked, dual-magnetic multi-walled carbon nanotube microextraction (USA-IL-LDMME), by the authors), the IL functions as the main extraction solvent, while magnetic CNTs are used to increase the extraction efficiency and to couple with the IL, facilitating the easier separation of the IL from the solution.

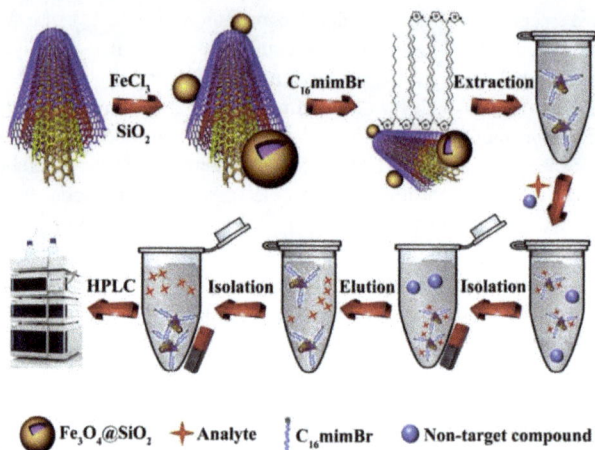

**Figure 3.** Preparation of surfactant-coated magnetic CNTs and the application to microextraction and preconcentration of analytes. Reproduced with permission from [51]. Copyright Elsevier, 2014.

## 6. Conclusions and Perspectives

Here, we have discussed a variety of IL-functionalized carbon-based materials, consisting predominantly of graphene and multi-walled carbon nanotubes, for microextraction purposes, and we have highlighted the emerging frontiers of this aspect of nanoscience and analytical chemistry. Increasing extraction yield, selectivity and sensitive detection are the most exciting and challenging aspects of this field.

Graphene-IL and CNT-IL hybrid materials have been synthesized and used as sorbents in microextraction procedures for various classes of organic compounds and metals in food and water samples. These materials proved to be advantageous and more efficient than bare graphene, CNTs or IL, underlining their combined capability via various interactions. Furthermore, it is noteworthy that the composite material can adsorb various classes of compounds by altering the anion of the IL, which is an elegant way to tune selectivity and applicability. Despite their successful synthesis and employment in microextraction, composite materials endowed with magnetic properties and outstanding characteristics were also synthesized.

Carbon-based nanomaterials, coated with IL, clearly contribute significantly towards the development of enhanced analytical methods due to their increased efficiency, recovery and selectivity, as adsorbents. Furthermore, in some cases, they proved to surpass commercially available sorbent materials, employed in microextraction procedures. In the future, large-scale synthesis of the materials should be examined, in order to use them as reliable commercial materials for microextraction purposes. Moreover, more research should be conducted focusing on other matrices or classes of compounds that can be extracted by the hybrid adsorbents. As "nanoparticles will be part of the heart and soul of analytical chemistry", the applications of composite nanomaterials, based on carbon structure-IL, in sample preparation, will continue to improve, and new analytical applications for the composite materials are expected to emerge, since their full potential is yet to be realized.

**Conflicts of Interest:** The authors report no conflict of interest.

## Abbreviations

| | |
|---|---|
| CNTs | carbon nanotubes |
| DI | direct-immersion |
| DSPE | dispersive solid-phase microextraction |
| EDOT | 3,4-ethylenedioxythiophene |
| EFs | enrichment factors |
| G | graphene |
| GO | graphene oxide |
| HS | headspace |
| IL | Ionic liquids |
| LODs | limits of detection |
| MSPE | magnetic solid-phase extraction |
| [NTf$_2$] | bis(trifluoromethanesulfonyl imide) |
| PAHs | polycyclic aromatic hydrocarbons |
| PANI | polyaniline |
| PEDOT | poly(3,4-ethylenedioxythiophene) |
| PILs | polymeric ionic liquids |
| PT-SPE | pipette-tip solid-phase extraction |
| QuEChERS | Quick, Easy, Cheap, Effective, Rugged and Safe |
| SPE | solid-phase extraction |
| SPME | solid-phase microextraction |
| USA-IL-LDMME | ultrasound-assisted, ionic liquid-linked, dual-magnetic multi-walled carbon nanotube microextraction |

## References

1. Stalikas, C.D.; Fiamegos, Y.C. Microextraction combined with derivatization. *TrAC Trends Anal. Chem.* **2008**, *27*, 533–542. [CrossRef]
2. Zhang, B.-T.; Zheng, X.; Li, H.-F.; Lin, J.-M. Application of carbon-based nanomaterials in sample preparation: A review. *Anal. Chim. Acta* **2013**, *784*, 1–17. [CrossRef] [PubMed]
3. Iijima, S. Helical microtubules of graphitic carbon. *Nature* **1991**, *354*, 56–58. [CrossRef]
4. Herrero Latorre, C.; Álvarez Méndez, J.; Barciela García, J.; García Martín, S.; Peña Crecente, R.M. Carbon nanotubes as solid-phase extraction sorbents prior to atomic spectrometric determination of metal species: A review. *Anal. Chim. Acta* **2012**, *749*, 16–35. [CrossRef] [PubMed]
5. Speltini, A.; Merli, D.; Profumo, A. Analytical application of carbon nanotubes, fullerenes and nanodiamonds in nanomaterials-based chromatographic stationary phases: A review. *Anal. Chim. Acta* **2013**, *783*, 1–16. [CrossRef] [PubMed]
6. Chen, D.; Feng, H.; Li, J. Graphene oxide: Preparation, functionalization, and electrochemical applications. *Chem. Rev.* **2012**, *112*, 6027–6053. [CrossRef] [PubMed]
7. Han, D.; Row, K.H. Recent applications of ionic liquids in separation technology. *Molecules* **2010**, *15*, 2405. [CrossRef] [PubMed]
8. Wasserscheid, P.; Welton, T. *Ionic Liquids in Synthesis*, 2nd ed.; Wiley-VCH: Berlin, Germany, 2003.
9. Vidal, L.; Riekkola, M.-L.; Canals, A. Ionic liquid-modified materials for solid-phase extraction and separation: A review. *Anal. Chim. Acta* **2012**, *715*, 19–41. [CrossRef] [PubMed]
10. Hou, X.; Guo, Y.; Liang, X.; Wang, X.; Wang, L.; Wang, L.; Liu, X. Bis(trifluoromethanesulfonyl)imide-based ionic liquids grafted on graphene oxide-coated solid-phase microextraction fiber for extraction and enrichment of polycyclic aromatic hydrocarbons in potatoes and phthalate esters in food-wrap. *Talanta* **2016**, *153*, 392–400. [CrossRef] [PubMed]
11. Sun, M.; Feng, J.; Bu, Y.; Duan, H.; Wang, X.; Luo, C. Development of a solid-phase microextraction fiber by the chemical binding of graphene oxide on a silver-coated stainless-steel wire with an ionic liquid as the crosslinking agent. *J. Sep. Sci.* **2014**, *37*, 3691–3698. [CrossRef] [PubMed]

12. Wu, M.; Ai, Y.; Zeng, B.; Zhao, F. In situ solvothermal growth of metal-organic framework-ionic liquid functionalized graphene nanocomposite for highly efficient enrichment of chloramphenicol and thiamphenicol. *J. Chromatogr. A* **2016**, *1427*, 1–7. [CrossRef] [PubMed]

13. Sun, M.; Bu, Y.; Feng, J.; Luo, C. Graphene oxide reinforced polymeric ionic liquid monolith solid-phase microextraction sorbent for high-performance liquid chromatography analysis of phenolic compounds in aqueous environmental samples. *J. Sep. Sci.* **2016**, *39*, 375–382. [CrossRef] [PubMed]

14. Wu, M.; Wang, L.; Zeng, B.; Zhao, F. Fabrication of poly(3,4-ethylenedioxythiophene)-ionic liquid functionalized graphene nanosheets composite coating for headspace solid-phase microextraction of benzene derivatives. *J. Chromatogr. A* **2014**, *1364*, 45–52. [CrossRef] [PubMed]

15. Serrano, M.; Chatzimitakos, T.; Gallego, M.; Stalikas, C.D. 1-butyl-3-aminopropyl imidazolium-functionalized graphene oxide as a nanoadsorbent for the simultaneous extraction of steroids and β-blockers via dispersive solid-phase microextraction. *J. Chromatogr. A* **2016**, *1436*, 9–18. [CrossRef] [PubMed]

16. Wu, M.; Wang, L.; Zhao, F.; Zeng, B. Ionic liquid polymer functionalized carbon nanotubes-coated polyaniline for the solid-phase microextraction of benzene derivatives. *RSC Adv.* **2015**, *5*, 99483–99490. [CrossRef]

17. Cordero-Vaca, M.; Trujillo-Rodríguez, M.J.; Zhang, C.; Pino, V.; Anderson, J.L.; Afonso, A.M. Automated direct-immersion solid-phase microextraction using crosslinked polymeric ionic liquid sorbent coatings for the determination of water pollutants by gas chromatography. *Anal. Bioanal. Chem.* **2015**, *407*, 4615–4627. [CrossRef] [PubMed]

18. Zhang, C.; Anderson, J.L. Polymeric ionic liquid bucky gels as sorbent coatings for solid-phase microextraction. *J. Chromatogr. A* **2014**, *1344*, 15–22. [CrossRef] [PubMed]

19. Feng, J.; Sun, M.; Bu, Y.; Luo, C. Facile modification of multi-walled carbon nanotubes-polymeric ionic liquids-coated solid-phase microextraction fibers by on-fiber anion exchange. *J. Chromatogr. A* **2015**, *1393*, 8–17. [CrossRef] [PubMed]

20. Feng, J.; Sun, M.; Li, L.; Wang, X.; Duan, H.; Luo, C. Multiwalled carbon nanotubes-doped polymeric ionic liquids coating for multiple headspace solid-phase microextraction. *Talanta* **2014**, *123*, 18–24. [CrossRef] [PubMed]

21. Narimani, O.; Dalali, N.; Rostamizadeh, K. Functionalized carbon nanotube/ionic liquid-coated wire as a new fiber assembly for determination of methamphetamine and ephedrine by gas chromatography-mass spectrometry. *Anal. Methods* **2014**, *6*, 8645–8653. [CrossRef]

22. Wu, M.; Wang, L.; Zeng, B.; Zhao, F. Ionic liquid polymer functionalized carbon nanotubes-doped poly(3,4-ethylenedioxythiophene) for highly-efficient solid-phase microextraction of carbamate pesticides. *J. Chromatogr. A* **2016**, *1444*, 42–49. [CrossRef] [PubMed]

23. Ai, Y.; Wu, M.; Li, L.; Zhao, F.; Zeng, B. Highly selective and effective solid phase microextraction of benzoic acid esters using ionic liquid functionalized multiwalled carbon nanotubes-doped polyaniline coating. *J. Chromatogr. A* **2016**, *1437*, 1–7. [CrossRef] [PubMed]

24. Xu, X.; Zhang, M.; Wang, L.; Zhang, S.; Liu, M.; Long, N.; Qi, X.; Cui, Z.; Zhang, L. Determination of rhodamine b in food using ionic liquid–coated multiwalled carbon nanotube-based ultrasound-assisted dispersive solid-phase microextraction followed by high-performance liquid chromatography. *Food Anal. Methods* **2016**, *9*, 1696–1705. [CrossRef]

25. Grijalba, A.C.; Escudero, L.B.; Wuilloud, R.G. Ionic liquid-assisted multiwalled carbon nanotube-dispersive micro-solid phase extraction for sensitive determination of inorganic as species in garlic samples by electrothermal atomic absorption spectrometry. *Spectrochim. Acta Part B* **2015**, *110*, 118–123. [CrossRef]

26. Alguacil, F.J.; García-Díaz, I.; López, F.A.; Rodríguez, O. Liquid-liquid extraction of cadmium(ii) by tioacl (tri-iso-octylammonium chloride) ionic liquid and its application to a tioacl impregnated carbon nanotubes system. *Rev. Met.* **2015**, *51*. [CrossRef]

27. Bazhenov, A.V.; Fursova, T.N.; Turanov, A.N.; Aronin, A.S.; Karandashev, V.K. Properties of a composite material based on multi-walled carbon nanotubes and an ionic liquid. *Phys. Solid State* **2014**, *56*, 572–579. [CrossRef]

28. Polo-Luque, M.L.; Simonet, B.M.; Valcárcel, M. Ionic liquid combined with carbon nanotubes: A soft material for the preconcentration of pahs. *Talanta* **2013**, *104*, 169–172. [CrossRef] [PubMed]

29. Polo-Luque, M.L.; Simonet, B.M.; Valcárcel, M. Solid-phase extraction of nitrophenols in water by using a combination of carbon nanotubes with an ionic liquid coupled in-line to ce. *Electrophoresis* **2013**, *34*, 304–308. [CrossRef] [PubMed]

30. Chen, H.; Yuan, Y.; Xiang, C.; Yan, H.; Han, Y.; Qiao, F. Graphene/multi-walled carbon nanotubes functionalized with an amine-terminated ionic liquid for determination of (z)-3-(chloromethylene)-6-fluorothiochroman-4-one in urine. *J. Chromatogr. A* **2016**, *1474*, 23–31. [CrossRef] [PubMed]
31. Tunckol, M.; Fantini, S.; Malbosc, F.; Durand, J.; Serp, P. Effect of the synthetic strategy on the non-covalent functionalization of multi-walled carbon nanotubes with polymerized ionic liquids. *Carbon* **2013**, *57*, 209–216. [CrossRef]
32. Stankovich, S.; Dikin, D.A.; Piner, R.D.; Kohlhaas, K.A.; Kleinhammes, A.; Jia, Y.; Wu, Y.; Nguyen, S.T.; Ruoff, R.S. Synthesis of graphene-based nanosheets via chemical reduction of exfoliated graphite oxide. *Carbon* **2007**, *45*, 1558–1565. [CrossRef]
33. Maidatsi, K.V.; Chatzimitakos, T.G.; Sakkas, V.A.; Stalikas, C.D. Octyl-modified magnetic graphene as a sorbent for the extraction and simultaneous determination of fragrance allergens, musks, and phthalates in aqueous samples by gas chromatography with mass spectrometry. *J. Sep. Sci.* **2015**, *38*, 3758–3765. [CrossRef] [PubMed]
34. Cai, M.-Q.; Su, J.; Hu, J.-Q.; Wang, Q.; Dong, C.-Y.; Pan, S.-D.; Jin, M.-C. Planar graphene oxide-based magnetic ionic liquid nanomaterial for extraction of chlorophenols from environmental water samples coupled with liquid chromatography–tandem mass spectrometry. *J. Chromatogr. A* **2016**, *1459*, 38–46. [CrossRef] [PubMed]
35. Huang, Y.; Wang, Y.; Wang, Y.; Pan, Q.; Ding, X.; Xu, K.; Li, N.; Wen, Q. Ionic liquid-coated fe3o4/aptes/graphene oxide nanocomposites: Synthesis, characterization and evaluation in protein extraction processes. *RSC Adv.* **2016**, *6*, 5718–5728. [CrossRef]
36. Ding, X.; Wang, Y.; Wang, Y.; Pan, Q.; Chen, J.; Huang, Y.; Xu, K. Preparation of magnetic chitosan and graphene oxide-functional guanidinium ionic liquid composite for the solid-phase extraction of protein. *Anal. Chim. Acta* **2015**, *861*, 36–46. [CrossRef] [PubMed]
37. Sun, W.; Li, L.; Luo, C.; Fan, L. Synthesis of magnetic graphene nanocomposites decorated with ionic liquids for fast lead ion removal. *Int. J. Biol. Macromol.* **2016**, *85*, 246–251. [CrossRef] [PubMed]
38. Wen, Q.; Wang, Y.; Xu, K.; Li, N.; Zhang, H.; Yang, Q.; Zhou, Y. Magnetic solid-phase extraction of protein by ionic liquid-coated fe@graphene oxide. *Talanta* **2016**, *160*, 481–488. [CrossRef] [PubMed]
39. Lotfi, Z.; Mousavi, H.Z.; Sajjadi, S.M. Covalently bonded double-charged ionic liquid on magnetic graphene oxide as a novel, efficient, magnetically separable and reusable sorbent for extraction of heavy metals from medicine capsules. *RSC Adv.* **2016**, *6*, 90360–90370. [CrossRef]
40. Wu, J.; Zhao, H.; Xiao, D.; Chuong, P.-H.; He, J.; He, H. Mixed hemimicelles solid-phase extraction of cephalosporins in biological samples with ionic liquid-coated magnetic graphene oxide nanoparticles coupled with high-performance liquid chromatographic analysis. *J. Chromatogr. A* **2016**, *1454*, 1–8. [CrossRef] [PubMed]
41. Kazemi, E.; Haji Shabani, A.M.; Dadfarnia, S.; Abbasi, A.; Rashidian Vaziri, M.R.; Behjat, A. Development of a novel mixed hemimicelles dispersive micro solid phase extraction using 1-hexadecyl-3-methylimidazolium bromide coated magnetic graphene for the separation and preconcentration of fluoxetine in different matrices before its determination by fiber optic linear array spectrophotometry and mode-mismatched thermal lens spectroscopy. *Anal. Chim. Acta* **2016**, *905*, 85–92. [PubMed]
42. Cao, X.; Shen, L.; Ye, X.; Zhang, F.; Chen, J.; Mo, W. Ultrasound-assisted magnetic solid-phase extraction based ionic liquid-coated fe3o4@graphene for the determination of nitrobenzene compounds in environmental water samples. *Analyst* **2014**, *139*, 1938–1944. [CrossRef] [PubMed]
43. Aliyari, E.; Alvand, M.; Shemirani, F. Modified surface-active ionic liquid-coated magnetic graphene oxide as a new magnetic solid phase extraction sorbent for preconcentration of trace nickel. *RSC Adv.* **2016**, *6*, 64193–64202. [CrossRef]
44. Farzin, L.; Shamsipur, M.; Sheibani, S. Solid phase extraction of hemin from serum of breast cancer patients using an ionic liquid coated fe3o4/graphene oxide nanocomposite, and its quantitation by using faas. *Microchim. Acta* **2016**, *183*, 2623–2631. [CrossRef]
45. Chen, Y.; Cao, S.; Zhang, L.; Xi, C.; Li, X.; Chen, Z.; Wang, G. Preparation of size-controlled magnetite nanoparticles with a graphene and polymeric ionic liquid coating for the quick, easy, cheap, effective, rugged and safe extraction of preservatives from vegetables. *J. Chromatogr. A* **2016**, *1448*, 9–19. [CrossRef] [PubMed]

46. Alvand, M.; Shemirani, F. Fabrication of fe3o4@graphene oxide core-shell nanospheres for ferrofluid-based dispersive solid phase extraction as exemplified for cd(ii) as a model analyte. *Microchim. Acta* **2016**, *183*, 1749–1757. [CrossRef]

47. Chen, F.; Song, Z.; Nie, J.; Yu, G.; Li, Z.; Lee, M. Ionic liquid-based carbon nanotube coated magnetic nanoparticles as adsorbent for the magnetic solid phase extraction of triazole fungicides from environmental water. *RSC Adv.* **2016**, *6*, 81877–81885. [CrossRef]

48. Luo, M.; Liu, D.; Zhao, L.; Han, J.; Liang, Y.; Wang, P.; Zhou, Z. A novel magnetic ionic liquid modified carbon nanotube for the simultaneous determination of aryloxyphenoxy-propionate herbicides and their metabolites in water. *Anal. Chim. Acta* **2014**, *852*, 88–96. [CrossRef] [PubMed]

49. Wen, Q.; Wang, Y.; Xu, K.; Li, N.; Zhang, H.; Yang, Q. A novel polymeric ionic liquid-coated magnetic multiwalled carbon nanotubes for the solid-phase extraction of cu, zn-superoxide dismutase. *Anal. Chim. Acta* **2016**, *939*, 54–63. [CrossRef] [PubMed]

50. Chen, J.; Wang, Y.; Huang, Y.; Xu, K.; Li, N.; Wen, Q.; Zhou, Y. Magnetic multiwall carbon nanotubes modified with dual hydroxy functional ionic liquid for the solid-phase extraction of protein. *Analyst* **2015**, *140*, 3474–3483. [CrossRef] [PubMed]

51. Xiao, D.; Yuan, D.; He, H.; Pham-Huy, C.; Dai, H.; Wang, C.; Zhang, C. Mixed hemimicelle solid-phase extraction based on magnetic carbon nanotubes and ionic liquids for the determination of flavonoids. *Carbon* **2014**, *72*, 274–286. [CrossRef]

52. Shirani, M.; Semnani, A.; Habibollahi, S.; Haddadi, H. Ultrasound-assisted, ionic liquid-linked, dual-magnetic multiwall carbon nanotube microextraction combined with electrothermal atomic absorption spectrometry for simultaneous determination of cadmium and arsenic in food samples. *J. Anal. At. Spectrom.* **2015**, *30*, 1057–1063. [CrossRef]

*separations*

MDPI

*Review*

# Fabric Sol–gel Phase Sorptive Extraction Technique: A Review

Viktoria Kazantzi and Aristidis Anthemidis *

Laboratory of Analytical Chemistry, Department of Chemistry, Aristotle University, Thessaloniki 54124, Greece; vicky—89@hotmail.com
* Correspondence: anthemid@chem.auth.gr; Tel.: +30-231-099-7826

Academic Editor: Janusz Pawliszyn
Received: 5 January 2017; Accepted: 18 May 2017; Published: 24 May 2017

**Abstract:** Since the introduction in 2014 of fabric phase sorptive extraction (FPSE) as a sample preparation technique, it has attracted the attention of many scientists working in the field of separation science. This novel sorbent extraction technique has successfully utilized the benefits of sol–gel derived hybrid sorbents and a plethora of fabric substrates, resulting in a highly efficient, sensitive and green sample pretreatment methodology. The proposed procedure is an easy and efficient pathway to extract target analytes from different matrices providing inherent advantages such as high sample loading capacity and short pretreatment time. The present review mainly focuses on the background and sol–gel chemistry for the preparation of new fabric sorbents as well as on the applications of FPSE for extracting target analytes, from the time that it was first introduced. New modes of FPSE including stir FPSE, stir-bar FPSE, dynamic FPSE, and automated on-line FPSE are also highlighted and commented upon in detail. FPSE has been effectively applied for the determination of various organic and inorganic analytes in different types of environmental and biological samples in high throughput analytical, environmental, and toxicological laboratories.

**Keywords:** fabric phase sorptive extraction; sample preparation; sol–gel; chromatography; automation; solid phase extraction; atomic spectrometry

## 1. Introduction

In modern analytical methods, sensitive techniques as well as new analytical instruments are used for the determination of various analytes in a plethora of different and complex matrices. However, the direct determination of very low concentrations has been broadly recognized as the Achilles heel of the analysis and is mainly associated with matrix interferences or inadequate sensitivity. On the other hand, it is well known that sample preparation is a key step of chemical analysis and somehow is considered as the bottleneck of the whole analytical cycle being the most tedious and time consuming stage and affecting significantly the precision as well as the accuracy of the overall analysis [1,2]. Thus, a preliminary step of preconcentration and separation of analytes from the original sample matrix or matrix simplification is usually required.

Among sample pretreatment techniques which are used prior to determination, solid–phase extraction (SPE) is recognized as an advantageous alternative to classic liquid–liquid extraction (LLE), which is undesirable due to the recent green analytical chemistry (GAC) regulations, regarding the use of organic solvents [3,4]. SPE has been widely accepted thanks to several inherent advantages like simplicity, the ability to extract polar compounds, no requirement of phase separation, lower volumes of organic solvent which can be reached even at microliter level (50–100 µL) in automated procedures, lower sample pretreatment time, as well as lower cost of analysis. Moreover, SPE facilitates miniaturization and automation [5,6].

Following the trends of analytical chemistry on miniaturization, several approaches of sorbent microextraction have been developed such as solid-phase microextraction (SPME) [7–9], stir-bar sorptive extraction (SBSE) [10], thin film microextraction (TFME) [11] and related techniques as well as magnetic solid-phase extraction (MSPE) [12]. Despite the fact that SPME is a well-established microextraction technique, in some applications, it may not provide the desired sensitivity due to the small sorbent mass and sample capacity as well as mechanical distortion resulting in poor precision and sensitivity [13].

In order to increase the volume of the extraction phase as well as the active surface area, Pawliszyn et al. proposed a TFME approach based on a thin sheet of a polydimethylsiloxane (PDMS) organic polymer membrane as an extraction phase [11]. The membrane was attached to a stainless-steel rod which was immersed into the sample solution containing the target analytes. The extraction procedure took place in both direct and headspace extraction mode. After the extraction procedure, the membrane was rolled around the rod and placed into a gas chromatography (GC) injector for thermal desorption. In comparison with SPME, it presents higher extraction efficiency as well as shorter equilibrium time, thanks to the larger extraction phase (25–125 times more than a fiber). In 2012, Kermani and Pawliszyn reported a modification of TFME based on the distribution of a polymeric sorbent consisting of a mixed carboxen/polydimethylsiloxane (CAR/PDMS) and polydimethylsiloxane/divinylbenzene (PDMS/DVB), spread onto a glass wool fabric support as the substrate [14]. The extraction procedure was similar to TFME for GC-MS analysis. The resulted samplers presented better stability and robustness thanks to the incorporation of the fabric substrate in their thin film structure. Recently, a carbon mesh support was presented by Grandy et al., as an alternative substrate for the TFME technique [15] for GC-toroidial ion trap MS (GC-TMS) analysis. The proposed carbon substrate seems to be more durable than other TFME designs.

In 2014, Kabir and Furton [16] developed a novel highly promising and versatile sample preparation sorbent extraction technique named as fabric phase sorptive extraction (FPSE). FPSE successfully combines the advantages of sol–gel derived sorbents used in microextraction and the wide variety of fabric substrates, resulting in a highly efficient and green sample pretreatment technique [13,17]. Two main limitations of sorptive extraction techniques have been addressed by FPSE, the low sorbent capacity and long sample preparation time. The inherent porous surface of cellulose or polyester used as fabric substrate together with the strength of sol–gel derived hybrid sorbents uniformly dispersed as an ultra-thin film within the fabric substrate, results in a plethora of sorbent materials with significant analyte retention capacity and very fast extraction equilibrium. In comparison with a typical SPME fiber, the sorbent loading in FPSE media is about 400-times higher. Also unlike SPME, the extraction sorbent is dispersed homogeneously on the surface of nanometer size polyester/cellulose micro-fibrils of FPSE [13]. Regarding the elution of the retained analytes from the FPSE media in organic solvents, this is also fast without the potential carryover risk. Highly acidic or basic chemical environments as well as any organic solvent can be used as eluents. The advantages of the FPSE technique include: (a) simplicity, low cost, minimal consumption of solvents; (b) sample preparation can be completed by directly introducing the FPSE media into the vessel containing the sample matrix; (c) enhanced efficiency by sonication, magnetic stirring; (d) a plethora of organic solvents can be used as eluent; (e) minimization of sample preparation steps, reducing potential sources of errors; (f) a variety of effective sol–gel coatings can be employed as sorbent; (g) high analyte preconcentration factors; (h) high chemical resistance of the FPSE media thanks to a strong chemical bonding between the sorbent phase and the substrate.

Typically, the FPSE procedure starts with the immersion of the FPSE medium in a solvent system to clean any unwanted residue, followed by subsequent rinsing with deionized water to remove residual organic solvents. An amount of sample containing the target analytes is taken into a screw-capped glass vial. The FPSE medium is inserted into the vial along with a clean Teflon-coated magnetic stir bar. The sample solution is stirred for a defined extraction time for the sorption of the analytes. The FPSE medium is then removed from the extraction vial and is inserted into another vial containing the eluting solvent, for 4–10 min. Finally, the eluent is centrifuged and filtered to remove any particulate matter prior to injection into HPLC or other systems [18]. The FPSE procedure is shown in Figure 1. The FPSE medium can be reused by washing with the solvent system or it can be left to dry on a watch glass and stored in an air-tight glass container for future use.

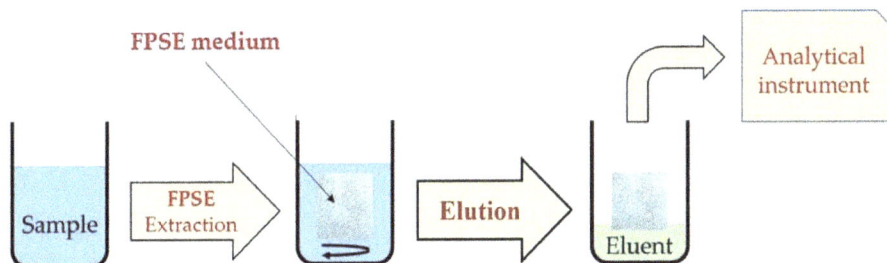

**Figure 1.** Schematic presentation of the main steps involved in FPSE process.

The recent trend of solid-phase extraction and microextraction is related to the development and characterization of new sorbent materials. The main targets in the search of novel sorbents depend on the extraction mode as well as on the analytes and samples, including better selectivity (or even specificity towards definite target species), improved sorptive or adsorptive capacity, as well as enhanced thermal, chemical or mechanical stability of the extractive media [19].

Sol–gel technology has many features in producing new materials of high purity and homogeneity, in forms of bulk, fibers, sheets, coating films as well as particles [20]. In analytical chemistry, the sol–gel process is commonly used in the synthesis of materials as sorbents for sample preparation techniques like SPE, SPME, and SBSE. The first demonstration of sol–gel technology for preparation of SPME fibers was presented by Malik and co-workers [21]. Since then, a wide variety of sol–gel sorbents with high selectivity, extraction sensitivity and FPSE applications have been presented.

In FPSE, a large pool of sol–gel sorbent materials with unique properties is available including a variety of polymers coated on either hydrophilic or hydrophobic substrates. The characteristics of the FPSE media used in the developed FPSE methods are given in Table 1.

**Table 1.** Characteristics of the FPSE media used in the developed methods.

| Sol–gel Coating | Sorbent Loading (mg·cm$^{-2}$) | Fabric Substrate | Polarity |
|---|---|---|---|
| PDMDPS | 1.93 | Cellulose/Polyester | Non-polar |
| C$_{18}$ | 2.4 | Cellulose | Non-polar |
| PDMS | 2.3 | Cellulose | Non-polar |
| PTHF | 3.96 | Cellulose | Medium polar |
| PEO-PPO-PEO | 5.68 | Cellulose | Polar |
| Graphene | 7.57 | Cellulose | Polar |
| PEG-PPG-PEG | 5.68 | Cellulose | Polar |
| PEG | 8.64 | Cellulose | Highly polar |

The objective of the present review article is to address, from both a presentative and critical point of view, the FPSE technique as well as its applications in diverse fields of analytical chemistry, either in batch or automated mode. In addition, the most important information regarding the principles of the FPSE technique and the sol–gel sorbent materials used for this purpose along with the process of preparing each substrate and the procedure of sol–gel coating are also presented.

## 2. Sol–gel Technology in Developing Microextraction Sorbents

Sol–gel technology is an interesting approach for the synthesis of inorganic polymers and organic–inorganic hybrid porous products of various sizes, shapes and formats like films, fibers, particles and monoliths, by employing mild reaction conditions [22]. By carefully modifying the synthesis process, the resulted materials are thermally and chemically stable with better homogeneity and purity, tunable porosity, and selectivity. The most important advantages of sol–gel technology for sorbent micro-extraction are the strong retention of the coating onto the substrate due to chemical bonding as well as the reduction of the extraction equilibrium time and the fast mass transfer thanks to the inherent porous structure [17].

A sol–gel process involves the catalytic hydrolysis of sol–gel precursor(s) and subsequent polycondensation of the hydrolyzed precursor(s), resulting in the transition of a liquid colloidal (solid particles with diameter of 1–100 nm) suspension known as "sol" into a 3D network of solid matrix "gel" with pores of sub-micrometer dimensions and polymeric chains whose average length is greater than a micrometer [23].

Initially, in the sol–gel process an inorganic or organically-modified inorganic precursor, for example methyltrimethoxysilane (MTMOS) is mixed with water and solvent in the presence of acid/base/fluoride catalyst. Afterwards, the hydrolyzed sol–gel precursor is condensed leading to the formation of a growing sol–gel network, into which a sol–gel active organic polymer, like hydroxy-terminated PDMS, can become chemically integrated. The reactions of the sol–gel process are schematically represented in Figure 2. As a result, a surface-bonded sol–gel hybrid organic-inorganic polymeric network is created and a new sol–gel hybrid material is synthesized.

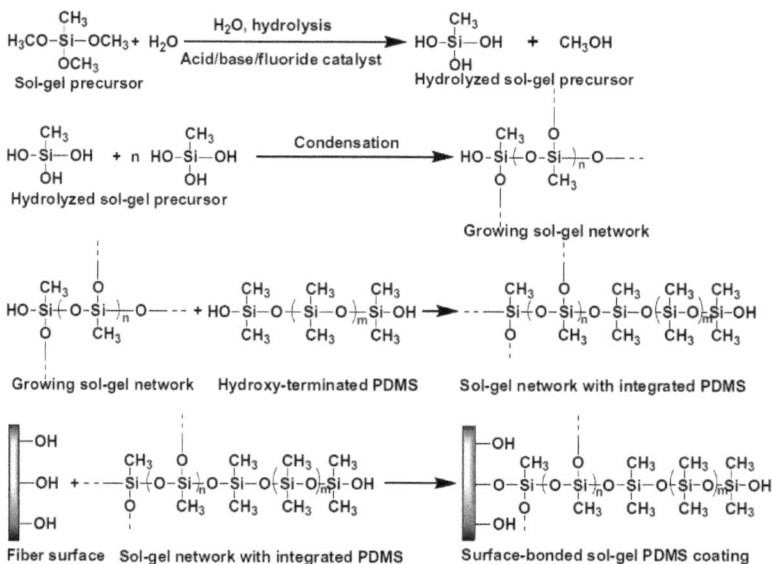

**Figure 2.** Schematic representation of chemical reactions involved in the synthesis of sol–gel hybrid organic–inorganic sorbents. Reproduced from [17] with permission of Elsevier.

A wide variety of available organically modified sol–gel precursors with different polarities can be used in the sol solution design to complement the overall polarity of the extraction sorbent [23]. The characteristics and the chemical properties of sol–gel hybrid organic-inorganic sorbents are affected by several factors, including the nature and type of precursors, the precursor to water ratio, the type of catalyst and its concentration, the pH of sol-solution, the organic solvent, the temperature and humidity during reactions, as well as the post-gelation aging conditions. The chemical structure of the produced sol–gel matrix depends on the type of the catalyst, used in the sol solution. The most commonly used precursors as well as different types of catalysts employed in sol–gel technology are given in Table 2.

**Table 2.** Types of precursors and catalysts used in a sol–gel process.

| Sol–gel Precursor | | Operation |
|---|---|---|
| Unchanged: | TEOS, Titanium isopropoxide, Zirconium butoxide, Tetramethoxygermane, Alumina | Inorganic component to the hybrid polymeric network. Offers active hydroxyl group to facilitate chemical bonding to the fiber/capillary surface. |
| Organically modified: | MTMS, $C_{18}$-MTMS, $C_8$-TMS, TMSPA, VTEOS, PTMOS | Forms inorganic backbone of the hybrid material. Organic moieties provide intermolecular interactions between analytes and the sorbent. |
| **Catalyst** | | **Operation** |
| Acid/Fluoride/Base: | GAA, HCl, $HNO_3$, Citric acid, TFA, HF, $NH_4OH$, NaOH, Basic amino acids | Catalyzes the hydrolysis and condensation reactions and controls the network structure and porosity. Acid catalyzed sol–gel materials possess weakly branched microporous structures. Base catalyzed sol–gel materials have highly branched particulate structures with large pore sizes. |

## 3. Preparation of FPSE Media

The preparation of FPSE media involves two main steps: (1) pretreatment of fabric substrates for sol–gel coating and (2) design and preparation of the sol solution for sol–gel coating process.

### 3.1. Pretreatment of Fabric Substrates

The segments of fabric substrates (e.g., cellulose, polyester) are first soaked with deionized water under sonication in order to become thoroughly wet. Fabric pieces are cleaned with a high amount of deionized water so that chemical residues are removed. Then, a process called mercerization follows, by treating the fabric with 1.0 mol·L$^{-1}$ NaOH under sonication and the mercerized fabric is washed several times with plenty of deionized water. The next step is treating the fabric with 0.1 mol·L$^{-1}$ HCl under sonication, washing again with deionized water, and finally drying overnight in an inert atmosphere. Dried fabric substrates are stored in clean glass airtight containers until they are coated with the appropriate sol–gel sorbent.

### 3.2. Preparation of the Sol Solution for the Sol–gel Coating Process

Designing the sol solution is the main step on the way to the development of a sol–gel sorbent due to the fact that its composition and the relative ratio of the constituents, define the porosity as well as the selectivity and specificity of the resulting sorbent [24]. For an effective sol–gel sorbent, the selection of the sol–gel active organic polymer, the inorganic or organically modified inorganic sol–gel precursor, the solvent/solvent system, the catalyst, the amount of water, as well as an appropriate relative molar ratio of the constituents must be considered. The composition of the sol solutions for the FPSE media preparation is given in details in Table 3.

The coating procedure is integrated by inserting gently the treated fabric pieces into the vial containing the sol solution. As a result, a three dimensional network is formed throughout the porous substrate matrix (cellulose, polyester etc.). After a predetermined coating time the fabric is taken away from the sol solution, dried, and placed in a desiccator overnight for solvent evaporation and for aging of the sol–gel coating. The objective of this step is to complete the condensation reaction and remove solvents and unreacted residuals from the sol–gel matrix, ensuring a clean, surface bonded sol–gel sorbent free of structural deformation and internal stress. The coated FPSE media is then rinsed with the appropriate solvent system under sonication for a few minutes in order to remove residual sol solution ingredients from the coated surface. The FPSE media is then cut into 2.5 cm × 2.0 cm pieces and stored in clean airtight containers, for future use.

*Separations* **2017**, *4*, 20

**Table 3.** Composition of the sol solutions for FPSE media preparation.

| FPSE Media | Substrate | Polymer | Precursor | Organic Solvent System | Catalyst | Reference |
|---|---|---|---|---|---|---|
| Sol–gel PDMDPS | Polyester | Poly(dimethyldiphenylsiloxane) | MTMS | Methylene chloride:acetone | TFA | [25–30] |
| Sol–gel PDMDPS | Polyester | Poly(dimethyldiphenylsiloxane) | 3-CPTEOS | Methylene chloride | TFA | [31] |
| Sol–gel PDMDPS | Cellulose | Poly(dimethyldiphenylsiloxane) | MTMS | Methylene chloride:acetone | TFA | [32] |
| Sol–gel PTHF | Cellulose | Poly(tetrahydrofuran) | MTMS | Methylene chloride:acetone | TFA | [13,24–26,28–30,32–35] |
| Sol–gel PTHF | Cellulose | Poly(tetrahydrofuran) | 3-CPTEOS | Methylene chloride | TFA | [31] |
| Sol–gel PEO-PPO-PEO | Cellulose | Poly(ethylene oxide)–poly(propylene oxide)–poly(ethylene oxide) triblock copolymer | MTMS | Methylene chloride: acetone | TFA | [36] |
| Sol–gel Graphene | Cellulose | Graphene | MTMS | Methylene chloride: acetone | TFA | [36] |
| Sol–gel PEG–PPG-PEG | Cellulose | Poly(ethyleneglycol)–block-poly(propyleneglycol)–block-Poly(ethyleneglycol) triblock copolymer | MTMS | Methylene chloride: acetone | TFA | [24], [29], [31], [37], |
| Sol–gel PEG | Cellulose | Poly(ethyleneglycol) | MTMS | Methylene chloride: acetone | TFA | [18,24–26,28–32,34,37] |
| Sol–gel C$_{18}$ | Cellulose | Octadecyl carbon chain | MTMS | Methylene chloride:acetone | TFA | [24], [37] |
| Sol–gel PDMS | Cellulose | Poly(dimethylsiloxane) | MTMS | Methylene chloride:acetone | TFA | [34] |

Images of the surface by scanning electron microscope (SEM) of (a) uncoated surface of cellulose fabric substrate at 100× magnification; (b) sol–gel poly-PEG coated surface of FPSE media at 100× magnification; (c) uncoated surface of cellulose fabric substrate surface at 500× magnification; and (d) sol–gel poly-PEG coated surface of FPSE media at 500× magnification are presented in Figure 3. Both low and high magnification SEM images of the sol–gel PEG coated FPSE media clearly demonstrate that the ultra-thin coating of the sol–gel sorbent is uniformly distributed throughout the substrate matrix without blocking the pores of the fabric. The easy access of the low viscosity sol solution into the fabric matrix helps in achieving this uniform ultrathin sol–gel PEG coating. The sol–gel PEG coating does not clog the through-pores of the fabric substrate, allowing an easy permeation of sample matrix through its body, which consequently helps to accomplish extraction equilibrium in a very short period of time [18].

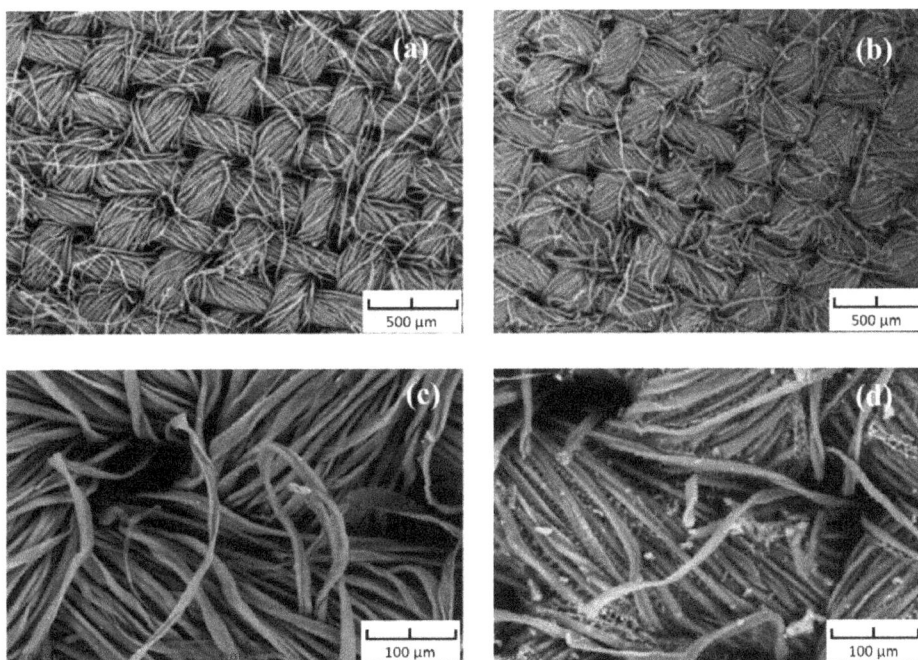

**Figure 3.** SEM images of (**a**) uncoated cellulose at 100× magnification; (**b**) PEG coated FPSE media at 100× magnification; (**c**) uncoated cellulose at 500× magnification; (**d**) PEG coated FPSE media at 500× magnification. Reproduced from [18] with permission of Elsevier.

## 4. Applications of Fabric Phase Sorptive Extraction

Since the presentation of FPSE in 2014 as a novel sample pretreatment technique prior to high performance liquid chromatography for the determination of selected estrogens, various applications for organic and inorganic analytes have been reported in the literature. All reported FPSE methods are presented briefly in Table 4, containing information about all FPSE procedures for sample preparation, the main parameters with the analytical performance characteristics, as well as the analytes and the types of samples. Most of the reported methods were applied for drugs and pharmaceuticals using HPLC or UPLC coupled with DAD, MS or MS/MS.

**Table 4.** Applications of the FPSE technique.

| Analytical Technique | Sol–gel Coating | Fabric Substrate | Elution Solvent | E.T. min | Sample | Analyte | Type of Analyte | EF | LOD*, ng·L⁻¹a, ng·g⁻¹ | LOQ**, ng·L⁻¹b, ng·g⁻¹ | CCα μg·kg⁻¹ | CCβ μg·kg⁻¹ | R(%),c Rapp (%) | Reference |
|---|---|---|---|---|---|---|---|---|---|---|---|---|---|---|
| FPSE-HPLC-FLD | PTHF | Cellulose | Methanol | 20 | Urine, Ground-River-Drinking water | BPA | Estrogens | 13.9 | 42 | 139 | – | – | 88.7–96.4 | [13] |
|  |  |  |  |  |  | E2 |  | 14.4 | 20 | 66 |  |  | 89.4–97.4 |  |
|  |  |  |  |  |  | EE2 |  | 14.7 | 36 | 119 |  |  | 89.0–98.0 |  |
| FPSE-HPLC-DAD | PEG | Cellulose | Methanol-acetonitrile | 20 | Blood serum | BRZ | Drugs Benzodiazepines | – | 10 | 30 | – | – | 91.6–97.6 | [26] |
|  |  |  |  |  |  | DZP |  |  |  |  |  |  | 90.0–102.2 |  |
|  |  |  |  |  |  | LRZ |  |  |  |  |  |  | 87.6–95.5 |  |
|  |  |  |  |  |  | APZ |  |  |  |  |  |  | 93.2–106 |  |
| FPSE-HPLC-DAD | PEG | Cellulose | Acetonitrile | 40 | Intact milk | PENG | Antibiotics Penicillin | – | 3.0ᵃ | 10.0ᵇ | 11.2 | 12.3 | 89.2–104.3 | [37] |
|  |  |  |  |  |  | CLO |  |  | 6.0ᵃ | 20.0ᵇ | 32.8 | 35.4 | 82.8–88.1 |  |
|  |  |  |  |  |  | DICLO |  |  | 7.5ᵃ | 25.0ᵇ | 33.2 | 36.1 | 80.8–92.4 |  |
|  |  |  |  |  |  | OXA |  |  | 9.0ᵃ | 30.0ᵇ | 33.0 | 36.7 | 82.6–90.5 |  |
| FPSE-HPLC-DAD | PEG | Cellulose | Methanol-acetonitrile | 30 | Raw milk | TAP | Antibiotics Amphenicols | – | – | – | 52.49 | 56.8 | 90.5–103.3 | [18] |
|  |  |  |  |  |  | FF |  |  |  |  | 55.23 | 58.99 | 92.3–103.3 |  |
|  |  |  |  |  |  | CAP |  |  |  |  | 53.8 | 55.9 | 97.0–106.6 |  |
| FPSE-HPLC-UV | PEG | Cellulose | Methanol-acetonitrile | 13 | Raw milk | SMTH | Antibiotics Sulfonamides | – | – | – | 116.5 | 120.4 | 94.7–107 | [24] |
|  |  |  |  |  |  | SIX |  |  |  |  | 114.4 | 118.5 | 93.0–104.6 |  |
|  |  |  |  |  |  | SDMX |  |  |  |  | 94.7 | 104.1 | 96.1–102.7 |  |
| FPSE-HPLC-UV | PTHF | Cellulose | Methanol | 25 | Ground-River water, Treated water, Soil, Sludge | 4-TBP | Chemicals Alkyl phenols | – | 182 | 601 | – | – | 90.1–95.0 | [33] |
|  |  |  |  |  |  | 4-SBP |  |  | 179 | 599 |  |  | 90.6–95.7 |  |
|  |  |  |  |  |  | 4-TAP |  |  | 192 | 640 |  |  | 89.0–96.1 |  |
|  |  |  |  |  |  | 4-CP |  |  | 161 | 531 |  |  | 91.1–96.0 |  |
| FPSE-UPLC-MS | PTHF | Cellulose | Acetonitrile | 20 | Food simulants | DEP | Chemicals non-volatile Migrants | 3.1 | 5.0ᵃ | 15ᵇ | – | – | 67.6 | [34] |
|  |  |  |  |  |  | TBC |  | 6.4 | 1.0ᵃ | 3ᵇ |  |  | 104.8 |  |
|  |  |  |  |  |  | DBM |  | 6.6 | 3.0ᵃ | 10ᵇ |  |  | 112.0 |  |
|  |  |  |  |  |  | TBoAC |  | 7.3 | 1.0ᵃ | 3ᵇ |  |  | 83.3 |  |
|  |  |  |  |  |  | TXIB |  | 5.1 | 1.0ᵃ | 3ᵇ |  |  | 87.4 |  |
|  |  |  |  |  |  | HAA C12 |  | – | 7.0ᵃ | 20ᵇ |  |  | 53.1 |  |
|  |  |  |  |  |  | DBP |  | 5.8 | 10ᵃ | 30ᵇ |  |  | 91.5 |  |
|  |  |  |  |  |  | TINU326 |  | 11.0 | 10ᵃ | 25ᵇ |  |  | 72.1 |  |
|  |  |  |  |  |  | CHIMA81 |  | 1.8 | 2.0ᵃ | 10ᵇ |  |  | 100.8 |  |
|  |  |  |  |  |  | TINU327 |  | 3.2 | 10ᵃ | 30ᵇ |  |  | 80.6 |  |
|  |  |  |  |  |  | 2EHAdip |  | – | 1.0ᵃ | 3ᵇ |  |  | 9.1 |  |
|  |  |  |  |  |  | 2EHSeb |  | 2.9 | 1.0ᵃ | 3ᵇ |  |  | 64.7 |  |
|  |  |  |  |  |  | CYA1084 |  | – | 12ᵃ | 30ᵇ |  |  | 86.5 |  |
|  |  |  |  |  |  | IRGA38 |  | – | 1.0ᵃ | 3ᵇ |  |  | 78.1 |  |
|  |  |  |  |  |  | TOPAC |  | 12.0 | 5.0ᵃ | 15ᵇ |  |  | 33.3 |  |
|  |  |  |  |  |  | IRGA1076 |  | – | 3.0ᵃ | 10ᵇ |  |  | 80.4 |  |
|  |  |  |  |  |  | IRGA168 |  | – | 3.0ᵃ | 10ᵇ |  |  | 45.7 |  |
|  |  |  |  |  |  | IRGA1010 |  | – | 3.0ᵃ | 10ᵇ |  |  | 67.6 |  |

Table 4. Cont.

| Analytical Technique | Sol–gel Coating | Fabric Substrate | Elution Solvent | E.T. min | Sample | Analyte | Type of Analyte | EF | LOD*, ng·L⁻¹ᵃ, ng·g⁻¹ | LOQ**, ng·L⁻¹ᵇ, ng·g⁻¹ | CCα µg·kg⁻¹ | CCβ µg·kg⁻¹ | R(%);ᶜ, Rapp (%) | Reference |
|---|---|---|---|---|---|---|---|---|---|---|---|---|---|---|
| FPSE-UHPLC-MS/MS | PTHF | Cellulose | Methanol-acetonitrile | 20 | Tap water, Osmosis effluent wastewater, Untreated effluent/ biological wastewater | NORET | Drugs Androgens & Progestogens | – | 33.5 | | – | – | 80.6–94.4 | [35] |
| | | | | | | NOR | | | 1.7 | | | | 94.1–103.5 | |
| | | | | | | MGA | | | 21.4 | | | | 102.2–121.2 | |
| | | | | | | PRO | | | 6.9 | | | | 79.8–84.2 | |
| | | | | | | BOL | | | 46.9 | | | | 66.6–76.2 | |
| | | | | | | NAN | | | 50.7 | | | | 82.8–102.4 | |
| | | | | | | ADTD | | | 19.4 | | | | 63.9–77.9 | |
| | | | | | | DHEA | | | 264 | | | | 77.6–87.4 | |
| | | | | | | TES | | | 2.2 | | | | 76.6–81.4 | |
| | | | | | | AND | | | 63.6 | | | | 92.2–98.9 | |
| FPSE-UHPLC-MS/MS | PTHF | Cellulose | Methanol-acetonitrile | 20 | Urine | NORET | Drugs Androgens & Progestogens | – | 33.5 | | – | – | – | [35] |
| | | | | | | NOR | | | 1.7 | | | | | |
| | | | | | | MGA | | | 11.1 | | | | | |
| | | | | | | PRO | | | 12.8 | | | | | |
| | | | | | | BOL | | | 37.9 | | | | | |
| | | | | | | NAN | | | 50.1 | | | | | |
| | | | | | | ADTD | | | 25.6 | | | | | |
| | | | | | | DHEA | | | 110.6 | | | | | |
| | | | | | | TES | | | 8.9 | | | | | |
| | | | | | | AND | | | 80.0 | | | | | |
| FPSE-UHPLC-MS/MS | PDMDPS | Polyester | Methanol | 60 | Seawater | UV P | UV Stabilizers Personal care | – | 5.63 | 18.8 | – | – | – | [28] |
| | | | | | | UV 329 | | | 4.33 | 14.5 | | | | |
| | | | | | | UV 326 | | | 8.96 | 29.9 | | | | |
| | | | | | | UV 328 | | | 1.63 | 5.44 | | | | |
| | | | | | | UV 327 | | | 1.06 | 3.54 | | | | |
| | | | | | | UV 360 | | | 2.72 | 9.08 | | | | |
| FPSE-UHPLC-MS/MS | PDMDPS | Polyester | Methanol | 60 | Sewage | UV P | UV Stabilizers Personal care | – | 12.8–25.3 | 42.7–84.3 | – | – | 82–96 | [27] |
| | | | | | | UV 329 | | | 12.2–19.8 | 40.7–66.0 | | | 48–61 | |
| | | | | | | UV 326 | | | 51.6–60.7 | 172–202 | | | 49–58 | |
| | | | | | | UV 328 | | | 9.44–18.1 | 31.5–60.3 | | | 43–59 | |
| | | | | | | UV 327 | | | 36.2–38.6 | 121–129 | | | 65–73 | |
| | | | | | | UV 571 | | | 40.0–44.3 | 133–148 | | | 49–53 | |
| | | | | | | UV 360 | | | 6.01–7.34 | 20.0–24.5 | | | 35–46 | |
| FPSE-LC-MS/MS | PEG | Cellulose | Methanol | 240 | River water, Effluent-Influent wastewater | MPB | Pharmaceuticals Personal care | – | 10 | 50 | – | – | 9–27ᶜ | [31] |
| | | | | | | CBZ | | | 10 | 50 | | | 20–92ᶜ | |
| | | | | | | PrPB | | | 2 | 20 | | | 41–65ᶜ | |
| | | | | | | DHB | | | 5 | 50 | | | 44–74ᶜ | |
| | | | | | | BzPB | | | 1 | 20 | | | 45–67ᶜ | |
| | | | | | | DHMB | | | 2 | 20 | | | 50–74ᶜ | |
| | | | | | | DICLO | | | 1 | 20 | | | 44–73ᶜ | |
| | | | | | | BP-3 | | | 2 | 20 | | | 59–93ᶜ | |
| | | | | | | TCC | | | 3 | 10 | | | 57–59ᶜ | |
| | | | | | | TCS | | | 50 | 200 | | | 43–54ᶜ | |
| FPSE-GC-MS | PEG | Cellulose | Ethyl acetate | 120 | River water, Wastewater | IBU | Drugs anti-inflammatory | 418 | 0.8 | 3 | – | – | 82–109 | [25] |
| | | | | | | NAP | | 263 | 2 | 3 | | | 93–111 | |
| | | | | | | KET | | 223 | 5 | 15 | | | 92–108 | |
| | | | | | | DIC | | 162 | 2 | 7 | | | 94–116 | |

**Table 4.** *Cont.*

| Analytical Technique | Sol-gel Coating | Fabric Substrate | Elution Solvent | E.T. min | Sample | Analyte | Type of Analyte | EF | LOD*, ng·L⁻¹a, ng·g⁻¹ | LOQ**, ng·L⁻¹b, ng·g⁻¹ | CCα µg·kg⁻¹ | CCβ µg·kg⁻¹ | R(%);c, Rapp (%) | Reference |
|---|---|---|---|---|---|---|---|---|---|---|---|---|---|---|
| Stir-FPSE-UPLC-DAD | PEG | Cellulose | Methanol | 60 | River water | Simazine | Herbicides | 444 | 140 | 460 | — | — | 84–124 | [30] |
|  |  |  |  |  |  | Atrazine |  | 729 | 240 | 790 |  |  | 75–126 |  |
|  |  |  |  |  |  | Secbumeton |  | 988 | 80 | 260 |  |  | 76–103 |  |
|  |  |  |  |  |  | Terbumeton |  | 1165 | 80 | 260 |  |  | 75–104 |  |
|  |  |  |  |  |  | Propazine |  | 996 | 110 | 360 |  |  | 75–97 |  |
|  |  |  |  |  |  | Prometryn |  | 1286 | 470 | 1500 |  |  | 78–111 |  |
|  |  |  |  |  |  | Terbutryn |  | 1411 | 80 | 260 |  |  | 78–99 |  |
| Stir-FPSE-HPLC-DAD | PTHF | Cellulose | Acetonitrile | 15 | Wastewater, Reservoir water | TBBPA | Flame Retardants | — | 30 | — | — | — | 93 | [32] |
|  |  |  |  |  |  | TBBPA-BAE |  |  | 20 |  |  |  | 95 |  |
|  |  |  |  |  |  | TBBPA-BDBPE |  |  | 40 |  |  |  | 92–99 |  |
| Stir-bar-FPSE-HPLC-DAD PTHF |  | Cellulose | Acetonitrile | 10 | Wastewater, Reservoir water | TBBPA | Flame Retardants | — | 10 | — | — | — | 92–95 | [32] |
|  |  |  |  |  |  | TBBPA-BAE |  |  | 50 |  |  |  | 90–97 |  |
|  |  |  |  |  |  | TBBPA-BDBPE |  |  | 10 |  |  |  | 91–98 |  |
| DFSE-LC-MS/MS | PEG | Cellulose | Ethyl acetate | 10 | River water, Influent-Effluent wastewater | MPB | Pharmaceuticals Personal care | — | 4 | 50 | — | — | 12–30 c | [29] |
|  |  |  |  |  |  | CBZ |  |  | 4 | 50 |  |  | 18–53 c |  |
|  |  |  |  |  |  | PrPB |  |  | 2 | 50 |  |  | 20–64 c |  |
|  |  |  |  |  |  | DHB |  |  | 2 | 50 |  |  | 21–68 c |  |
|  |  |  |  |  |  | BzPB |  |  | 2 | 50 |  |  | 33–70 c |  |
|  |  |  |  |  |  | DHMB |  |  | 2 | 20 |  |  | 39–76 c |  |
|  |  |  |  |  |  | DICLO |  |  | 2 | 50 |  |  | 23–50 c |  |
|  |  |  |  |  |  | BP-3 |  |  | 2 | 100 |  |  | 45–52 c |  |
|  |  |  |  |  |  | TCC |  |  | 8 | 50 |  |  | 15–49 c |  |
|  |  |  |  |  |  | TCS |  |  | 20 | 100 |  |  | 22–43 c |  |
| FDSE-FI-FAAS | PDMDPS | Polyester | Methyl isobutyl ketone | 1.5 | River-Coastal-Ditch water | Lead | Toxic Metals | 140 | 1.8 µg·L⁻¹ | 6.0 µg·L⁻¹ | — | — | 95.0–101.0 | [36] |
|  |  |  |  |  |  | Cadmium |  | 38 | 0.4 µg·L⁻¹ | 1.2 µg·L⁻¹ |  |  | 94.0–98.0 |  |

*; LOD calculated by S/N = 3, **; LOQ calculated by S/N = 10, E.T.; extraction time, EF; enrichment factor, a,b, LOD and LOQ values given in ng·g⁻¹; c, Rapp (%); Apparent recovery including the extraction recovery and the matrix effect.

In the first FPSE application, Kumar et al. [13] developed a FPSE-HPLC-FLD method for selected estrogen (BPA, E2 and EE2) determination in urine and environmental (ground, river and drinking) water samples by HPLC coupled with an FLD detection system. The estrogens which are medium polar compounds, were extracted using a medium polar FPSE medium prepared by sol–gel PTHF coated on a cellulose substrate. The FPSE procedure took place in a glass vial containing the target analytes along with the coated FPSE medium and followed by magnetic stirring for an efficient adsorption. A volume of 500 μL of methanol was used to elute the analytes from the fabric. After 5 min centrifugation and filtration of the eluent, it was injected into the HPLC system. The proposed method presented good analytical characteristics as shown in Table 4.

Samanidou et al. [26] used the FPSE technique for the determination of selected benzodiazepines, APZ, BRZ, DZP, and LRZ in blood serum by HPLC-DAD. Benzodiazepines are commonly prescribed drugs with sedative, anti-depressive, tranquilizing, hypnotic, and anticonvulsant properties. Their determination in biological fluids is of great importance not only in clinical assays, but also in forensics and toxicological studies. Three different FPSE sorbents were examined; the sol–gel PEG and sol–gel PTHF coated on cellulose substrates and the sol–gel PDMDPS coated on a polyester substrate. The optimization experiments showed that sol–gel PEG was the most appropriate FPSE medium to extract benzodiazepines. After conditioning the FPSE medium, it was inserted into a glass vial containing the sample solution along with a Teflon-coated magnetic stirrer. The sample was stirred for about 20 min and then, elution followed by immersing the FPSE medium into another vial containing an elution solvent system of acetonitrile: methanol (50:50 $v/v$). The elution time was 10 min. The extracted benzodiazepines were analyzed directly in the HPLC instrument with photodiode array detector (DAD) operated at 240 nm. In this work, the FPSE sol–gel PEG medium was introduced directly into the matrix without prior deproteinization, resulting in the minimization of sample preparation steps as well as in the elimination of probable errors during the sample pretreatment procedure. However, it is not mentioned if the FPSE medium is replaced for each analytical cycle as it could become clogged from complex matrices like blood, milk, etc.

Guedes-Alonso et al. [35] developed a simple, fast and sensitive method for the quantification of natural and synthetic steroid hormones, androgens and progestogens by coupling sol–gel PTHF coated FPSE medium with UHPLC-MS/MS. Steroid hormones have been widely used in both human and veterinary medicine and their effective determination in water samples is of crucial importance for the assessment of the concentration levels and their related ecological risk. A UHPLC system coupled to a triple quadrupole detector was used for the quantification of selected androgens and progestogens, NORET, NOR, MGA, PRO, BOL, NAN, ADTD, DHEA, TES, and AND, in tap water and wastewater treated with different techniques as well as in urine samples. The FPSE procedure took place in glass vials with a Teflon coated magnetic stirrer. After submerging the FPSE medium into the sample solution (water or urine samples), it was stirred at 1000 rpm, with an extraction efficiency maximum achieved in 20 min and elution in 3 min. Methanol in a small quantity, approximately 0.75 mL, was used as the elution solvent. The recovery of the developed method ranged from 65.9% to 121.2%. Both repeatability values (intra-and inter-day) were in all type of samples lower than 20%.

The FPSE technique was also applied for the determination of four non-steroidal anti-inflammatory drugs, namely: IBU, NAP, KET, and DIC, in environmental water samples in combination with GC-MS [25]. Up to now, this is the only method that associates FPSE with the GC-MS analytical technique. Three different sorbents, sol–gel PDMDPS on a polyester substrate, sol–gel PTHF and sol–gel PEG on cellulose substrates, were investigated. The sol–gel PEG coated FPSE medium was found to be the most efficient one for the NSAIDs analytes independently of the pH and the ionic strength. The FPSE extraction was performed in a glass screw-cap vial by immersing the FPSE medium into the sample solution with the help of a tweezers. The solution was magnetically stirred at 500 rpm for 2 h. Ethyl acetate was used to elute the analytes with an elution time of 15 min. For the elution of the compounds, the FPSE medium was introduced into a 1 mL sample of ethyl acetate for 15 min. Subsequently, the FPSE medium was removed from the vial and the extract was evaporated to dryness by nitrogen stream.

The residue was reconstituted into a 100 μL insert with 40 μL of ethyl acetate and 10 μL of MTBSTFA. The derivatization reaction took place in an oven at 60 °C for 60 min. After cooling to room temperature, the extract was ready for GC-MS analysis. The total runtime was 38.2 min. The instrument was a gas chromatograph equipped with a capillary column, combined with a quadrupole mass spectrometer. Recoveries were in the range of 82%–116% and RSDs between 3.5% and 18%.

Samanidou et al. [18] reported a facile sol–gel synthesis to incorporate short-chain PEG into the sol–gel matrix and anchor the growing sol–gel PEG network to a flexible hydrophilic cellulose substrate, resulting in a homogenous and ultra-thin, highly polar FPSE medium. The sol–gel PEG FPSE medium was applied for the first time to assess the concentration of amphenicol drugs in raw milk samples. The extraction of amphenicols residues, TAP, FF, and CAP from raw milk followed by HPLC-DAD was investigated. The sol–gel PEG FPSE medium together with an aliquot of milk sample (0.5 g) with 500 μL DI water (or 0.5 g milk spiked with 500 μL of amphenicols standard solution) was magnetically stirred for 30 min. Methanol was used as the elution solvent and elution was performed for 10 min. The extract was centrifuged for 15 min and filtered prior to HPLC. The FPSE medium could be reused by washing with 2 mL acetonitrile: methanol (50:50 *v/v*) for 5 min. The proposed method showed good linearity and sensitivity. The precision was evaluated within a day and between days and ranged from 1.0%–10.7% and 7.6%–14.0% for TAP, FF, and CAP respectively. In this study, no protein precipitation or solvent evaporation and sample reconstitution were necessary, thanks to the FPSE technique which reduces the sample preparation steps, ending up in a short extraction equilibrium time pretreatment technique.

Karageorgou et al. [24] developed a simple FPSE-HPLC-UV method for the isolation of three sulfonamides, SMTH, SIX, and SDMX, from untreated milk samples. The highly polar sol–gel PEG coated on a cellulose substrate was used for the extraction of the selected sulfonamides. The FPSE procedure was the same as the previous method [18]. After the extraction, the FPSE medium was first inserted into a clean vial with 250 μL MeOH for 8 min and then into another vial with 250 μL ACN for a further 5 min for the elution of the analytes. The resulting solution was filtered prior to HPLC injection to remove any particulate matter. The same medium could be reused up to 30 times without any significant loss in extraction performance. The precision of the method was estimated for within-day repeatability and between-day repeatability with RSDs ranging from 5.6% to 6.7% respectively. The flexibility of the FPSE medium facilitates the easy insertion of the fabric into the sample solution, resulting in the fast extraction of the analytes. In addition, neither prior pretreatment of the sample was carried out including protein precipitation, nor solvent evaporation was performed followed by sample reconstitution. Another application from the same group, using the sol–gel PEG FPSE medium on cellulose substrate and offering the same advantages regarding the reduction of sample preparation steps, was reported for the extraction of four penicillin antibiotic residues (PENG, CLO, DICLO, and OXA) from cows' milk [37]. Sol–gel PEG FPSE medium was adopted for method optimization and validation among sol–gel $C_{18}$ and sol–gel PEG-PPG-PEG all coated on cellulose substrates, as the first one provided better absolute recovery values ranging between 22% and 58%, than 5% provided by the two other sorbent FPSE media.

FPSE has been also employed for the preconcentration of four endocrine disruptor alkylphenol molecules, namely, 4-TBP, 4-SBP, 4-TAP, and 4-CP in aqueous and soil samples followed by HPLC-UV detection [33]. The sol–gel PTHF on a cellulose substrate was selected for the extraction procedure. The sample was stirred at a speed of 1000 rpm for an extraction time of 25 min and then elution with methanol followed, for 6 min. The eluent was centrifuged for 5 min, filtered with a syringe filter and finally it was injected into the HPLC system. Relative recoveries were satisfactory and ranged from 91% to 97% in aqueous samples, while for soil and sludge samples they were lower between 89% and 91%, probably due to the complicated sample matrix.

Another interesting application of FPSE using UPLC-MS(QqQ) is for migration analysis of several non-volatile additives in food packaging materials [34]. Three different sol–gel coated FPSE media with diverse polarities were investigated: sol–gel PDMS (non-polar), sol–gel PTHF (medium polar) and sol–gel PEG (highly polar) all coated on cellulose substrates as well as several extraction parameters were optimized. The proposed method was applied for 18 plastic additives: eight plasticizers (DEP, TBC, DBM, TBoAC, TXIB, DBP, 2EHAdip, 2EHSeb), five antioxidants (IRGA38, TOPAC, IRGA1076, IRGA168 and IRGA1010), four UV absorbers (TINU326, CHIMA81, TINU327, CYA1084) and one antistatic agent (HAA C12). Three different food simulants with concentration of 300 μg·L$^{-1}$ of the target analytes: A, ethanol 10%; B, acetic acid 3%, and D1, ethanol 50% were prepared to optimize the FPSE protocol. Better retention was achieved using sol–gel PTHF and sol–gel PEG coated FPSE media with analytes dissolved in ethanol 10% or acetic acid 3%. For these simulants, the retention efficiency was over 75%. On the other hand, they were slightly retained using ethanol 50% as simulant. The extraction of these compounds with the low polar sol–gel PDMS was, as expected, lower than the other FPSE media, especially for DEP. Generally, compounds with low logP values proved to have higher enrichment factors, especially with sol–gel PTHF and sol–gel PEG FPSE media. The use of sol–gel PDMS improved the enrichment capacity, in the case of compounds with high log*P* values. Sample extraction assisted by magnetic stirring at 700 rpm was optimized at 20 min and 10 min for solvent elution assisted by ultrasound. Acetonitrile was adopted as the elution solvent since recoveries were higher than 70% for 13 out of the 18 selected compounds in all FPSE media. The best extraction recovery values were achieved when analytes were dissolved in 3% aqueous acetic acid solution, where 17 out of 18 compounds showed improved sensitivity and 10 of them obtained enrichment factors higher than 3 for all tested FPSE media. When FPSE eluents were evaporated using nitrogen, 11 out of 18 compounds reached EFs higher than 100. This significant improvement of the sensitivity was based on the combination of FPSE technique with nitrogen evaporation allowing the determination of such analytes at very low concentrations in various types of samples.

Recently, Montesdeoca-Esponda et al., applied the FPSE methodology followed by UHPLC-MS/MS detection for the determination of benzotriazole UV stabilizers (BUVSs) in sewage samples [27]. BUVSs are classified as emerging pollutants used in different personal care products such as sunscreens, soaps, shampoos, lip gloss, hair dyes or makeup that can affect the aqueous environmental ecosystem in various ways. The target analytes were: UV P, UV 329, UV 326, UV 328, UV 327, UV 571, and UV 360. The non-polar sol–gel PDMDPS coated on a polyester substrate was used to extract the analytes. A sample of volume 10 mL together with the fabric were stirred at 1000 rpm for 60 min and the analytes were eluted with 1.0 mL methanol for 5 min. Under these conditions, the preconcentration factor was 10 times. A UPLC system coupled with a triple quadropole detector with an ESI interface was used to determine the target analytes. The proposed method was applied to sewage samples with recoveries ranging from 42% to 99%.

Garcia-Guerra et al. [28] reported a FPSE-UHPLC-MS/MS method for benzotriazole UV stabilizers' (UV P, UV 329, UV 326, UV 328, UV 327, and UV 360) determination in seawater samples collected from beaches used by tourists where the direct input of these compounds may be significant. Three different coated FPSE media; sol–gel PDMDPS, sol–gel PTHF, and sol–gel PEG were evaluated using aqueous solutions at pH 6 as initial conditions. Among them, sol–gel PDMDPS coated on a polyester substrate, showed the greatest capability to extract the non-polar target analytes, so it was selected for further experiments. For 60 min extraction time and 10 min elution in methanol the developed method provided absolute recoveries in the range from 40.9% to 44.3%, except for UV P and UV 329 whose values were between 9.30% and 20.6%.

Lakade et al. [31] presented a comparative study of four FPSE media: non-polar sol–gel PDMDPS, medium-polar sol–gel PTHF, polar sol–gel PEG-PPG-PEG triblock, and polar sol–gel PEG to extract a group of PPCPs (log$K_{ow}$ values range from −0.6 to 6.1) from environmental water samples. The FPSE sol–gel PEG media was selected as the most appropriate sorbent material for the target PPCPs (MPB, CBZ, PrPB, DHB, BzPB, DHMB, DICLO, BP-3, TCC, and TCS) in the developed method. The analytes were determined in river and wastewater samples by LC-MS/MS. A portion of 1.0 mL of methanol was used as the elution solvent with a time of 5 min and the total extraction time was extremely long, for about 4 h.

### 5. Stir Fabric Phase Sorptive Extraction

In order to increase extraction kinetics, various interesting approaches have been reported by increasing the contact surface area of micro-extraction devices. Pawliszyn et al. [11] proposed the thin-film microextraction technique where a thin membrane of PDMS was used as the extracting surface by enhancing the interaction between the sample and the active sorptive phase. In this context, as FPSE is a typical diffusion extraction process, it can be improved by the stirring of the whole extraction system simultaneously including the fabric sorbent medium as well as the sample.

The potential combination of FPSE media with the advantages of stir membrane extraction (SME) [38] was investigated by Roldan-Pijuan et al. [30]. The proposed stir fabric phase sorptive extraction (SFPSE), which integrates sol–gel hybrid organic–inorganic coated FPSE media with a magnetic stirring mechanism, is presented for the first time and demonstrated for determination of seven triazine herbicides. Two flexible fabric substrates, cellulose and polyester were used as the host matrix for three different sorbents, sol–gel PTHF, sol–gel PEG, and sol–gel PDMDPS. Results showed that the analytes were better extracted by sol–gel PEG, so it was selected for further studies.

The SFPSE unit was constructed using a section of a polypropylene SPE cartridge, an FPSE medium, an external element cut from a pipette tip, and an iron wire to allow magnetic stirring of the unit. The configuration of the extraction device as well as the extraction procedure, are depicted in Figure 4. The extraction time was fixed at 60 min, while elution time with 1.0 mL methanol at 5 min. The absolute recovery values were in the range 22.2%–70.5%. In this new approach, following the SME model, fast analyte diffusion as well as high contact surface area are provided thanks to the device design, leading to the enhancement of the extraction efficiency as well as to the reduction of total extraction time.

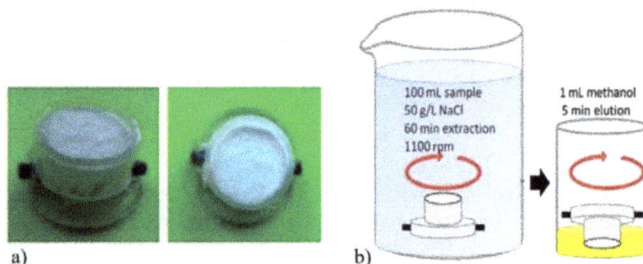

**Figure 4.** Configuration of stir FPSE device (**a**) along with the extraction procedure (**b**). Reproduced from [30] with permission of Elsevier.

Huang et al. [32] presented a stir FPSE system similar to that presented previously [30] to extract brominated flame retardants: TBBPA, TBBPA-BAE, and TBBPA-BDBPE, from wastewater and reservoir water samples, followed by HPLC-DAD analysis. In the same work [32] and in the context of stirring of the fabric, Huang et al., presented an alternative extraction device called as stir-bar fabric phase sorptive extraction (stir bar-FPSE). Briefly, the FPSE media was cut into a house shape, clamped, and fixed by using a stir bar. Schematic diagram of the two extraction procedures stir-bar FPSE and magnetic stir-FPSE are shown in Figure 5.

Three FPSE sorbents: sol–gel PTHF, sol–gel PEG, and sol–gel PDMDPS, were prepared on cellulose fabric substrates. Based on the medium polarity of three BFRs, the sol–gel PTHF fabric was selected for further studies. In both analytical procedures 300 µL acetonitrile was used to elute the BFRs with an elution time of 15 min. Due to the large sorbent loading capacity and unique stirring performance, both techniques possessed high extraction capability and fast extraction equilibrium. Due to the low solvent consumption, the proposed methods could meet the green analytical criteria.

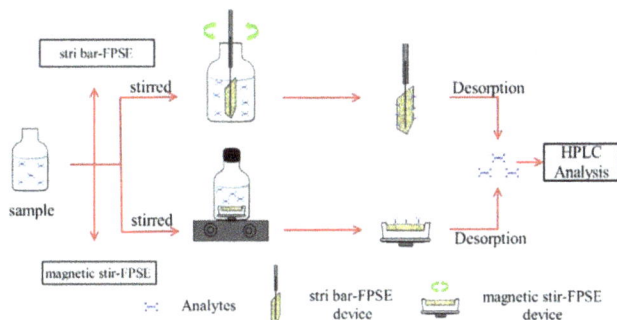

**Figure 5.** Schematic presentation of stir-bar FPSE and magnetic stir-FPSE procedure. Reproduced from [32] with permission of Elsevier.

## 6. Dynamic Fabric Phase Sorptive Extraction

As already mentioned, FPSE technique has been applied for the extraction of various analytes from different types of samples providing satisfactory results by achieving better extraction recoveries in several cases than in other sample preparation techniques. However, the main drawback of FPSE is the extraction time, which takes up to four hours (see Table 4) for extraction equilibrium.

In order to overcome this drawback of the long extraction time, Lakade et al. [29] proposed a new mode of FPSE, called dynamic phase sorbent extraction (DFPSE). DFPSE uses 47 mm circular disks of FPSE media in a filtration assembly instead of 25 mm × 20 mm fabric media introduced directly into the sample solution. The retained analytes on the FPSE disks are eluted by passing a volume of elution solvent through them. This configuration decreases the extraction time extremely as the interfacial area is highly increased. The performance efficiency of the DFPSE technique was evaluated for the extraction of pharmaceuticals and personal care products (PPCPs), MPB, CBZ, PrPB, DHB, BzPB, DHMB, DICLO, BP-3, TCC, and TCS, from river and wastewater samples, followed by LC-MS/MS detection. Taking into account that the majority of the target PPCPs are either highly polar (PARA, CAFF, APy, MPB, CBZ) or medium polar (PROP, PrPB, DHB, BzPB), a hydrophilic substrate such as cellulose, would be a suitable choice. In addition, a polar polymer PEG was selected as organic polymer from a large number of polymer candidates.

Initially, three FPSE disks were placed into the filtration assembly conditioned by passing MeOH followed by ultrapure water and then they were dried by applying vacuum. For the extraction procedure, 50 mL of sample solution was loaded into the filtration assembly and left for 10 min in contact with the FPSE disks for the adsorption of analytes. After that, the sample was passed entirely through the FPSE disks by vacuum. Subsequently, the FPSE disks were dried by an air flow generated by vacuum. The retained analytes were eluted by passing 10 mL of EtOAc, as it took less time to be evaporated in comparison with other solvents that were examined. The analysis was performed by HPLC-MS/MS in MRM mode in positive or negative ionization mode. Recovery values were better than those provided by static FPSE except for PARA and CAFF whose recoveries were lower than 38%. The extraction time was significantly reduced from 240 min to 10 min [31], proving that the proposed dynamic mode provides promising results and that it can be used for the determination of various target analytes in different kinds of samples, with shorter equilibrium time and higher retention than static FPSE.

## 7. Automated Fabric Phase Sorptive Extraction

Current trends in analytical chemistry, are mainly focused on three significant objectives namely miniaturization, simplification, and automation. During recent years, noteworthy progress has been made in order to enhance the quality of analytical results and follow the concept of green analytical chemistry. The implementation of flow-based sample pretreatment methodologies used for fluidic manipulation and on-line sample/reagent pretreatment holds many advantages in contrast with the batch mode of sample preparation. These include low consumption of solvents and reagents and thus low cost of total analysis as well as a significant improvement of the repeatability of the extraction procedure. The combination of flow injection analysis (FIA), sequential injection analysis (SIA), and related techniques with atomic spectrometry (AS) [6,39] provide unique capabilities and enhanced performance of the developed methods. Despite the numerous advantages of the FPSE technique, there is a need for automation in order to reduce sample preparation time extremely and considerably improve the analytical characteristics of each developed method.

Very recently, Anthemidis et al. [36], demonstrated and successfully evaluated for the first time, an automated platform using the FPSE technique in an on-line column preconcentration system. The novel automated on-line flow injection fabric disk sorptive extraction system (FI-FDSE) coupled with flame atomic absorption spectrometry (FAAS), was applied for the preconcentration and determination of lead and cadmium in environmental water samples. Generally, the automation of the FPSE technique is based on the effective packing of a minicolumn with FPSE sorbent media in a shape of disks packed in a series and fixed appropriately. The minicolumn was incorporated onto the FIA system as shown in Figure 6. Four different FPSE sorbent media, sol–gel PDMDPS coated on a polyester substrate, sol–gel PTHF, sol–gel PEO-PPO-PEO triblock copolymer, and sol–gel graphene coated on cellulose substrates, were examined through the developed FDSE platform. Sol–gel PDMDPS presented the highest extraction sensitivity and very good reproducibility in comparison to the other FPSE media, so it was selected for further experiments. The way of FDSE media preparation as well as the way of the minicolumn packing are given elsewhere [36].

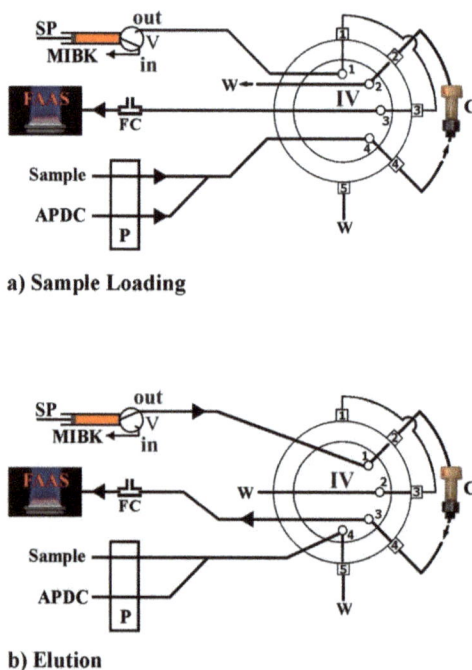

a) Sample Loading

b) Elution

**Figure 6.** Schematic diagram of the FI-FDSE-FAAS manifold for metal preconcentration and determination by FAAS. APDC, aqueous solution 0.2% (m/v) ammonium pyrrolidine dithiocarbamate; MIBK, methyl isobutyl ketone; W, waste; P, peristaltic pump; SP, syringe pump; IV, injection valve in the load or elution position; V, two-position valve; FC, flow compensation unit; C, FDSE minicolumn. (**a**) Sample loading step, (**b**) elution step. Reproduced from [36] with permission of Elsevier.

Briefly, the FDSE minicolumn was created using the body of a polyethylene syringe (4.0 mm i.d.) shortened to 1.5 cm and packed with 38–40 disks FPSE disks in a row as shown in Figure 7. No frits or glass wool were used at either end of the column to block fabric disks.

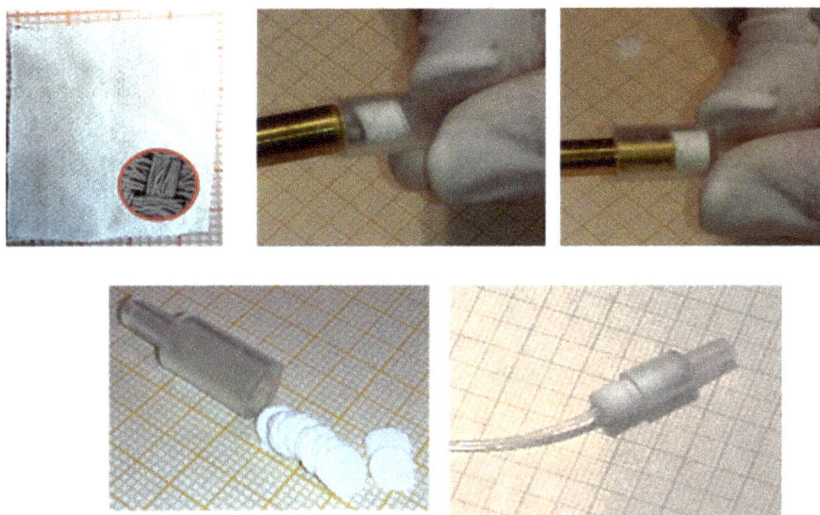

**Figure 7.** FDSE technique: The FPSE fabric (**up–left**); the construction of the minicolumn packed with the FPSE media in the form of disks in a row; the minicolumn with FPSE disks ready for use (**down–right**). Reproduced from [36] with permission of Elsevier.

This configuration of the minicolumn provides limited backpressure due to the easy permeation of the incoming flow through the pores of the FPSE substrate. Thus, high loading flow rates can be applied, resulting in higher extraction efficiency and lower time of analysis. Enrichment factors of 140 and 38 and detection limits of 1.8 and 0.4 $\mu g \cdot L^{-1}$ were achieved for lead and cadmium determination, respectively, with a sampling frequency of 30 $h^{-1}$ for 90 s preconcentration time. The precision as relative standard deviation (RSD) was 3.1% and 3.3% for lead and cadmium respectively. The FDSE minicolumn was efficient and stable for at least 500 sorption/elution cycles.

## 8. Conclusions and Future Outlook

Fabric phase sorptive extraction is a newly developed technique used for isolation and preconcentration of different analytes, from various matrices where even untreated samples demonstrate high extraction efficiency, operational flexibility, simplicity, and a shortened sample pretreatment scheme. FPSE has successfully eliminated inherent errors of conventional sample preparation as well as excessive time consumption. Submerging the FPSE media directly into the sample solution for analyte extraction offers great flexibility and simplicity, decreasing drastically potential analyte loss. The FPSE is easy to use with chromatographic techniques (e.g., liquid chromatography and gas chromatography) coupled with various detection systems such as spectrometry, mass spectrometry and atomic absorption spectrometry.

The inherent porosity of sol–gel sorbent and characteristic permeability of flexible cellulose or polyester fabric substrate result in rapid extraction of target analytes and complete the extraction equilibrium in a short time. Small volumes of organic solvent for elution purposes, elimination of solvent evaporation, and a sample reconstitution step, make the technique environmentally friendly and cost effective in accordance with Green Analytical Chemistry requirements. This review of the literature dealing with FPSE and its applications clearly shows that the technique has achieved great importance. Most of the applications have been developed for the selective extraction of drugs, pharmaceuticals, and various chemical compounds in water, and in biological and food samples.

In the future, research will have the challenge to develop new sol–gel coatings and materials to determine any type of analyte in complex sample matrices. The new automated approach opens up the possibility of sample preparation and total analysis without human intervention, increasing throughput, and improving several analytical performance characteristics.

**Author Contributions:** Viktoria Kazantzi wrote the review article and Aristidis Anthemidis reviewed and revised the article. All authors have read and approved the final manuscript.

**Conflicts of Interest:** The authors declare no conflict of interest.

## Abbreviations

2EHAdip, Bis(2-ethylhexyl) adipate; 2EHSeb, Bis(2-ethylhexyl) sebacate; 3-CPTEOS, 3-cyanopropyltriethoxysilane; 4-CP, 4-Cumylphenol; 4-TAP, 4-tert-amylphenol; 4-SBP, 4-sec-butylphenol; 4-TBP, 4-tert-butylphenol; ADTD, Androstenedione; AND, Androsterone; APDC, Ammonium pyrrolidinedithiocarbamate; APy, Antipyrine; APZ, Alprazolam; BOL, Boldenone; BPA, Bisphenol A; BP-3, Benzophenone; BRZ, Bromazepam; BzPB, Benzylparaben; $C_{18}$-MTMS, Octadecyltrimethoxysilane; $C_8$-TMS, Octyltrimethoxysilane; CAFF, Caffeine; CAP, Chloramphenicol; CBZ, Carbamazepine; CHIMA81, Chimassorb 81; CLO, Cloxacillin; CYA1084, Cyassorb 1084; DBM, Dibutyl maleate; DBP, Dibutyl phthalate; DEP, Diethyl phthalate; DFPSE, Dynamic fabric phase sorptive extraction; DHB, 2,4-dihydroxybenzophenone; DHEA, Dehydroepiaandrosterone; DHMB, 2,2-dihydroxy-4-4-methoxybenzophenone; DIC, Diclofenac; DICLO, Dicloxacillin; DZP, Diazepam; E2, 17β- Estradiol; EE2, 17α-Ethynylestradiol; FF, Florfenicol; FDSE, Fabric disk sorptive extraction; FPSE, Fabric phase sorptive extraction; GAA, Glacial acetic acid; HAA C12, N-Lauryldiethanolamine; IBU, Ibuprofen; IRGA38, Irgafos38; IRGA168, Irgafos 168; IRGA1010, Irganox 1010; IRGA1076, Irganox 1076; KET, Ketoprofen; LRZ, Lorazepam; MGA, Megestrol acetate; MIBK, Methyl isobutyl ketone; MPB, Methylparaben; MTBSTFA, N-Methyl-N-(tert-butyldimethylsilyl)trifluoroacetamide; MTMS, Methyltrimethoxysilane; NAN, Nandrolone; NAP, Naproxen; NOR, Norgestrel; NORET, Norethisterone; OXA, Oxacillin; PARA, Paracetamol; PDMDPS, Poly(dimethyldiphenylsiloxane); PDMS, Poly(dimethylsiloxane); PDPS, Poly(diphenylsiloxane); PEG, Poly(ethyleneglycol); PEG-PPG-PEG, Poly(ethyleneglycol)–block-poly (propyleneglycol) –block-poly(ethyleneglycol); PENG, Penicillin-G; PEO-PPO-PEO, Poly(ethylene oxide)–poly (propylene oxide)–poly(ethylene oxide) triblock copolymer; PPCPs, Pharmaceuticals and personal care products; PRO, Progesterone; PROP, Propranololhydrochloride; PrPB, Propylparaben; PTHF, Poly(tetrahydrofuran); PTMOS, Phenyltrimethoxysilane; SDMX, Sulfadimethoxine; SFPSE, Stir fabric phase sorptive extraction; SIX, Sulfisoxazole; SMTH, Sulfamethazine; TAP, Thiamphenicol; TBBPA, Tetrabromobisphenol A; TBBPA-BAE, Tetrabromobisphenol A bisallylether; TBBPA-BDBPE, Tetrabromobisphenol A bis(2,3-dibromopropyl)ether; TBC, Tributyl citrate; TBoAC, Tributyl-o-acetyl citrate; TCC, Triclocarban; TCS, Triclosan; TEOS, Tetraethoxysilane; TES, Testosterone; TFA, Trifluoroacetic acid; TINU326, Tinuvin 326; TINU327, Tinuvin 327; TMSPA, (Trimethoxysilyl)propylamine; TOPAC, Topanol CA; TXIB, 2,2,4-trimethyl-1,3-pentanediol diisobutyrate; UV 326, 2-tert-butyl-6-(5-chlorobenzotriazol-2-yl)- 4-methylphenol; UV 327, 2,4-ditert-butyl-6-(5-chlorobenzotriazol-2-yl) phenol; UV 328, 2-(benzotriazol-2-yl)-4,6-bis(2-methybutan-2-yl) phenol; UV 329, 2-(benzotriazol-2-yl)-4-(2,4,4-trimethylpentan-2-yl) phenol; UV 360, 2-(benzotriazol-2-yl)-6-[[3-(benzotriazol-2-yl)-2-hydroxy-5-(2,4,4-trimethy lpentan-2-yl)phenyl]methyl]-4-(2,4,4-trimethylpentan-2-yl)phenol; UV 571, 2-(benzotriazol-2-yl)-6-dodecyl-4-methyl phenol; UV P, 2-(benzotriazol-2-yl)-4-methylphenol; VTEOS, Vinyltriethoxysilane.

## References

1.  Płotka-Wasylka, J.; Szczepańska, N.; Guardia de la, M.; Namieśnik, J. Modern trends in solid phase extraction: New sorbent media. *TrAC–Trends Anal. Chem.* **2016**, *77*, 23–43. [CrossRef]

2.  Płotka-Wasylka, J.; Szczepańska, N.; Guardia de la, M.; Namieśnik, J. Miniaturized solid-phase extraction techniques. *TrAC–Trends Anal. Chem.* **2015**, *73*, 19–38. [CrossRef]

3.  Armenta, S.; Garrigues, S.; de la Guardia, M. Green analytical chemistry. *Trends Anal. Chem.* **2008**, *27*, 497–511. [CrossRef]

4.  Gałuszka, A.; Migaszewski, Z.; Namiesnik, J. The 12 principles of green analytical chemistry and the significance mnemonic of green analytical practices. *Trends Anal. Chem.* **2013**, *50*, 78–84. [CrossRef]

5.  Anthemidis, A.N.; Miró, M. Recent developments in flow injection/sequential injection liquid-liquid extraction for atomic spectrometric determination of metals and metalloids. *Appl. Spectrosc. Rev.* **2009**, *44*, 140–167. [CrossRef]

6.  Miró, M.; Hansen, E.H. On-line sample processing involving microextraction techniques as a front-end to atomic spectrometric detection for trace metal assays: a review. *Anal. Chim. Acta* **2013**, *782*, 1–11. [CrossRef] [PubMed]

7. Malik, A.K.; Kaur, V.; Verma, N. A review on solid phase microextraction—High performance liquid chromatography as a novel tool for the analysis of toxic metal ions. *Talanta* **2006**, *68*, 842–849. [CrossRef] [PubMed]

8. Duan, C.; Shen, Z.; Wu, D.; Guan, Y. Recent developments in solid-phase microextraction for on-site sampling and sample preparation. *Trends Anal. Chem.* **2011**, *30*, 1568–1574. [CrossRef]

9. Arthur, C.L.; Pawliszyn, J. Solid phase microextraction with thermal desorption using fused silica optical fibers. *Anal. Chem.* **1990**, *62*, 2145–2148. [CrossRef]

10. Baltussen, E.; Sandra, P.; David, F.; Cramers, C. Stir bar sorptive extraction (SBSE), a novel extraction technique for aquatus samples: Theory and principles. *J. Microcolumn Sep.* **1999**, *10*, 737–747. [CrossRef]

11. Bruheim, I.; Liu, X.; Pawliszyn, J. Thin-film microextraction. *Anal. Chem.* **2003**, *75*, 1002–1010. [CrossRef] [PubMed]

12. Giakisikli, G.; Anthemidis, A.N. Magnetic materials as sorbents for metal/metalloid preconcentration and/or separation. A review. *Anal. Chim. Acta* **2013**, *789*, 1–16. [CrossRef] [PubMed]

13. Kumar, R.; Gaurav; Heena; Malik, A.K.; Kabir, A.; Furton, K.G. Efficient analysis of selected estrogens using fabric phase sorptive extraction and high performance liquid chromatography-fluorescence detection. *J. Chromatogr. A* **2014**, *1359*, 16–25. [CrossRef] [PubMed]

14. Kermani, F.R.; Pawliszyn, J. Sorbent Coated Glass Wool Fabric as a Thin Film Microextraction Device. *Anal. Chem.* **2012**, *84*, 8990–8995. [CrossRef] [PubMed]

15. Grandy, J.J.; Pawliszyn, J. Development of a Carbon Mesh Supported Thin Film Microextraction Membrane As a Means to Lower the Detection Limits of Benchtop and Portable GC/MS Instrumentation. *Anal. Chem.* **2016**, *88*, 1760–1767. [CrossRef] [PubMed]

16. Kabir, A.; Furton, K.G. Fabric Phase Sorptive Extractor (FPSE). U.S. Patent and Trademark Office 14,216,121, 17 March 2014.

17. Kabir, A.; Furton, K.G.; Malik, A. Innovations in sol–gel microextraction phases for solvent-free sample preparation in analytical chemistry. *Trends Anal. Chem.* **2013**, *45*, 197–218. [CrossRef]

18. Samanidou, V.; Galanopoulos, L.D.; Kabir, A.; Furton, K.G. Fast extraction of amphenicols residues from raw milk using novel fabric phase sorptive extraction followed by high-performance liquid chromatography-diode array detection. *Anal. Chim. Acta* **2015**, *855*, 41–50. [CrossRef] [PubMed]

19. Augusto, F.; Carasek, E.; Silva, R.G.C.; Rivellino, S.R.; Batista, A.D.; Martendal, E. New sorbents for extraction and microextraction techniques. *J. Chromatogr. A* **2010**, *1217*, 2533–2542. [CrossRef] [PubMed]

20. Fumes, B.H.; Silva, M.R.; Andrade, F.N.; Nazario, C.E.D.; Lanças, F.M. Recent advances and future trends in new materials for sample preparation. *TrAC–Trends Anal. Chem.* **2015**, *71*, 9–25. [CrossRef]

21. Chong, S.L.; Wang, D.; Hayes, J.D.; Wilhite, B.W.; Malik, A. Sol–gel coating technology for the preparation of solid-phase microextraction fibers of enhanced thermal stability. *Anal. Chem.* **1997**, *69*, 3889–3898. [CrossRef] [PubMed]

22. Kumar, A.; Gaurav; Malik, A.K.; Tewary, D.K.; Singh, B. A review on development of solid phase microextraction fibers by sol–gel methods and their applications. *Anal. Chim. Acta* **2008**, *610*, 1–14. [CrossRef] [PubMed]

23. Valcárcel, M.; Cárdenas, S.; Lucena, R. Novel Sol–gel Sorbents in Sorptive Microextraction Analytical Microextraction Techniques. In *Analytical Microextraction Techniques*; Bentham Science: Sharjah, United Arab Emirates, 2016; pp. 28–69.

24. Karageorgou, E.; Manousi, N.; Samanidou, V.; Kabir, A.; Furton, K.G. Fabric phase sorptive extraction for the fast isolation of sulfonamides residues from raw milk followed by high performance liquid chromatography with ultraviolet detection. *Food Chem.* **2016**, *196*, 428–436. [CrossRef] [PubMed]

25. Racamonde, I.; Rodil, R.; Quintana, J.B.; Sieira, B.J.; Kabir, A.; Furton, K.G.; Cela, R. Fabric phase sorptive extraction: A new sorptive microextraction technique for the determination of non-steroidal anti-inflammatory drugs from environmental water samples. *Anal. Chim. Acta* **2015**, *865*, 22–30. [CrossRef] [PubMed]

26. Samanidou, V.; Kaltzi, I.; Kabir, A.; Furton, K.G. Simplifying sample preparation using fabric phase sorptive extraction technique for the determination of benzodiazepines in blood serum by high-performance liquid chromatography. *Biomed. Chromatogr.* **2015**, *30*, 829–836. [CrossRef] [PubMed]

27. Montesdeoca-Esponda, S.; Sosa-Ferrera, Z.; Kabir, A.; Furton, K.G.; Santana-Rodríguez, J.J. Fabric phase sorptive extraction followed by UHPLC-MS/MS for the analysis of benzotriazole UV stabilizers in sewage samples. *Anal. Bioanal. Chem.* **2015**, *407*, 8137–8150. [CrossRef] [PubMed]

28. Garcia-Guerra, R.B.; Montesdeoca-Esponda, S.; Sosa-Ferrera, Z.; Kabir, A.; Furton, K.G.; Santana-Rodríguez, J.J. Rapid monitoring of residual UV-stabilizers in seawater samples from beaches using fabric phase sorptive extraction and UHPLC-MS/MS. *Chemosphere* **2016**, *164*, 201–207. [CrossRef] [PubMed]

29. Lakade, S.S.; Borrull, F.; Furton, K.G.; Kabir, A.; Marcé, R.M.; Fontanals, N. Dynamic fabric phase sorptive extraction for a group of pharmaceuticals and personal care products from environmental waters. *J. Chromatogr. A* **2016**, *1456*, 19–26. [CrossRef] [PubMed]

30. Roldán-Pijuán, M.; Lucena, R.; Cárdenas, S.; Valcárcel, M.; Kabir, A.; Furton, K.G. Stir fabric phase sorptive extraction for the determination of triazine herbicides in environmental waters by liquid chromatography. *J. Chromatogr. A* **2015**, *1376*, 35–45. [CrossRef] [PubMed]

31. Lakade, S.S.; Borrull, F.; Furton, K.G.; Kabir, A.; Fontanals, N.; Marce, R.M. Comparative study of different fabric phase sorptive extraction sorbents to determine emerging contaminants from environmental water using liquid chromatography-tandem mass spectrometry. *Talanta* **2015**, *144*, 1342–1351. [CrossRef] [PubMed]

32. Huang, G.; Dong, S.; Zhang, M.; Zhang, H.; Huang, T. Fabric phase sorptive extraction: Two practical sample pretreatment techniques for brominated flame retardants in water. *Water Res.* **2016**, *101*, 547–554. [CrossRef] [PubMed]

33. Kumar, R.; Gaurav; Kabir, A.; Furton, K.G.; Malik, A.K. Development of a fabric phase sorptive extraction with high-performance liquid chromatography and ultraviolet detection method for the analysis of alkyl phenols in environmental samples. *J. Sep. Sci.* **2015**, *38*, 3228–3238. [CrossRef] [PubMed]

34. Aznar, M.; Alfaro, P.; Nerin, C.; Kabir, A.; Furton, K.G. Fabric phase sorptive extraction: An innovative sample preparation approach applied to the analysis of specific migration from food packaging. *Anal. Chim. Acta* **2016**, *936*, 97–107. [CrossRef] [PubMed]

35. Guedes-Alonso, R.; Ciofi, L.; Sosa-Ferrera, Z.; Santana-Rodríguez, J.J.; Del Bubba, M.; Kabir, A.; Furton, K.G. Determination of androgens and progestogens in environmental and biological samples using fabric phase sorptive extraction coupled to ultra-high performance liquid chromatography tandem mass spectrometry. *J. Chromatogr. A* **2016**, *1437*, 116–126. [CrossRef] [PubMed]

36. Anthemidis, A.; Kazantzi, V.; Samanidou, V.; Kabir, A.; Furton, K.G. An automated flow injection system for metal determination by flame atomic absorption spectrometry involving on-line fabric disk sorptive extraction technique. *Talanta* **2016**, *156–157*, 64–70. [CrossRef] [PubMed]

37. Samanidou, V.; Michaelidou, K.; Kabir, A.; Furton, K.G. Fabric phase sorptive extraction of selected penicillin antibiotic residues from intact milk followed by high performance liquid chromatography with diode array detection. *Food Chem.* **2017**, *224*, 131–138. [CrossRef] [PubMed]

38. Alcudia-León, M.C.; Lucena, R.; Cárdenas, S.; Valcárcel, M. Stir membrane extraction: A useful approach for liquid simple preparation. *Anal. Chem.* **2009**, *81*, 8957–8961. [CrossRef] [PubMed]

39. Yu, H.L.; Wang, J.H. Recent advances in flow-based sample pretreatment for the determination of metal species by atomic spectrometry. *Chi. Sci. Bull.* **2013**, *58*, 1992–2002. [CrossRef]

*separations*

MDPI

*Article*

# Fabric Phase Sorptive Extraction Explained

**Abuzar Kabir \*, Rodolfo Mesa, Jessica Jurmain and Kenneth G. Furton**

International Forensic Research Institute, Department of Chemistry and Biochemistry,
Florida International University, 11200 SW 8th Street, Miami, FL 33199, USA; rmesa004@fiu.edu (R.M.);
jurmainjl@mymail.vcu.edu (J.J.); furtonk@fiu.edu (K.G.F.)
\* Correspondence: akabir@fiu.edu; Tel.: +1-305-348-2396; Fax: +1-304-348-4172

Academic Editor: Victoria F. Samanidou
Received: 17 January 2017; Accepted: 12 May 2017; Published: 2 June 2017

**Abstract:** The theory and working principle of fabric phase sorptive extraction (FPSE) is presented. FPSE innovatively integrates the benefits of sol–gel coating technology and the rich surface chemistry of cellulose/polyester/fiberglass fabrics, resulting in a microextraction device with very high sorbent loading in the form of an ultra-thin coating. This porous sorbent coating and the permeable substrate synergistically facilitate fast extraction equilibrium. The flexibility of the FPSE device allows its direct insertion into original, unmodified samples of different origin. Strong chemical bonding between the sol–gel sorbent and the fabric substrate permits the exposure of FPSE devices to any organic solvent for analyte back-extraction/elution. As a representative sorbent, sol–gel poly(ethylene glycol) coating was generated on cellulose substrates. Five (cm$^2$) segments of these coated fabrics were used as the FPSE devices for sample preparation using direct immersion mode. An important class of environmental pollutants—substituted phenols—was used as model compounds to evaluate the extraction performance of FPSE. The high primary contact surface area (PCSA) of the FPSE device and porous structure of the sol–gel coatings resulted in very high sample capacities and incredible extraction sensitivities in a relatively short period of time. Different extraction parameters were evaluated and optimized. The new extraction devices demonstrated part per trillion level detection limits for substitute phenols, a wide range of detection linearity, and good performance reproducibility.

**Keywords:** fabric phase sorptive extraction (FPSE); sol–gel; phenols; environmental pollution; sample preparation; microextraction; green analytical chemistry (GAC)

---

## 1. Introduction

Sample preparation is an important but often neglected step in chemical analysis [1]. The importance of an efficient sample preparation technique becomes more inevitable when dealing with trace and ultra-trace levels of target analytes dispersed in complex sample matrices e.g., environmental, pharmaceutical, food, and biological samples. These samples are not generally suitable for direct injection into the analytical instrument. Three main factors may explain this unsuitability of direct instrumental injection. First, the matrix ingredients may exert detrimental effect on the performance of the analytical instrument, or they may interfere with the analysis of target analytes; second, the concentration of the target analytes in the sample matrix may be below the detection limit of the analytical instrument. Third, the sample matrix may be incompatible with the analytical instrument. As such, the primary objective of sample preparation is to isolate and concentrate the target analytes from various sample matrices to a new solvent/solvent system and to minimize matrix interference so that the cleaner analytes solution can be introduced into the analytical instrument for separation, identification, and quantification.

Classical sample preparation techniques such as liquid-liquid extraction (LLE) and solid phase extraction (SPE) are still among popular choices for analytical sample preparation [2–4]. However, these procedures are time consuming, laborious, multi-step, generally utilize large volume of toxic and hazardous organic solvents, and often involve lengthy and error-prone post-extraction steps such as solvent evaporation and sample reconstitution in a suitable solvent. In order to mitigate some of these problems, solid-phase microextraction (SPME) was introduced by Pawliszyn and co-workers in 1987 [5] as a solvent-free/solvent-minimized microextraction technique. Due to the substantial advantages over conventional sample preparation techniques, SPME has gained enormous popularity within a very short period of time. The broad spectrum applications of SPME have been extensively reviewed in a number of recent articles [6–8].

However, some major shortcomings of SPME are yet to be addressed. Among these, one major shortcoming of SPME (fiber format) is the miniscule amount (typically ~0.5 µL) of sorbent loading that often results in poor extraction sensitivity [9]. The low extraction sensitivity of fiber-SPME prompted the invention of a number of microextraction techniques with higher sorbent loading including in-tube SPME [10], SBSE [11], MEPS [12], rotating-disk sorptive extraction (RDSE) [13], and thin film microextraction (TFME) [14]. A comprehensive list of different modifications of sorbent based sorptive microextraction techniques can be found elsewhere [1].

SPME and its different formats, modifications, and implementations are generally governed by two principle criteria: (1) thermodynamics and (2) kinetics [15]. Thermodynamic properties determine the maximum amount of analytes that can be extracted by unit mass of sorbent under a specific set of extraction conditions. Since higher sorbent loading allows the accumulation of a larger amount of target analytes by the sorbent when adequate time is allowed to reach the extraction equilibrium, sorbent loading is directly related to extraction efficiency. On the other hand, kinetics controls the rate of extraction and hence the time required to reach the extraction equilibrium. The faster the extraction equilibrium, the higher the throughput in the analytical lab. As a result, there is a pressing demand for developing new microextraction techniques that can simultaneously satisfy the required sensitivity (by increasing sorbent loading) and reduce the sample preparation time to its lowest level (by minimizing the extraction equilibrium time).

A critical evaluation of different microextraction systems revealed that the shortcomings of all contemporary microextraction systems originate from (1) coating technology used for immobilizing the sorbent on the substrate surface [16] and (2) the physical format of the extraction system [14] that determines the primary contact surface area (PCSA) of the device and consequently the extraction kinetics. PCSA is defined as the surface area of the extraction medium that can be accessed directly by the sample matrix containing the analytes during the extraction process. Therefore, if a sample preparation technique is to be highly sensitive as well as fast, both the coating technology and the PCSA have to be augmented.

One major challenging area of microextraction devices is the sorbent coating technology that often use a dilute solution of pristine polymer to form a thin surface film on the substrate followed by a free radical cross-linking reaction to immobilize the film [17,18]. The weak physical adhesion of the polymer to the substrate results in a number of the unwanted phenomena such as high phase bleeding, washing away of the polymer with organic solvent, long extraction equilibrium time, limited selectivity, poor extraction reproducibility, and swelling of the sorbent when exposed to organic solvents. The lack of chemical bonding between the polymeric sorbent and the substrate is believed to be the primary cause of these coating-related problems. A number of alternative coating techniques have also been proposed including physical deposition [19,20], electrochemical deposition of conducting polymers [21,22], gluing with adhesives [23], and sol–gel column technology [1,16]. Nonetheless, sol–gel column technology pioneered by Malik and co-workers [1,16] have been proven to be the most convenient and versatile [1]. In addition to the suitability of the coating process, sol–gel technology opens up the possibility of making multi-component materials that can be conveniently used to customize the surface morphology, selectivity, and affinity of the sorbent toward the target analytes. The sorbent coating created by sol–gel technology is highly porous and chemically bonded to the substrate. As an outcome, such coatings demonstrate remarkable thermal, solvent, and chemical stability. Due to its inherent porosity, a thin film of sol–gel coating can extend equivalent or higher sensitivity than commercially available thick SPME coatings. The high porosity of the sol–gel coating also makes it possible to reach extraction equilibrium in a fraction of the time that is often required by commercial SPME fibers.

Although tremendous efforts have been made to increase the sensitivity of the microextraction systems by merely increasing the sorbent loading on the same substrate (fused silica fiber/glass tube), little work has been done to increase the primary contact surface area (PCSA) of the extraction device. The increase in PCSA of the extraction device not only allows higher sorbent loading without increasing the coating thickness, but may also considerably reduce the extraction equilibrium time. TFME, SBSE, RDSE, etc. were developed to increase the PCSA, but the use of conventional sorbent immobilization approach did not offer much benefit in boosting the extraction sensitivity.

In addition to improving the coating technology and enhancing the PCSA of the microextraction device, some other important factors that may potentially improve the quality of sample preparation to a new height need further consideration: (1) the ability to preconcentrate target analytes directly from the unmodified samples without any clean-up exercises; (2) resistance to harsh chemical environments (i.e., highly acidic and basic) so that matrix pH can be adjusted to wider pH values; (3) the ability to use any organic solvent to elute the extracted analytes so that the final solution can be injected simultaneously into gas chromatograph (GC), high performance liquid chromatograph (HPLC), and/or capillary electrophoresis (CE) to obtain complementary information depending on the analytical need; (4) equal effectiveness in field sampling and sample preparation to eliminate the burden of sample collection, transportation, storage, and sample preparation in the laboratory; (5) ability to achieve a high preconcentration factor during the extraction so that solvent evaporation and sample reconstitution may be avoided; and (6) ability to reach extraction equilibrium fast enough so that field sampling and sample preparation do not become an inconvenient task.

Taking all of these challenges into consideration, a new green sample preparation approach fabric phase sorptive extraction (FPSE) has been developed [24,25], which creatively addresses the majority of the problems often encountered in contemporary sample preparation practices. In addition to the advanced material properties of sol–gel derived hybrid organic-inorganic sorbents, FPSE has successfully utilized flexibility, permeability, and the rich surface chemistry of natural/synthetic fabric substrates, resulting in a microextraction sorbent chemically bonded to the substrate with a very high, readily accessible active extraction surface for fast and high efficiency analytes extraction. FPSE has introduced major advantages in solvent-less/solvent-minimized sample preparation techniques including an extraordinarily high primary contact surface area (PCSA), and the ability to directly preconcentrate the target analytes even from excessively complicated sample matrices containing debris, cells, proteins, particulates, etc.

Although FPSE has emerged as a major analytical sample preparation technique [25–36], the theory and principle of this simple, innovative and environment friendly technique is yet to be fully studied and understood. As such, the primary objective of the current project is to thoroughly study the theory and principle of fabric phase sorptive extraction. In order to investigate different factors involved in fabric phase sorptive extraction, a polar sorbent sol–gel poly(ethylene glycol) (sol–gel PEG) coated on a hydrophilic cellulose substrate was used as the extraction device and a number of substituted phenols were used as the representative environmental pollutants. Substituted phenols, a class of important industrial raw materials, are known as highly toxic environmental pollutants. They are frequently found in the wastewater generated by wood processing and pharmaceutical industries, organic synthesis, and oil refineries [37]. Substituted phenols have severe health implications for human kidneys, heart, lungs, and the central nervous system [38,39]. As a result, a rapid and efficient sample preparation strategy is warranted to monitor these pollutants.

## 2. Materials and Methods

### 2.1. Chemicals and Materials

All chemicals used in the study were of analytical grade or superior. 4-chlorophenol (≥99%), 3,5-dimethylphenol (98%), 2,6-dichlorophenol (99%), 2,4,6-trichlorophenol (98%), 2,4-diisopropyl phenol (98%), acetone (HPLC grade), dichloromethane (anhydrous), methyltrimethoxysilane (MTMS) (98%), and trifluoroacetic acid (TFA) (99%) were purchased from Sigma-Aldrich (St. Louis, MO, USA). Sodium hydroxide and hydrochloric acid were purchased from Thermo Fisher Scientific (Milwaukee, WI, USA). Polyethylene glycol was purchased from Alfa Aesar (Ward Hill, MA, USA). Fabric phase sorptive extraction vials (20 mL) and HPLC sample vials (2 mL) were purchased from Supelco (St. Lois, MO, USA). HPLC grade methanol and water were purchased from Fisher Scientific (Pittsburg, PA, USA). Unbleached Muslin 100% cotton cellulose fabrics were purchased from Jo-Ann Fabric (Miami, FL, USA).

### 2.2. Instrumentation

An Agilent 1100 series HPLC-UV system (Agilent Technologies, Santa Clara, CA, USA) equipped with G1311A quaternary pump, G1313A ALS auto sampler, tray holder, G1322A vacuum degasser, G1316A thermostated column compartment, G1314A variable wavelength detector was used for the separation, identification and quantification of substituted phenol compounds. The separation of substituted phenols was performed on a reversed phase Zorbax Extend-C18 HPLC column (5 µm, 150 mm, 4.6 mm; Agilent Technologies, Santa Clara, CA, USA). Extraction conditions: 5 cm$^2$ FPSE media, sample volume 10 mL, stirring speed 1100 rpm, salt content 15%, extraction time 40 min, desorption time 5 min, and desorption solvent MeOH:ACN (50/50) followed by: high-performance liquid chromatographic analysis. HPLC conditions: isocratic elution 69/31 $H_2O$/ACN, 1 mL/min flow rate, UV detection at 200 nm, ambient temperature. Gas chromatography-mass spectrometry (GC-MS) analysis was carried out in an Agilent 5973 GC-MS. GC-MS conditions: DB-5MS column, 30 m × 0.25 mm, 0.5 µm, splitless injection, Injector temperature 280 °C, GC oven temperature programmed from 40 °C for 1 min, increased temperature to 100 °C at 7 °C/min, stayed for 0 min, then increased the temperature to 250° at 25 °C/min. MS ion source temperature 230 °C, MS transfer line temperature 280 °C, Injection vol. 1 µL. Centrifugation of different solutions to obtain particle free solutions was carried out in an Eppendorf Centrifuge Model 5415 R (Eppendorf North America Inc., Hauppauge, NY, USA). A Fisher Scientific Digital Vortex Mixer (Fisher Scientific, Pittsburg, PA, USA) was employed to thoroughly mixing different solutions. On-line data collection and processing of chromatographic data was done using ChemStation software (Revision A.08.03) for Windows (Agilent Technologies, Santa Clara, CA, USA). A Philips XL30 scanning electron microscope equipped with an EDAX detector was used to obtain SEM images presented in the article. A Perkin Elmer Spectrum 100 FT-IR Spectrometer equipped with Universal ATR Sampling Accessory (Santa Clara, CA, USA) was used to perform FT-IR characterization of the substrates and FPSE media coated with sol–gel sorbents. A Barnstead NANOPure Diamond (model D11911) deionized water system (Dubuque, IA, USA) was employed to obtain high purity deionized water (18.2 MΩ) used in sol–gel synthesis and aqueous sample preparation for fabric phase sorptive extraction.

### 2.3. Preparation of Sol–gel PEG Coated FPSE Media

#### 2.3.1. Surface Cleaning and Activation of Fabric Substrates

A 100 cm$^2$ segment of the cellulose fabric was cleaned with a copious amount of deionized water, followed by soaking in 1 M NaOH solution for 1 h under continuous sonication. The base-treated fabrics were then washed several times with deionized water, followed by treating with 0.1 M HCl solution for 1 h under sonication. The treated fabric was then washed with copious amount of deionized water and finally dried in home-made drying chamber with continuous helium gas flow at

50 °C overnight. The dried fabric was stored in clean airtight glass container until they are coated with sol–gel sorbents.

### 2.3.2. Preparation of the Sol Solutions for Coating on the Substrate

The sol solutions for creating the sol–gel PEG coatings were prepared using a modified version of a previously described formulation [25,40]. Briefly, the sol solution for sol–gel PEG coating was prepared by dissolving 10 g of poly(ethylene glycol) polymer, 10 mL methyltrimethoxysilane sol–gel precursor (MTMS), 20 mL methylene chloride as the organic solvent, 4 mL trifluoroacetic acid (5% water) as the sol–gel catalyst. The mixture was then vortexed for 3 min, centrifuged for 5 min, and finally the clear supernatant of the sol solution was transferred to a clean 100 mL amber colored glass reaction bottle.

### 2.3.3. Creation of Sol–gel PEG Coatings on the Substrate

For the sol–gel PEG coatings, cellulose fabric was used as the substrate. The clean and treated fabrics were gently inserted into the reaction bottle containing the sol solution so that a three-dimensional network of sol–gel PEG could be formed on the surface of the substrate as well as throughout the porous matrix. The fabrics were kept inside the sol solution for 4 h. After completing the coating period, the sol solution was expelled from the reaction bottle and the coated fabrics were dried and aged in a home-made conditioning device built inside a gas chromatography oven with continuous helium gas flow at 50 °C for 24 h. Before using for extraction, the sol–gel PEG coated fabrics were rinsed sequentially with methylene chloride and methanol followed by drying at 50 °C under an inert atmosphere for 1 h. The fabric phase sorptive extraction media coated with sol–gel PEG were then cut into 2.5 cm × 2.0 cm pieces (area of each side, 5 cm$^2$) and stored in a closed glass container to prevent contamination.

### *2.4. Fabric Phase Sorptive Extraction and Back-Extraction Procedure*

### 2.4.1. Preparation of Standard Solutions for Fabric Phase Sorptive Extraction

All primary stock solutions (substituted phenols) were prepared by dissolving 100 mg of each analyte in 10 mL methanol in a 22 mL amber glass vial to obtain a solution concentration of 10 mg/mL. The intermediate stock solution was prepared by diluting the primary stock solution to 1.0 mg/mL in methanol. All working solutions were prepared by diluting intermediate stock solutions in HPLC grade water to reach the desired concentrations. FPSE method development and validation exercises were carried out using HPLC grade water impregnated with substituted phenols at different concentration levels.

### 2.4.2. Fabric phase Sorptive Extraction

A 5 cm$^2$ piece (2.5 cm × 2.0 cm) of sol–gel PEG coated FPSE device was immersed in 1 mL methanol: acetonitrile (50:50, *v*/*v*) for 5 min to ensure cleanliness. The device was air dried to remove the residual organic solvent. The clean FPSE device was then immersed into the sampling vial (10 mL) containing the sample impregnated with the analytes of interest. A small Teflon coated magnet was placed into the vial. Finally, the vial was placed on top of a magnetic stirrer to promote diffusion of the analytes throughout sample for a predetermined period of time. After that, the FPSE device was removed from the sampling vial and was shaken off or dried with a Kim wipe. This is particularly important if the prepared sample is to be analyzed in gas chromatography.

### 2.4.3. Back-Extraction/Solvent Desorption

The analytes extracted on the FPSE device were back-extracted into a suitable solvent system. 500 µL of the solvent system was transferred into a 1 mL glass vial. The dry FPSE device containing the extracted analytes was immersed into the solvent mixture. The back-extraction was carried out by simply keeping the FPSE device immersed into solvent system for a predetermined period and no external diffusion mechanism (stirring/sonication) was imposed. The back-extraction solution containing the extracted analytes was then transferred into an Eppendorf tube for centrifugation to compel any remaining particulates to precipitate. Finally, the particle free preconcentrated solution of analytes was transferred into a HPLC/GC sample vial for chromatographic analysis. In order to reuse the FPSE medium in future, it was cleaned with 1 mL methanol: acetonitrile (50:50, *v/v*) for 5 min, dried on a watch-glass for 5 min and then stored in a clean glass container. A series of photographic images presented in Figure 1 to demonstrate different steps involve in fabric phase sorptive extraction.

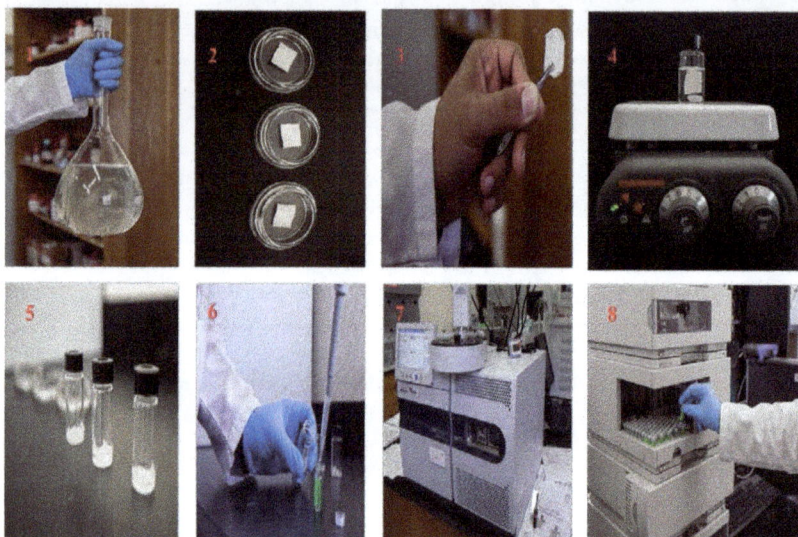

**Figure 1.** Steps in fabric phase sorptive extraction (FPSE): (**1**) unfiltered, dirty environmental sample; (**2**) FPSE media on watch glass; (**3**) an FPSE medium on a tweezer; (**4**) fabric phase sorptive extraction; (**5**) back-extraction in organic solvent; (**6**) transferring preconcentrated eluent into sample vial; (**7**) analysis in GC; (**8**) analysis in HPLC.

## 3. Results and Discussion

### 3.1. Theoretical Considerations

The principle of FPSE, similar to SPME and related equilibrium-driven sorptive microextraction techniques [14,41,42], is based on the interactions between the analytes and the extraction sorbent. Under equilibrium extraction conditions, the amount of analyte extracted ($n$) by FPSE medium is proportional to the partition coefficient between the extraction phase and the sample matrix ($K_{es}$), volume of the extracting phase ($V_e$), volume of the sample ($V_s$), equilibrium extraction amount ($n$) of the analytes, and the original concentration of the analyte ($C_0$) as expressed in Equation (1):

$$n = \frac{K_{es}V_eV_sC_0}{K_{es}V_e + V_s} \tag{1}$$

When the sample volume is large compared to the volume of the extraction sorbent, $K_{es}V_e \ll V_s$, Equation (1) can be simplified to,

$$n = K_{es}V_eC_0 \tag{2}$$

As Equation (2) suggests, the amount of the extracted analyte(s) is directly proportional to the volume of the extraction sorbent and hence can be increased by expanding the volume of the extraction sorbent if the analyte(s) concentration is kept constant.

However, according to the kinetic theory of extraction, expanding the volume of extraction sorbent by increasing the coating thickness of sorbent may result in long extraction equilibrium time as expressed in Equation (3). When the extraction sorbent is dispersed as a thin film on the substrate, the diffusion of analytes through the boundary layer controls the extraction kinetic (rate of extraction). As such, the time required to extract 95% of the equilibrium extraction amount of the analytes, $t_{e,95\%}$ can be calculated as,

$$t_{e,95\%} = \frac{B\delta b K_{es}}{D_s} \qquad (3)$$

where $b$ is the extraction sorbent thickness, $\delta$ is the boundary layer thickness, $K_{es}$ is the distribution constant of analyte between the extraction sorbent and the sample matrix, $D_s$ is the diffusion coefficient of the analyte in the sample matrix, and $B$ is a geometric factor referring to the geometry on which the extraction sorbent is coated.

As is evident from Equation (3), the equilibrium extraction time can be reduced by reducing the coating thickness ($b$), by increasing primary contact surface area of the extraction medium (smaller $B$ value), or by increasing the analyte diffusion in the sample matrix by applying external stimuli e.g., stirring, sonication, shaking, etc.

The rate of extraction in sorptive microextraction is not linear. The extraction rate is very fast at the beginning of the extraction process and slows down as the extraction progresses toward equilibrium. The initial rate $\left(\frac{dn}{dt}\right)$ of extraction is proportional to the surface area of the extraction phase, A as shown in Equation (4):

$$dn/dt = (DsA/\delta)C_0 \qquad (4)$$

Therefore, it is obvious that in order to increase the sensitivity of the microextraction system, the loading of the sorbent has to be increased. On the other hand, to reduce the extraction equilibrium time, surface area of the extraction device has to be increased.

### 3.2. Significance of Porous Substrates for Microextraction Sorbent

The permeability of the microextraction substrate plays a pivotal role in the mass transfer rate from the sample matrix to the microextraction sorbent. A plausible model explaining the events that occur during the extraction process for both impermeable and permeable substrate is presented in Figure 2. It is worthy to mention that among all contemporary sorbent based sorptive microextraction techniques, FPSE is the only technique that utilizes permeable substrate to expedite the extraction equilibrium.

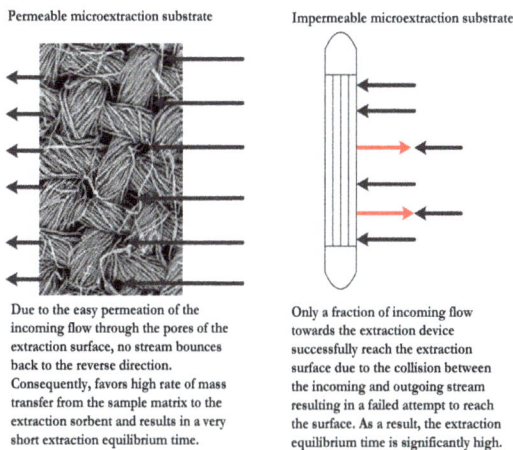

Permeable microextraction substrate — Impermeable microextraction substrate

Due to the easy permeation of the incoming flow through the pores of the extraction surface, no stream bounces back to the reverse direction. Consequently, favors high rate of mass transfer from the sample matrix to the extraction sorbent and results in a very short extraction equilibrium time.

Only a fraction of incoming flow towards the extraction device successfully reach the extraction surface due to the collision between the incoming and outgoing stream resulting in a failed attempt to reach the surface. As a result, the extraction equilibrium time is significantly high.

**Figure 2.** Comparison of liquid flow behavior through a permeable and impermeable substrate.

Under the influence of external stimuli (stirring, sonication, etc.), a continuous flow of the sample moves toward the substrate holding the microextraction sorbent for interaction. When an impermeable substrate such as fused silica fiber is used, the flow bounces back and collides with the incoming flow, redirecting some of the incoming flow away from the substrate surface. As a consequence, only a fraction of the target analytes may interact with the sorbent. On the contrary, when the substrate is permeable, majority of the incoming flow of the sample penetrates the surface without being redirected. As a result, the mass transfer between the sorbent and sample occurs rapidly, within a fraction of the time required for similar interactions with impermeable substrates. This is consistent with the experimental result obtained by Alcudia-León et al. [43]. In this study, the authors compared the extraction behavior of stir bars coated with impermeable PTFE of varying surface areas to a porous PTFE stir-membrane unit. The highest extraction sensitivity was obtained by the porous stir-membrane among all the test devices even though it had a low surface area relative to the impermeable units. This observation was a clear manifestation of the role of porous substrates as the hosts for microextraction sorbents.

### 3.3. Surface Chemistry of Cellulose Substrate

Cellulose is a hydrophilic linear polymer of β-D-glucopyranose, whose structure is shown in Figure 3. Each dimer of cellulose contains three hydroxyl functional groups in positions 2, 3, and 6 that can participate in polycondensation during the sol–gel coating process at a varying degree of reactivity [44]. As such, cellulose seems to be an excellent candidate as a potential substrate for sol–gel sorbent coating.

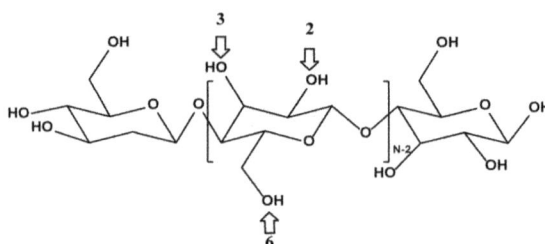

**Figure 3.** Chemical structure of cellulose substrate demonstrating available hydroxyl functional groups for anchoring sol–gel inorganic-organic network.

The cellulose fabric used in this experiment was activated by treating it with 1 M NaOH solution for an hour under sonication. Swelling of cellulose, also known as mercerization, is an important treatment that improves its chemical reactivity and significantly increases the availability of all the hydroxyl groups for chemical reactions [45].

### 3.4. Sol–gel Coating Process on Fabric Phase Sorptive Extraction Media

Sol–gel process in creating the sorbent coatings on cellulose substrate involves the following reactions:

(i)    Catalytic hydrolysis of sol–gel precursor;
(ii)   Polycondensation of hydrolyzed precursors leading to a growing sol–gel network;
(iii)  Random incorporation of sol–gel active organic polymer into the growing sol–gel network;
(iv)  Immobilization of the growing sol–gel network on the substrate surface via polycondensation.

Under the appropriate experimental conditions, sol–gel precursor MTMS undergoes hydrolysis in the presence of sol–gel catalyst, trifluoroacetic acid (TFA). The hydrolyzed MTMS molecules then participate in polycondensation to create a sol–gel network. Sol–gel active organic polymer, poly(ethylene glycol) present in the respective sol solution enters randomly into the growing sol–gel network during this step. Polycondensation of the sol–gel network with the available hydroxyl functional groups of cellulose results in a thin, chemically bonded layer of sorbent wrapped on cellulose microfibrils.

A tentative reaction scheme of the sol–gel process is given in Figure 4.

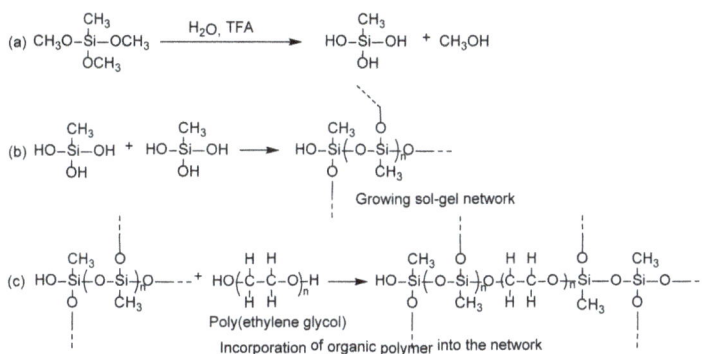

**Figure 4.** Chemical reactions involved in the synthesis of surface-bonded sol–gel PEG coatings on fabric substrate (**a**) hydrolysis of methyl trimethoxysilane (MTMS) under acid catalyst; (**b**) condensation of hydrolyzed MTMS to generate a rapidly growing sol–gel network; (**c**) random incorporation of poly(ethylene glycol) into the growing sol–gel network.

Figure 5 demonstrates the schematic representation of sol–gel PEG coated FPSE media. Although the sol–gel PEG coating on the fabric substrate can't be seen in naked eye, the structure of the sorbent coating can be visualized at higher magnification in scanning electron microscopy or similar imaging instrument.

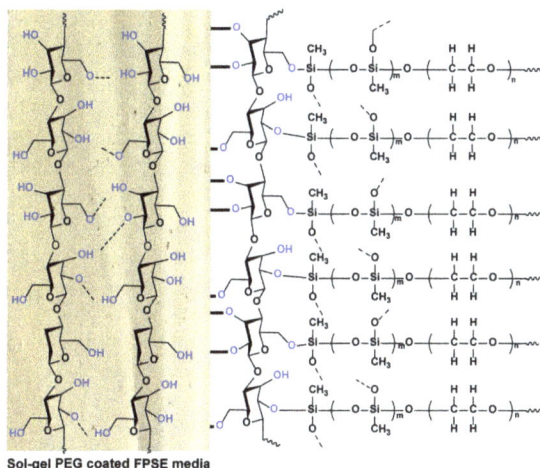

Sol-gel PEG coated FPSE media

**Figure 5.** Schematic representation of surface bonded sol–gel PEG coating on cellulose fabric surface.

The sol–gel coating process results in a highly porous and uniformly distributed ultra-thin three-dimensional network of hybrid inorganic-organic sorbent chemically bonded to the substrate.

As the coating occurs on individual cellulose microfibrils, the sorbent loading is relatively high with low coating thickness.

### 3.5. Characterization of FPSE Media

#### 3.5.1. Characterization of FPSE Media Using FT-IR

Figure 6 shows FT-IR spectra of uncoated cellulose, MTMS, poly(ethylene glycol), and sol–gel PEG coated FPSE media.

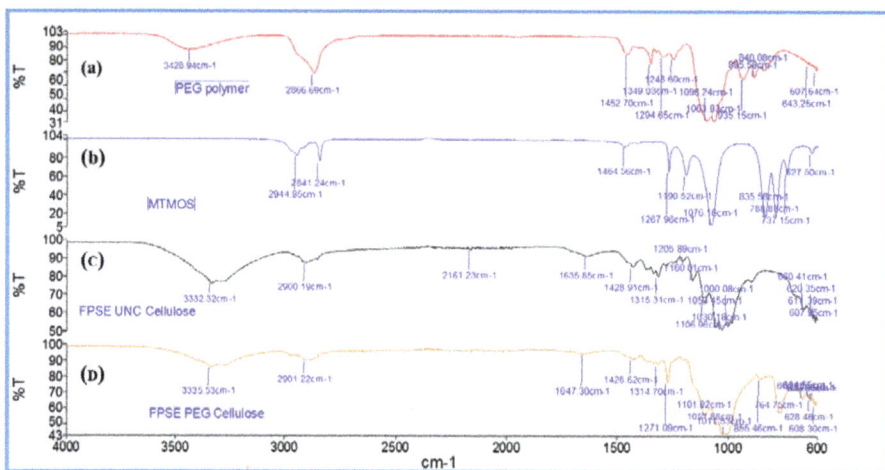

**Figure 6.** FT-IR spectra of different components in (**a**) poly(ethylene glycol) polymer; (**b**) methyltrimethoxysilane (MTMS); (**c**) uncoated cellulose substrate; (**d**) sol–gel PEG coated FPSE media.

As can be seen from the FT-IR spectra, MTMS [46] demonstrates characteristic Si–O–C–H band at 2944 cm$^{-1}$. Uncoated cellulose demonstrated characteristics peaks at 3332 cm$^{-1}$ and 2900 cm$^{-1}$, 1315 cm$^{-1}$, and at 1030 cm$^{-1}$ corresponding to O–H, C–H and C–O stretching and C–H bending vibration, respectively. In the fingerprint region between 1428 cm$^{-1}$ and 850 cm$^{-1}$, the characteristic peaks were recorded at 1105 cm$^{-1}$ (C–O–C) bridges, glycoside bonds; 1029 cm$^{-1}$ C–OH primary alcohols, 1052 cm$^{-1}$ C–OH secondary alcohols, 1000 cm$^{-1}$, 983 cm$^{-1}$ –CH-bonds [47]. The characteristic peak of poly(ethylene glycol) include 2866 cm$^{-1}$, 1349 cm$^{-1}$ which represent different vibration modes of C–H bonds. The peak at 1294 cm$^{-1}$ and 1248 cm$^{-1}$ belong to C–C double bond [48]. For sol–gel PEG coated FPSE devices, no additional signals of deposited sol–gel network were clearly visible due to the overlapping with characteristic absorption bands of uncoated fabrics. However, the sol–gel coated FPSE media demonstrated a decrease in the absorption bands, which is considered to be clear manifestation of the presence of the coating on the substrate [49].

#### 3.5.2. Sorbent Loading on FPSE Media and Coating Reproducibility

Reproducibility of the sol–gel coating process is an important factor that potentially causes significant impact on the overall error profile of FPSE process. In order to verify the reproducibility of sol–gel coating as well as to determine sorbent loading per square centimeter of FPSE media, four conditioned fabrics were coated under identical reaction conditions. The fabrics (with known area) were weighed before and after the sol–gel coating followed by cleaning and drying to determine the amount of sol–gel sorbent immobilized on the surface. FPSE media were found to host high mass of sol–gel sorbent. Average mass of sol–gel PEG sorbent on FPSE media were calculated as 8.63 mg/cm$^2$ (~43.2 mg/FPSE unit).

### 3.5.3. Scanning Electron Microscopy of FPSE Media

Figure 7 represents scanning electron micrographs (SEM) of uncoated, surface activated, and sol–gel PEG coated FPSE media.

As is evident from these micrographs, cellulose substrate is made up of microfibrils within the individual thread that are weaved to form the fabric. In addition to the inherent pores, capillaries, voids, and interstices, cellulose fibers are woven to contain well defined pores which are dependent upon their end usage. Low viscosity sol solution can easily permeate through the microfibrils and pores to create a uniform ultra-thin layer of sorbent coating on the surface. As such, sol–gel coating on FPSE media is not limited to only its outer surface. Instead, the coating is bonded throughout the three dimensional matrix of the substrate, resulting in a comparatively larger extraction surface with enhanced sorption capacity for the target analytes.

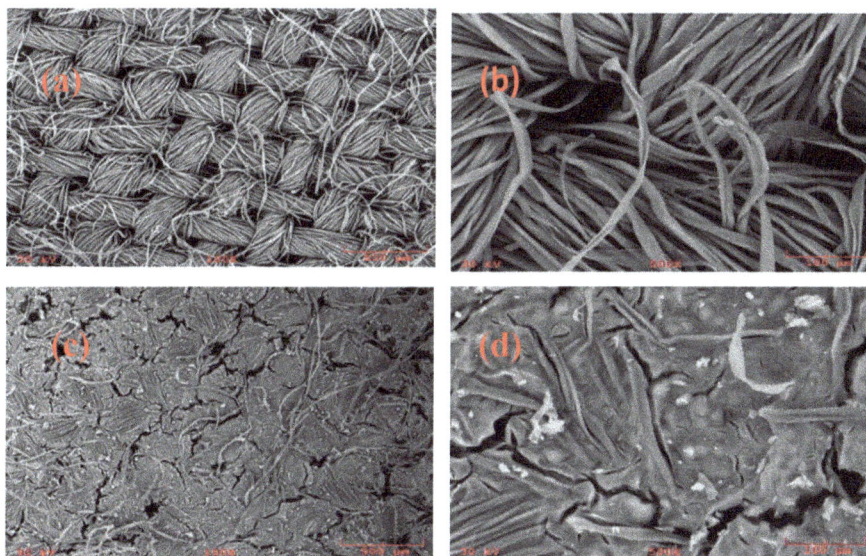

**Figure 7.** Scanning electron microscopy images of (**a**) uncoated cellulose fabric surface at 100× magnifications; (**b**) uncoated cellulose fabric surface at 500× magnifications; (**c**) sol–gel PEG coated cellulose fabric surface at 100× magnifications; (**d**) sol–gel PEG coated cellulose fabric surface at 500× magnifications.

### 3.6. Optimization of Fabric Phase Sorptive Extraction (FPSE) Conditions

In order to maximize the extraction efficiency of FPSE as a novel new-generation microextraction technique, a number of factors that can impact extraction performance were evaluated including sample volume, extraction time, stirring speed, back-extraction solvent type and volume, and ionic strength.

### 3.6.1. Sample Volume

Although FPSE follows the general principle of sorption-based microextraction techniques, the high sorbent loading and extended primary contact surface area (10 cm$^2$ for a typical 5 cm$^2$ FPSE unit) have made this system different from other microextraction techniques. Therefore, it is important to study the impact of sample volume on the extraction behavior. The amount of analyte(s) extracted in any microextraction system is expected to increase with the increase in the sample volume up to a certain point, at which no further increase is expected. This optimum sample volume was determined by extracting substituted phenols at 5 ng/mL concentration from 10 mL, 15 mL, and 20 mL aqueous solutions for 40 min. The extraction results revealed that there is no statistically significant difference ($p \geq 0.05$) in the extraction of 3,5-dimethyl phenol, 2,4,6-trichlorophenol, and 2,4-diisopropyl phenol between 10 mL, 15 mL, and 20 mL. However, a slight change (statistically significant, but practically insignificant) was observed for 4-chlorophenol and 2,6-dichlorophenol with different volumes. Based on these results, a 10 mL sample volume was used in all the subsequent experiments.

### 3.6.2. Extraction Time

Extraction time is one of the most important factors that influence extraction efficiency. In this work, the effect of extraction time on substituted phenols were investigated in over the span of 10–40 min. Experimental data (Figure 8) demonstrate that the extraction equilibrium for all five phenols was reached in 40 min.

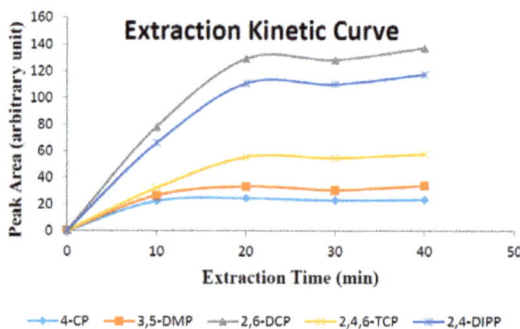

**Figure 8.** Extraction profiles of substituted phenols for the sol–gel PEG coated FPSE device.

### 3.6.3. Stirring Speed

Diffusion of the analytes through the sample matrix increases mass transfer rate to the extraction phase, reduces the time required for reaching extraction equilibrium, and augments the extraction kinetics [50,51]. Diffusion of the analytes can be increased in a number of ways including sonication, heating the sample matrix, and stirring [52,53]. In the current study, different stirring speeds (0–1100 rpm) were studied to determine the impact of stirring on extraction sensitivity. Extraction sensitivity of all 5 analytes was increased with higher stirring speed. Based on the results obtained, a stirring speed of 1100 rpm was selected as optimum.

### 3.6.4. Effect of Salt Content

Adding salt to the sample matrix generally decreases the solubility of organic analytes in water and thus results in higher extraction efficiency. On the other hand, the presence of salt in the sample matrix increases the viscosity of the solution and consequently reduces the rate of diffusion of the analytes. Since both factors counter-balance each other, experimental determination of the impact of salt addition is recommended [53]. The influence of salt was studied by adding different amounts of sodium chloride ranging from 0–20% ($w/v$) to the sample. Addition of salt in substituted phenols solution generally increased sensitivity up to 15% NaCl ($w/v$) (Figure 9). Therefore, NaCl concentration at 15% ($w/v$) is considered optimum for extracting substituted phenols from aqueous solutions.

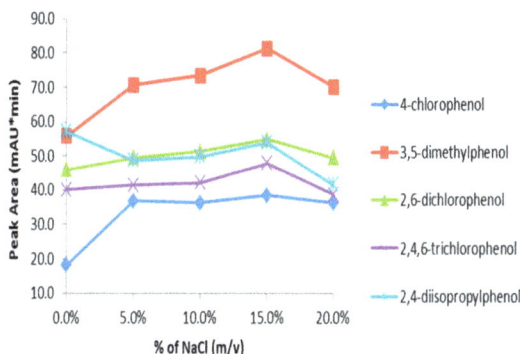

**Figure 9.** Graphs demonstrating the impact of salt addition to the FPSE efficiency for extracting substituted phenols.

### 3.6.5. Optimization of Elution Solvent, Volume of Solvent, and Elution Time

Due to the polarity differences among all the substituted phenols ($\log K_{ow}$ 2.4–3.8), all back-extraction solvents may not be equally effective for quantitative solvent mediated desorption. Considering the high to medium polarity of the substituted phenols, 500 μL of methanol (MeOH) (relative polarity 0.762), acetonitrile (ACN) (relative polarity 0.46) or methanol-acetonitrile mixture (MeOH: ACN) (50:50 $v/v$) were used for elution/back-extraction of substituted phenols from FPSE media. MeOH was found better than MeOH: ACN mixture and ACN for back-extraction of 4-chlorophenol, while MeOH: ACN mixture performed better than MeOH or ACN for back-extraction of 2,6-dichlorophenol. The trend changes with the polarity of the analyte relative to the back-extraction solvents. The back-extraction of high polarity compound 4-chlorophenol ($\log K_{ow}$ 2.4) was favored by highest polarity solvent MeOH. 2,6-dichlorophenol ($\log K_{ow}$ 2.7), being relatively less polar, was better extracted by MeOH:ACN. However, other substituted phenols did not show preference for any specific solvents. Because the MeOH:ACN mixture provided better reproducibility in solvent back-extraction, it was considered optimum and was used as the back-extraction solvent in all subsequent experiments.

Due to the ultra-thin coating of sol–gel sorbent as well as well-preserved pores in the fabric substrate, it was expected that low viscosity organic solvent would easily penetrate the sorbent to quantitatively desorb the extracted analytes back very fast. Two desorption times, 5 min and 10 min, were investigated with the MeOH:ACN (50:50 $v/v$) mixture. No discernible difference in extraction sensitivity was observed between 5 min and 10 min desorption. Therefore, a 5 min back extraction was accepted as the optimum desorption time.

These experiments showed that 500 μL of the ACN: MeOH mixture is the minimum volume needed for complete immersion of the 5 cm$^2$ FPSE unit and quantitative desorption of analytes without any carry over problem. The ultra-thin coating of highly porous sol–gel sorbent is easily accessible to organic solvent without any external stimuli (e.g., stirring, sonication), resulting in complete desorption within a very short period of time. A low volume of back-extraction solvent maintains the preconcentration achieved from the extraction, and therefore, reduces the time required for typically laborious and time consuming solvent elution and sample reconstitution steps.

### 3.7. Model Correlating Octanol-Water Coefficient and Absolute Recovery in FPSE

Mathematical model was created for sol–gel PEG coated FPSE media that can be used as a predictive tool to estimate approximate absolute recovery of a given compound on the specific FPSE media using its octanol-water coefficient value. In order to develop this model, a series of substituted phenol solutions (10 ppm, 20 ppm, 40 ppm, 60 ppm, and 80 ppm) were prepared in DI water. For each compound, a calibration curve was developed using concentration and corresponding chromatographic peak area. All samples were analyzed in HPLC-UV before and after the fabric phase sorptive extraction in triplicate. The single and average absolute recovery of individual compounds across the concentration range was calculated. The average for an individual compound was plotted against its $logK_{ow}$ value. Based on the data, absolute recovery of any analyte in sol-gel PEG can be calculated using its $logK_{ow}$ value into the following equation:

Absolute % recovery in sol-gel PEG coated FPSE medium= $-77.61822 + 44.931167 \times logK_{ow}$.

Figure 10 shows that substituted phenols demonstrates excellent linear relationship between average absolute recovery and logarithm of octanol-water coefficient. This graphs can be used as a predictive tool to estimate absolute recovery of a compound using its $logK_{ow}$ value and may assist in selecting the appropriate FPSE media suitable for a particular analytical problem.

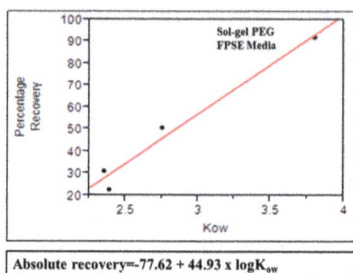

**Figure 10.** Graphs showing relationship between absolute percentage recovery and $logK_{ow}$ for sol–gel PEG coated FPSE media.

## 3.8. Analyte Sorption Capacity of FPSE Media

As seen in Section 3.5.2, FPSE media contained high volume of sorbents. This high sorbent loading should be translated into high sample loading capacity. In order to establish the maximum amount of analytes FPSE media can retain, the FPSE media were exposed to a series of increasingly higher concentration of analytes. An aliquot of each solution was injected before and after the extraction. Extraction of phenols up to 80 ppm (Figure 11) without competing against each other. As such, a 5 cm$^2$ unit of sol–gel PEG coated FPSE media was able to retain approximately 2000 µg of phenols (combined), which is about 10% of the sorbent loading. This high sample capacity may open up a new application of FPSE media as a scavenger of environmental pollutants.

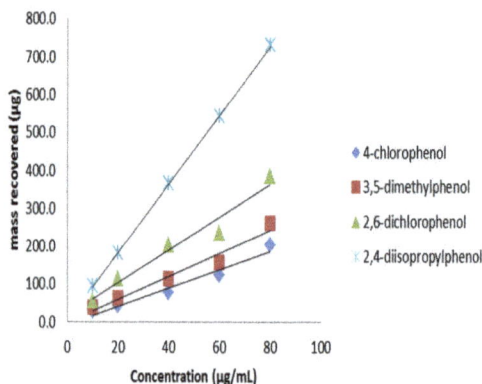

**Figure 11.** Analyte sorption capacity study for sol–gel PEG coated FPSE device.

### 3.9. Operational Stability of Sol–gel Coated FPSE Media

Exposing FPSE media to different organic solvents for solvent mediated desorption is an integral part of the FPSE process if analytes are to be transferred to a suitable organic solvent for subsequent analysis. Unlike commercially available SPME fibers, due to the strong chemical bonding between the sorbent and the substrate of FPSE media, they can be exposed to any organic solvent. To verify that, a set of 3 FPSE media were used to extract phenols before and after exposing to acetone, methylene chloride, and acetone: methylene chloride (50:50 *v/v*) (for 8 h at room temperature). No significant changes were observed between the extraction sensitivity before and after the treatments.

Adjusting the pH of the sample matrix to maximize the ionizable analytes is a common practice routinely used by analytical chemists. The chemical stability of FPSE media were evaluated at both acidic (pH 1) and basic (pH 12) environment. Extractions were performed in triplicate before and after an 8 h exposure to an acidic or basic environment. Extraction sensitivity was similar for all fabrics before and after the acid/base treatment. Although both silica-based sorbent and cellulose substrate are vulnerable to highly acidic/basic environments, these experiment suggests that short exposure of FPSE media to highly acidic and basic environment does not have an impact on their performances. Thus, FPSE media coated with sol–gel PEG sorbents have demonstrated remarkable operational stability.

### 3.10. Analytical Performance

The analytical parameters including linear range, correlation coefficient ($r^2$), intra-day and inter-day repeatability, and limit of detection of substituted phenols were investigated under the optimized extraction conditions and are presented in Table 1 (sample volume 10 mL, stirring speed 1100 rpm, salt content 15%, extraction time 40 min, desorption time 5 min, and desorption solvent MeOH: ACN (50/50 *v/v*).

**Table 1.** Analytical figures of merit for the proposed FPSE-HPLC-UV method for substituted phenols analysis.

| Compound Class | Compound Name | Linear Range ($\mu g \cdot L^{-1}$) | Determination Co-Efficient ($r^2$) | Repeatability (RSD%) [a] Intra-Day [b] | Inter-Day [b] | LOD [c] ($\mu g \cdot L^{-1}$) |
|---|---|---|---|---|---|---|
| Substituted Phenols | 4-chlorophenol (4-CP) | 1–5000 | 0.9970 | 1.8 | 25.5 | 0.03 |
| | 3,5-dimethylphenol (3,5-DMP) | 1–5000 | 0.9997 | 1.6 | 9.8 | 0.01 |
| | 2,6-dichlorophenol (2,6-DCP) | 1–5000 | 0.9983 | 1.3 | 0.5 | 0.04 |
| | 2,4,6-trichlorophenol (2,4,6-TCP) | 1–5000 | 0.9952 | 1.6 | 6.0 | 0.02 |
| | 2,4-diisopropylphenol (2,4-DIPP) | 1–5000 | 0.9981 | 5.1 | 4.9 | 0.02 |

[a] Expressed as relative standard deviation; [b] Intra- and inter-day repeatability was calculated by analyzing water samples spiked with substituted phenols at 50 $\mu g \cdot L^{-1}$ within one day (*n* = 5) and over a period of three days (*n* = 3); [c] Limit of detection (LOD) calculated as S/N = 3.

As can be seen in Table 1, the FPSE-HPLC-UV method demonstrates a wide linear range for all the tested analytes with correlation coefficients of $r^2 > 0.99$. The repeatability (%RSD) of substituted phenols at 50 ppb level concentration within a day were between 1.3 and 5.1, whereas, the between days repeatability were found to be between 0.5–26.0. As such, the method can be conveniently used in routine environmental monitoring laboratories.

### 3.11. Analysis of Prepared Sample Using Multiple Analytical Systems

As different chromatographic systems have different advantages, e.g., GC favors the analysis of highly volatile, low molecular weight compounds, whereas the HPLC is suitable for medium to high molecular weight compounds. In some cases, the whole ranges of analytes are of interest. Samples are prepared differently for both the system which inflicts additional burden to the analyst. FPSE has ingeniously solved this problem. Selection of a solvent equally compatible to both the GC and HPLC as the FPSE back-extraction solution and injection of the eluent simultaneously into the GC and HPLC may offer complimentary information that cannot be obtained from either of the systems alone.

Figure 12 presents GC and HPLC chromatograms obtained from analyzing the same sample prepared by FPSE.

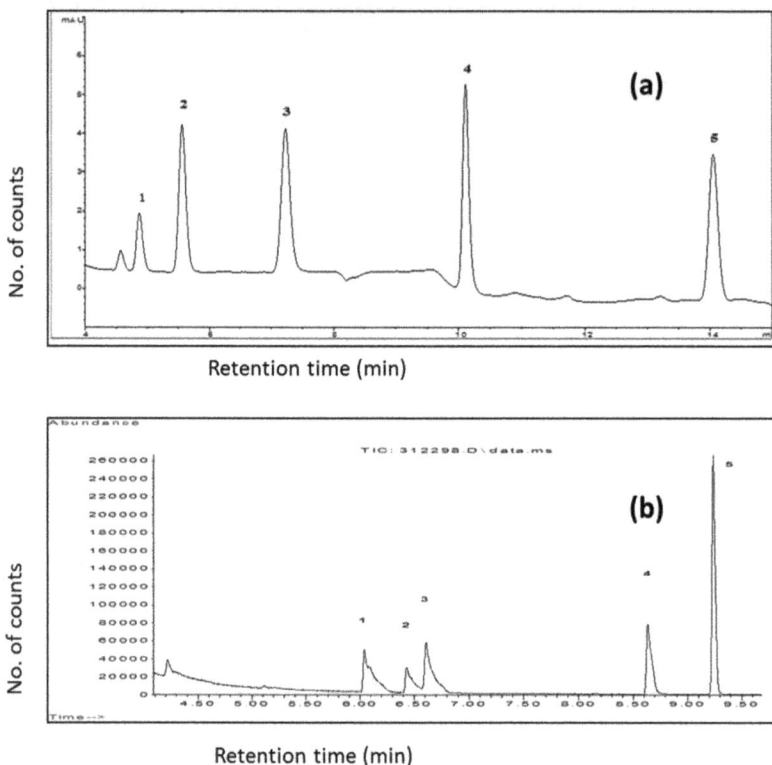

**Figure 12.** Fabric phase sorptive extraction of substituted phenols using a sol–gel PEG coated FPSE media (**a**) Chromatogram obtained from HPLC analysis: peaks: (1) 4-chlorophenol, (2) 3,5-dimethyl phenol, (3) 2,6-dichlorophenol, (4) 2,4,6-trichlorophenol, (5) 2,4-diisopropyl phenol; (**b**) Chromatogram obtained from GC-MS analysis: peaks: (1) 3,5-dimethyl phenol, (2) 4-chlorophenol, (3) 2,6-dichlorophenol, (4) 2,4,6-trichlorophenol, (5) 2,4-diisopropyl phenol.

### 3.12. Method Comparison

A literature review of selected methods used for the determination of substituted phenols is presented in Table 2. As evident from the Table 2, the new FPSE method has unequivocally demonstrated superior extraction sensitivity, higher pre-concentration factor, precision and simplicity. Considering only a fraction of the eluent was injected for analysis, the sensitivity of FPSE method can be easily increased even further by evaporating 500 μL sample solution to dryness and reconstituting the sample in a smaller volume of organic solvent or by increasing injection volume.

**Table 2.** Comparison of limit of detection of the proposed FPSE-HPLC-UV method with other reported methods for substituted phenols analysis.

| Compound [a] | Sample Preparation Method | Sorbent Chemistry | Analytical Instrument | Limit of Detection ($\mu g \cdot L^{-1}$) | Reference |
|---|---|---|---|---|---|
| 2,4-DCP; 2,4,6-TCP | SPME | Vinyl-SBA-15 mesoporous organosilica | GC-MS | 0.002–0.004 | [54] |
| 4-CP; 2,6-DCP; 2,4,6-TCP | HS-SPME (after derivatization) | PDMS/DVB | GC-ECD | 0.0001–0.122 | [55] |
| 4-CP;2,4,6-TCP | HS-SPME | polyaniline | GC-FID | 1.3–58 | [22] |
| 2,6-DCP; 2,4,6-TCP | HS-SPME | Polyacrylate | GC-MS | 0.0500 | [56] |
| 2,4,6-TCP | SPME | Polyacrylate | HPLC-UV; HPLC-ED | 2.0 (UV); 0.017 (ED) | [57] |
| 4CP; 3,5DMP; 2,6-DCP; 2,4,6-TCP; 2,4-DIPP | FPSE | sol–gel PEG | HPLC-UV | 0.01–0.03 | Current study |

[a] Abbreviations: 4-Chlorophenol (4-CP); 2,6-Dichlorophenol (2,6-DCP); 2,4,6-Trichlorophenol (2,4,6-TCP); 3,5-Dimethylphenol (3,5-DMP); 2,4-Diisopropylphenol (2,4-DIPP).

### 3.13. Applications of the FPSE-HPLC-UV Methods for Environmental Sample Analysis

The optimized FPSE-HPLC-UV methods for substituted phenols were applied for the determination of these compounds in tap water, pond water and reclaimed water released from a sewerage treatment plant. The pond water and reclaimed water samples were visibly cloudy with high content of suspended soil particles and biomass. However, no filtration or other clean-up procedure was applied to these samples prior to FPSE procedure. Extraction recoveries of sol–gel PEG coated FPSE media were calculated by comparing the extraction efficiency obtained by extracting spiked environmental blank samples with 50 µg/L of substituted phenols compounds to that of the deionized water at the same concentration. The relative percentage recovery data of substituted phenols are shown in Table 3.

**Table 3.** Analytical results of substituted phenols in real samples for FPSE-HPLC-UV method.

| Compound Name | Sample Matrix | Amount Added (ng) | Amount Recovered (ng) | Relative Recovery (%) | %RSD ($n = 3$) |
|---|---|---|---|---|---|
| 4-chlorophenol | | | 91.0 | 91.0 | 3.6 |
| 3,5-dimethylphenol | | | 73.2 | 73.2 | 1.3 |
| 2,6-dichlorophenol | Tap water | 100 | 19.7 | 19.7 | 11.4 |
| 2,4,6-trichlorophenol | | | 26.9 | 26.9 | 2.0 |
| 2,4-diisopropylphenol | | | 35.8 | 35.8 | 15.9 |
| 4-chlorophenol | | | 109.5 | 109.5 | 0.9 |
| 3,5-dimethylphenol | | | 88.0 | 88.0 | 3.3 |
| 2,6-dichlorophenol | Pond water | 100 | 21.9 | 21.9 | 7.8 |
| 2,4,6-trichlorophenol | | | 34.5 | 34.5 | 5.7 |
| 2,4-diisopropylphenol | | | 77.2 | 77.2 | 6.3 |
| 4-chlorophenol | | | 106.8 | 106.8 | 4.4 |
| 3,5-dimethylphenol | | | 90.7 | 90.7 | 1.2 |
| 2,6-dichlorophenol | Reclaimed water | 100 | 40.7 | 40.7 | 7.2 |
| 2,4,6-trichlorophenol | | | 57.0 | 57.0 | 7.3 |
| 2,4-diisopropylphenol | | | 90.8 | 90.8 | 9.9 |

The percentage recovery of substituted phenols in tap water, pond water, and reclaimed water were in the range of 19.7–91.0, 21.9–109.5, and 40.7–106.8, respectively. The probable cause of the low recovery of substituted phenols in environmental water may be attributed to the formation of complexes with numerous treatment chemicals and/or particulates. A holistic investigation is needed to understand the real cause behind the low recovery of substituted phenols from environmental water samples.

## 4. Conclusions

The theory and working principle of FPSE that candidly explains the uniquely superior performance of the new and emerging microextraction system have been presented. In a very short period of time, FPSE has found applications in a wide variety of sample matrices including pharmaceutical, environmental, food, biological, and toxicological samples. In all these applications, FPSE has established itself as an efficient, green, new-generation sample preparation technique. Unique integration of the sol–gel coating technology into highly porous fabric substrate resulted in an efficient sample preparation device capable of extracting target analytes from complex environmental samples with high efficiency. Strong chemical bonding between the sol–gel extraction sorbents and the host substrates provide excellent solvent and chemical stability to FPSE device. As a result, utilization of a back-extraction solvent equally compatible with both gas chromatography and liquid chromatography opens up the possibility of analyzing the same preconcentrated sample simultaneously in GC and LC to obtain complimentary information, if necessary. High chemical stability of FPSE media extends the ability of matrix pH adjustment between pH 1 to pH 12 that is often important to efficiently extract ionizable compounds (organic acids/bases) from an aqueous solution. High sorbent loading in FPSE media provided very high sample capacity. The inherently porous structure of the sol–gel sorbent and the permeability of fabric have significantly reduced the equilibrium extraction time from hours to minutes, making FPSE an excellent field-deployable sample preparation technique.

Parts per trillion level detection sensitivities were achieved for substituted phenols. Eliminating the necessity of sample clean-up prior to the extraction, FPSE presents a new possibility in analyzing complex samples such as environmental water, biological fluids, foods, and pharmaceuticals with high matrix interference leading to high quality analytical data that truly represents the sample of interest.

**Acknowledgments:** The authors would like to thank Tatiana Trejos, Trace Evidence Analysis Facility and Thomas Beasley, Florida Center for Analytical Electron Microscopy at Florida International University for their help in obtaining SEM images. Help extended by Linda Maiben as an undergraduate research assistant is also highly appreciated.

**Author Contributions:** Rodolfo Mesa and Jessica Jurmain performed the laboratory experiments reported in the current study. Abuzar Kabir wrote the manuscript. Kenneth G. Furton reviewed the manuscript. All authors have read and approved the final manuscript.

**Conflicts of Interest:** The authors declare no conflict of interest.

## References

1. Kabir, A.; Furton, K.G.; Malik, A. Innovations in sol–gel microextraction phases for solvent-free sample preparation in analytical chemistry. *TRAC Trends Anal. Chem.* **2013**, *45*, 197–218.
2. Wang, H.; Yan, H.; Qiu, M.; Qiao, J.; Yang, G. Determination of dicofol in aquatic products using molecularly imprinted solid-phase extraction coupled with GC-ECD detection. *Talanta* **2011**, *85*, 2100–2105. [CrossRef] [PubMed]
3. Varga, R.; Somogyvári, I.; Eke, Z.; Torkos, K. Determination of antihypertensive and anti-ulcer agents from surface water with solid-phase extraction–liquid chromatography–electrospray ionization tandem mass spectrometry. *Talanta* **2011**, *83*, 1447–1454. [CrossRef] [PubMed]
4. Ribeiro, C.; Tiritan, M.E.; Rocha, E.; Rocha, M.J. Development and Validation of a HPLC-DAD Method for Determination of Several Endocrine Disrupting Compounds in Estuarine Water. *J. Liquid Chromatogr. Relat. Technol.* **2007**, *30*, 2729–2746. [CrossRef]
5. Pawliszyn, J.; Liu, S. Sample introduction for capillary gas chromatography with laser desorption and optical fibers. *Anal. Chem.* **1987**, *59*, 1475–1478. [CrossRef]
6. Ribeiro, C.; Ribeiro, A.R.; Maia, A.S.; Goncalves, V.M.F.; Tiritan, M.E. New trends in sample preparation techniques for environmental analysis. *Crit. Rev. Anal. Chem.* **2014**, *44*, 142–185. [CrossRef] [PubMed]
7. Spietelun, A.; Marcinkowski, L.; de la Guardia, M.; Namiesnik, J. Recent developments and future trends in solid phase microextraction techniques towards green analytical chemistry. *J. Chromatogr. A* **2013**, *1321*, 1–13. [CrossRef] [PubMed]
8. Risticevic, S.; Niri, V.H.; Vuckovic, D.; Pawliszyn, J. Recent developments in solid-phase microextraction. *Anal. Bioanal. Chem.* **2009**, *393*, 781–795. [CrossRef] [PubMed]
9. Bigham, S.; Medlar, J.; Kabir, A.; Shende, C.; Alli, A.; Malik, A. Sol−Gel Capillary Microextraction. *Anal. Chem.* **2002**, *74*, 752–761. [CrossRef] [PubMed]

10. Eisert, R.; Pawliszyn, J. Automated In-Tube Solid-Phase Microextraction Coupled to High-Performance Liquid Chromatography. *Anal. Chem.* **1997**, *69*, 3140–3147. [CrossRef]

11. Baltussen, E.; Sandra, P.; David, F.; Cramers, C. Stir bar sorptive extraction (SBSE), a novel extraction technique for aqueous samples: Theory and principles. *J. Microcolumn Sep.* **1999**, *11*, 737–747. [CrossRef]

12. Abdel-Rehim, M. New trend in sample preparation: On-line microextraction in packed syringe for liquid and gas chromatography applications: I. Determination of local anaesthetics in human plasma samples using gas chromatography–mass spectrometry. *J. Chromatogr. B* **2004**, *801*, 317–321. [CrossRef]

13. Richter, P.; Leiva, C.; Choque, C.; Giordano, A.; Sepulveda, B. Rotating-disk sorptive extraction of nonylphenol from water samples. *J. Chromatogr. A* **2009**, *1216*, 8598–8602. [CrossRef] [PubMed]

14. Bruheim, I.; Liu, X.C.; Pawliszyn, J. Thin-film microextraction. *Anal. Chem.* **2003**, *75*, 1002–1010. [CrossRef] [PubMed]

15. Lucena, R. Extraction and stirring integrated techniques: Examples and recent advances. *Anal. Bioanal. Chem.* **2012**, *403*, 2213–2223. [CrossRef] [PubMed]

16. Chong, S.L.; Wang, D.X.; Hayes, J.D.; Wilhite, B.W.; Malik, A. Sol−Gel Coating Technology for the Preparation of Solid-Phase Microextraction Fibers of Enhanced Thermal Stability. *Anal. Chem.* **1997**, *69*, 3889–3898. [CrossRef] [PubMed]

17. Wright, B.W.; Peaden, P.A.; Lee, M.L.; Stark, T.J. Free radical cross-linking in the preparation of non-extractable stationay phases for capillary gas chromatography. *J. Chromatogr. A* **1982**, *248*, 17–34. [CrossRef]

18. Blomberg, L.G. Current aspects of stationary phase immobilization in open tubular column chromatography. *J. Microcolumn Sep.* **1990**, *2*, 62–68. [CrossRef]

19. Graham, C.M.; Meng, Y.J.; Ho, T.; Anderson, J.L. Sorbent coatings for solid-phase microextraction based on mixtures of polymeric ionic liquids. *J. Sep. Sci.* **2011**, *34*, 340–346. [CrossRef] [PubMed]

20. Musteata, F.M.; Walles, M.; Pawliszyn, J. Fast assay of angiotensin 1 from whole blood by cation-exchange restricted-access solid-phase microextraction. *Anal. Chim. Acta* **2005**, *537*, 231–237. [CrossRef]

21. Mohammadi, A.; Yamini, Y.; Alizadeh, N. Dodecylsulfate-doped polypyrrole film prepared by electrochemical fiber coating technique for headspace solid-phase microextraction of polycyclic aromatic hydrocarbons. *J. Chromatogr. A* **2005**, *1063*, 1–8. [CrossRef] [PubMed]

22. Bagheri, H.; Mir, A.; Babanezhad, E. An electropolymerized aniline-based fiber coating for solid phase microextraction of phenols from water. *Anal. Chim. Acta* **2005**, *532*, 89–95. [CrossRef]

23. Liu, Y.; Shen, Y.F.; Lee, M.L. Porous layer solid phase microextraction using silica bonded phases. *Anal. Chem.* **1997**, *69*, 190–195. [CrossRef]

24. Kabir, A.; Furton, K.G. Fabric Phase Sorptive Extractors. U.S. Patent 9,283,544 B2, 18 March 2014.

25. Kumar, R.; Gaurav; Heena; Malik, A.K.; Kabir, A.; Furton, K.G. Efficient analysis of selected estrogens using fabric phase sorptive extraction and high performance liquid chromatography-fluorescence detection. *J. Chromatogr. A* **2014**, *1359*, 16–25. [CrossRef] [PubMed]

26. Roldan-Pijuan, R.L.M.; Cardenas, S.; Valcarcel, M.; Kabir, A.; Kenneth, G. Stir fabric phase sorptive extraction for the determination of triazine herbicides in environmental water by using ultra-high performance liquid chromatography-UV detection. *Furton J. Chromatogr. A* **2015**, *1376*, 35–45. [CrossRef] [PubMed]

27. Kumar, R.; Gaurav; Kabir, A.; Furton, K.G.; Malik, A.K. Development of a fabric phase sorptive extraction with high-performance liquid chromatography and ultraviolet detection method for the analysis of alkyl phenols in environmental samples. *J. Sep. Sci.* **2015**, *38*, 3228–3238. [CrossRef] [PubMed]

28. Lakade, S.S.; Borrull, F.; Furton, K.G.; Kabir, A.; Fontanals, N.; Marcé, R.M. Comparative study of different fabric phase sorptive extraction sorbents to determine emerging contaminants from environmental water using liquid chromatography–tandem mass spectrometry. *Talanta* **2015**, *144*, 1342–1351. [CrossRef] [PubMed]

29. Lakade, S.S.; Borrull, F.; Furton, K.G.; Kabir, A.; Marce, R.M.; Fontanals, N. Dynamic fabric phase sorptive extraction for a group of pharmaceuticals and personal care products from environmental waters. *J. Chromatogr. A* **2015**, *1456*, 19–26. [CrossRef] [PubMed]

30. Racamonde, I.; Rodil, R.; Quintana, J.B.; Sieira, B.J.; Kabir, A.; Furton, K.G.; Cela, R. Fabric phase sorptive extraction: A new sorptive microextraction technique for the determination of non-steroidal anti-inflammatory drugs from environmental water samples. *Anal. Chim. Acta* **2015**, *865*, 22–30. [PubMed]

31. Aznar, M.; Alfaro, P.; Nerin, C.; Kabir, A.; Furton, K.G. Fabric phase sorptive extraction: An innovative sample preparation approach applied to the analysis of specific migration from food packaging. *Anal. Chim. Acta* **2016**, *936*, 97–107. [CrossRef] [PubMed]

32. García-Guerra, R.B.; Montesdeoca-Esponda, S.; Sosa-Ferrera, Z.; Kabir, A.; Furton, K.G.; Santana-Rodríguez, J.J. Rapid monitoring of residual UV-stabilizers in seawater samples from beaches using fabric phase sorptive extraction and UHPLC-MS/MS. *Chemosphere* **2016**, *164*, 201–207. [CrossRef] [PubMed]

33. Guedes-Alonso, R.; Ciofi, L.; Sosa-Ferrera, Z.; Santana-Rodriguez, J.J.; del Bubba, M.; Kabir, A.; Furton, K.G. Determination of androgens and progestogens in environmental and biological samples using fabric phase sorptive extraction coupled to ultra-high performance liquid chromatography tandem mass spectrometry. *J. Chromatogr. A* **2016**, *1437*, 116–126. [CrossRef] [PubMed]

34. Samanidou, V.; Kaltzi, I.; Kabir, A.; Furton, K.G. Simplifying Sample Preparation Using Fabric Phase Sorptive Extraction Technique for the Determination of Benzodiazepines in Blood Serum by High-Performance Liquid Chromatography. *Biomed. Chromatogr.* **2016**, *30*, 829–836. [CrossRef] [PubMed]

35. Aznar, M.; Úbeda, S.; Nerin, C.; Kabir, A.; Furton, K.G. Fabric phase sorptive extraction as a reliable tool for rapid screening and detection of freshness markers in oranges. *J. Chromatogr. A* **2017**, *1500*, 32–42. [CrossRef] [PubMed]

36. Locatelli, M.; Kabir, A.; Innosa, D.; Lopatriello, T.; Furton, K.G. A fabric phase sorptive extraction-High performance liquid chromatography-Photo diode array detection method for the determination of twelve azole antimicrobial drug residues in human plasma and urine. *J. Chromatogr. B* **2017**, *1040*, 192–198. [CrossRef] [PubMed]

37. Gruzdev, I.V.; Kuzivanov, I.M.; Zenkevich, I.G.; Kondratenok, B.M. Determination of methyl-substituted phenols in water by gas chromatography with preliminary iodination. *J. Anal. Chem.* **2013**, *68*, 161–169. [CrossRef]

38. Sarafraz-Yazdi, A.; Amiri, A.; Rounaghi, G.; Hosseini, H.E. A novel solid-phase microextraction using coated fiber based sol–gel technique using poly(ethylene glycol) grafted multi-walled carbon nanotubes for determination of benzene, toluene, ethylbenzene and *o*-xylene in water samples with gas chromatography-flam ionization detector. *J. Chromatogr. A* **2011**, *1218*, 5757–5764. [PubMed]

39. Zhang, W.-H.; Zhang, D.; Zhang, R.-J.; Xia, F.; Liu, Y.-F. Flow injection analysis of volatile phenols in environmental water samples using CdTe/ZnSe nanocrystals as a fluorescent probe. *Anal. Bioanal. Chem.* **2012**, *402*, 895–901. [CrossRef] [PubMed]

40. Kabir, A.; Hamlet, C.; Malik, A. Parts per quadrillion level ultra-trace determination of polar and nonpolar compounds via solvent-free capillary microextraction on surface-bonded sol–gel polytetrahydrofuran coating and gas chromatography–flame ionization detection. *J. Chromatogr. A* **2004**, *1047*, 1–13. [CrossRef] [PubMed]

41. Jiang, R.F.; Pawliszyn, J. TRAC Thin-film microextraction offers another geometry for solid-phase microextraction. *Trends Anal. Chem.* **2012**, *39*, 245–253. [CrossRef]

42. Cudjoe, E.; Vuckovic, D.; Hein, D.; Pawliszyn, J. Investigation of the effect of the extraction phase geometry on the performance of automated solid-phase microextraction. *Anal. Chem.* **2009**, *81*, 4226–4232. [CrossRef] [PubMed]

43. Alcudia-Leon, M.C.; Lucena, R.; Cardenas, S.; Valcarcel, M. Stir membrane extraction: A useful approach for liquid sample pretreatment. *Anal. Chem.* **2009**, *81*, 8957–8961. [CrossRef] [PubMed]

44. Klemm, D. *Comprehensive Cellulose Chemistry: Fundamentals and Analytical Methods*; Wiley-VCH: Weinheim, Germany, 1998.

45. Wakelyn, P.J.; Bertoniere, N.R.; French, A.D.; Thibodeaux, D.P.; Triplett, B.A.; Rousselle, M.A.; Goynes, W.R.; Edwards, J.V.; Hunter, L.; McAlister, D.D. *Cotton Fiber Chemistry and Technology*; CRC Press: Boca Raton, FL, USA, 2006.

46. Aegerter, M.A.; Koebel, M.M.; Leventis, N. *Aerogels Handbook*; Springer: New York, NY, USA, 2011.

47. Tarbuk, A.; Grancaric, A.M.; Leskovac, M. Novel cotton cellulose by cationisation during the mercerisation process—part 1: Chemical and morphological changes. *Cellulose* **2014**, *21*, 2167–2179. [CrossRef]

48. Loghmani, S.K.; Farrokhi-Rad, M.; Shahrabi, T. Effect of polyethylene glycol on the electrophoretic deposition of hydroxyapatite nanoparticles in isopropanol. *Ceramics Int.* **2013**, *39*, 7043–7051. [CrossRef]

49. Montarsolo, A.; Periolatto, M.; Zerbola, M.; Mossotti, R.; Ferrero, F. Hydrophobic sol–gel finishing for textiles: Improvement by plasma pre-treatment. *Text. Res. J.* **2013**, *83*, 1190–1200. [CrossRef]

50. Pedersen-Bjergaard, S.; Rasmussen, K.E.; Halvorsen, T.G. Liquid–liquid extraction procedures for sample enrichment in capillary zone electrophoresis. *J. Chromatogr. A* **2000**, *902*, 91–105. [CrossRef]
51. Esrafili, A.; Yamini, Y.; Shariati, S. Hollow fiber-based liquid phase microextraction combined with high-performance liquid chromatography for extraction and determination of some antidepressant drugs in biological fluids. *Anal. Chim. Acta* **2007**, *604*, 127–133. [CrossRef] [PubMed]
52. Pawliszyn, J.; Lord, H.L. *Handbook of Sample Preparation*; John Wiley & Sons, Inc.: Hoboken, NJ, USA, 2011.
53. Risticevic, S.; Lord, H.; Gorecki, T.; Arthur, C.L.; Pawliszyn, J. Protocol for solid-phase microextraction method development. *Nat. Protoc.* **2010**, *5*, 122–139. [CrossRef] [PubMed]
54. Zhu, F.; Liang, Y.J.; Xia, L.Y.; Rong, M.Z.; Su, C.Y.; Lai, R.; Li, R.Y.; Ouyang, G.F. Preparation and characterization of vinyl-functionalized mesoporous organosilica-coated solid-phase microextraction fiber. *J. Chromatogr. A* **2012**, *1247*, 42–48. [CrossRef] [PubMed]
55. De Morais, P.; Stoichev, T.; Basto, M.C.; Carvalho, P.; Vasconcelos, M.T.D. A headspace SPME-GC-ECD method suitable for determination of chlorophenols in water samples. *Anal. Bioanal.Chem.* **2011**, *399*, 2531–2538. [CrossRef] [PubMed]
56. Ribeiro, A.; Neves, M.H.; Almeida, M.F.; Alves, A.; Santos, L. Direct determination of chlorophenols in landfill leachates by solid-phase micro-extraction–gas chromatography–mass spectrometry. *J. Chromatogr. A* **2002**, *975*, 267–274. [CrossRef]
57. Penalver, A.; Pocurull, E.; Borrull, F.; Marce, R.M. Solid-phase microextraction coupled to high-performance liquid chromatography to determine phenolic compounds in water samples. *J. Chromatogr. A* **2002**, *953*, 79–87. [CrossRef]

![separations logo] *separations*

MDPI

*Article*

# A Novel Protocol to Monitor Trace Levels of Selected Polycyclic Aromatic Hydrocarbons in Environmental Water Using Fabric Phase Sorptive Extraction Followed by High Performance Liquid Chromatography-Fluorescence Detection

Shivender Singh Saini [1], Abuzar Kabir [2,*], Avasarala Lakshmi Jagannadha Rao [1], Ashok Kumar Malik [1,*] and Kenneth G. Furton [2]

[1]  Department of Chemistry, Punjabi University, Patiala 147002, India; shiv_algal2007@yahoo.com (S.S.S.); aljrao@gmail.com (A.L.J.R.)
[2]  International Forensic Research Institute, Department of Chemistry and Biochemistry, Florida International University, Miami, FL 33193, USA; furtonk@fiu.edu
*   Correspondence: akabir@fiu.edu (A.K.); malik_chem2002@yahoo.co.uk (A.K.M.); Tel.: +1-305-348-2396 (A.K.)

Academic Editor: Victoria F. Samanidou
Received: 21 March 2017; Accepted: 25 May 2017; Published: 8 June 2017

**Abstract:** Fabric phase sorptive extraction (FPSE) combines the advanced material properties of sol–gel derived microextraction sorbents and the flexibility and permeability of fabric to create a robust, simple and green sample preparation device. It simultaneously improves the extraction sensitivity and the speed of the extraction by incorporating high volumes of sponge-like, porous sol–gel hybrid inorganic–organic sorbents into permeable fabric substrates that are capable of extracting target analytes directly from both simple and complex aqueous sample matrices. For the first time, this technique was applied to the trace-level determination of selected polycyclic aromatic hydrocarbons (PAHs) in environmental water samples using a non-polar sol–gel C18 coated FPSE media. Several extraction parameters were optimized to improve the extraction efficiency and to achieve a high detection sensitivity. Validation tests of spiked samples showed good linearity for four selected PAHs ($R^2$ = 0.9983–0.9997) over a wide range of concentrations (0.010–10 ng/mL). Limits of detection (LODs) and quantification (LOQs) were measured at pg/mL levels; 0.1–1 pg/mL and 0.3–3 pg/mL, respectively. Inter- and intra-day precision tests showed variations of 1.1%–4.1% for four selected PAHs. Average absolute recovery values were in the range of 88.1%–90.5% with relative standard deviations below 5%, surpassing the values predicted by the recovery prediction model. Finally, the developed FPSE-HPLC-FLD protocol was applied to analyze 8 environmental water samples. Out of four selected PAHs, fluoranthene (Flu) and phenanthrene (Phen) were the most frequently detected in four samples, at concentrations of 5.6–7.7 ng/mL and 4.1–11 ng/mL, respectively, followed by anthracene (Anth) and pyrene (Pyr) in two samples. The newly developed FPSE-HPLC-FLD protocol is simple, green, fast and economical, with adequate sensitivity for trace levels of four selected PAHs and seems to be promising for routine monitoring of water quality and safety.

**Keywords:** fabric phase sorptive extraction (FPSE); polycyclic aromatic hydrocarbons (PAHs); persistent pollutants; green analytical chemistry (GAC); environmental water; sorptive microextraction

## 1. Introduction

Polycyclic aromatic hydrocarbons are toxic, carcinogenic and mutagenic organic compounds, and can exert immunologic and reproductive effects. They are introduced into the environment from both natural and anthropogenic sources, typically formed as unintentional byproducts of the incomplete combustion of organic matter. PAHs are ubiquitous in nature and the most damaging pollutants with regard to the ecosystem [1–5]. Although they have low water solubility and exist in the environment at very low concentrations, PAHs are strongly bioaccumulative (e.g., by fish) and can, therefore, pass up the food chain to top predators, including humans [6,7]. Natural inputs such as forest and prairie fires, volcanic eruptions and anthropogenic inputs, such as oil spills, waste incineration, coke and asphalt production, oil refining, aluminum production, urban runoff, emissions from combustion and industrial processes are the main sources of PAHs in the environment [8–10]. The elevated concentrations of PAHs in the environment, together with their ecological toxicity and health risk for humans, have spawned numerous environmental studies [11–14]. Due to their environmental concern, PAHs are included in the USEPA (United States Environmental Protection Agency) and in the EU (European Union) priority lists of pollutants [15]. Human exposure to PAHs occurs mainly by direct inhalation of polluted air and tobacco smoke, dietary intake of smoked food stuffs, direct intake of and contact with contaminated water, direct contact with contaminated soil and dermal contact with soot, tars and oils [16]. PAHs enter the aquatic environment through run off from contaminated roads or sealed parking lots [17], urban and industrial waste water discharge, direct spillage and wet and dry deposition of atmospheric contaminants [18]. Therefore, trace determination of PAHs in water matrices [19] is considered to be a valuable tool in risk assessment, remediation and management of PAHs in the environment.

In this study, four selected small PAHS including anthracene (a linear three ring PAH), phenanthrene (a fused three ring PAH), pyrene (a fused four ring PAH), and fluoranthene (a non-alternant four ring PAH) (Figure 1).

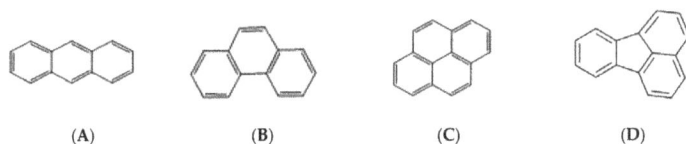

|       |       |       |       |
| :---: | :---: | :---: | :---: |
| (A)   | (B)   | (C)   | (D)   |

**Figure 1.** Chemical structures of four selected polycyclic aromatic hydrocarbons (PAHs): (**A**) Anthracene; (**B**) Phenanthrene; (**C**) Pyrene; (**D**) Fluoranthene.

Anthracene is used as a raw material in the industrial production of hydrogen peroxide and in the manufacturing of alizarin and anthraquinone dyes for cotton fibers [20]. Phenanthrene, a tricyclic aromatic hydrocarbon, has a "K-region" and a "bay-region", where the main carcinogenic species can be formed and is therefore commonly used as a model substrate for carcinogenic studies [21]. Fluoranthene, a non-alternant PAH, contains a five-member ring and is a persistent environmental organic pollutant structurally similar to other environmentally important compounds such as fluorene, dibenzothiophene, acenaphthylene, carbazole, dibenzofuran, and dibenzodioxin, i.e., compounds of [22,23]. Pyrene occurs at relatively high concentrations in PAH mixtures and is one of the most highly concentrated PAHs detected in drinking water [24].

Until now, numerous protocols comprising gas chromatography (GC) and high-performance liquid chromatography (HPLC) have been successfully developed for the determination of PAHs in the environment [25–29]. However, since PAHs are generally present at trace levels in the environment and are accompanied by diversified matrices, they cannot be directly handled by analytical instruments. This means that sample pre-treatment is required, with the aim of concentrating the PAHs, as well as eliminating or decreasing interference [30]. For the isolation of PAHs in environmental matrices, various sample pre-treatment techniques have been proposed, including solid-phase extraction (SPE) [31,32], liquid-liquid extraction (LLE) [33], miniaturized homogeneous liquid-liquid extraction (MHLLE) [34], solid-phase microextraction (SPME) [35,36] and single drop microextraction (SDME) [37,38]. Although these techniques are very useful, they possess some serious drawbacks. For example, LLE and SPE are laborious, time consuming, utilize high volumes of organic solvents and sample solutions, often incur significant loss of analytes and have poor reproducibility. The shortcomings of SPME include high cost, fragility of the fiber, low thermal and chemical stability, swelling if exposed to organic solvents, limited lifetime and sample carry-over problems [39,40]. Breaking of the organic drop due to fast stirring, reduced extraction rate due to air bubble formation, time-consumption and non-equilibrium are the major drawbacks associated with SDME [41].

Nowadays, a number of unique materials have been synthesized and applied as sorbents for PAH extraction, including multi-walled carbon nanotubes [32], magnetic nanoparticles [42], and metal-organic frameworks [43]. However, the synthesis process of such materials is rather complex and often consumes large amounts of organic solvents and time. Therefore, new, versatile and high-performance adsorbents with a simple preparation process are still highly desirable.

Kabir et.al., in 2014 integrated the rich surface chemistry of cellulose cotton fabric substrates and sol–gel technology to develop a novel sorptive microextraction technique called fabric phase sorptive extraction (FPSE) The inherent advantages of the synthesized sorptive material include (1) a flexible hydrophobic/hydrophilic fabric substrate that can be bent, twisted and squeezed to insert directly into unmodified samples; (2) the ability to extract target analytes directly from a raw sample matrix i.e., environmental water, whole milk, whole blood, urine, saliva containing proteins, lipids, particulates, biomasses or debris without any sample pre-treatment; (3) a high loading capacity with unique and tunable selectivity; (4) the possibility of using any organic or aqueous-organic solvent mixture for elution/solvent back-extraction; (5) the capability to extract polar, nonpolar, acidic, and basic compounds; (6) the absence of a requirement for solvent evaporation and analyte reconstitution; (7) the fact that its operational simplicity meets green analytical chemistry and economic criteria as well. One major advantage of FPSE is its ability to immobilize highly polar polymers in a sol–gel hybrid organic–inorganic network, chemically bonded to the fabric substrate, resulting in a microextraction material that can efficiently extract both polar and nonpolar analytes directly from an aqueous sample matrix. This has been used in the analysis of a wide variety of analytes in environmental and biological samples [44–47].

PAHs are inherently nonpolar, hydrophobic compounds which do not ionize in water. Intuitively, the selective extraction of highly nonpolar analytes, such as PAHs, from an aqueous sample matrix would be facilitated by a hydrophobic sorbent as the extraction phase material [48] and consequently sol–gel C18 coated FPSE media containing long hydrophobic C18 chains has been evaluated and successfully applied in this case.

Herein, we describe the design and preparation of sol–gel C18 coated FPSE media and the development of a novel application protocol for the efficient extraction of trace amounts of four selected small PAHs of high environmental interest from aqueous solutions. The hydrophilic cellulose fabric substrate incorporated in the core of the adsorbent aids in the extraction kinetics by attracting water molecules containing PAHs towards its surface for a successful sorbent-analyte interaction, resulting in the trapping of the analyte on the FPSE media. Our strategy offers the following advantages in water analysis: (1) highly efficient analysis of PAHs (to ppt level concentrations); (2) simultaneous determination of four different PAHs in water using just one adsorbent; (3) facile regeneration of the adsorbent; (4) no solvent evaporation or analyte reconstitution is needed and (5) the FPSE strategy can be extended to design a range of adsorbents for sensitive determination of related pollutant compounds. This study serves the analytical and environmental community by providing a better and superior pathway for effective extraction and determination of trace PAHs in water matrices of domestic and environmental interest. In addition, HPLC-FLD was used for chromatographic separation of four PAHs with the advantage of a shorter analysis time and superior resolution and sensitivity. Finally, and most importantly, this is the first manuscript presenting the application of FPSE for the trace analysis of polycyclic aromatic hydrocarbons in aqueous media.

## 2. Materials and Methods

### 2.1. Choice of Target PAHs

The four selected small, low molecular weight PAHs are in the list of 16 priority PAHs designated by USEPA [1] and are source markers in water with diagnostic ratios for discriminating petrogenic from pyrolytic sources (fluoranthene/pyrene) [27], a model substrate for cancer studies (phenanthrene) [21], and the PAHs most detected in the highest concentrations in drinking water (fluoranthene, phenanthrene, pyrene, and anthracene) [49]. In addition to the feasibility of HPLC-FLD analysis, the commercial availability of PAHs for analytical standards was also considered.

### 2.2. Chemicals and Materials

The Muslin 100% cotton cellulose substrate for the creation of sol–gel C18 coated FPSE media was purchased from Jo-Ann Fabric (Miami, FL, USA). The precursors for sol–gel synthesis, octadecyl trimethoxysilane, tetramethyl orthosilicate, trifluoroacetic acid (TFA) and organic solvents such as acetone and dichloromethane were purchased from Sigma-Aldrich (St. Louis, MO, USA). Sodium hydroxide and hydrochloric acid were purchased from Thermo Fisher Scientific (Milwaukee, WI, USA).

Certified analytical standards of anthracene (Anth), fluoranthene (Flu), pyrene (Pyr), and phenanthrene (Phen) were purchased from Sigma-Aldrich (Darmstadt, Germany). HPLC grade methanol, acetonitrile and water were purchased from Sigma-Aldrich (Darmstadt, Germany) and filtered through a 0.22 μm filter before use. Stock solutions of each PAH were made at a concentration of 1 mg/mL, and a standard mixture of all four PAHs (1 mg/mL each) was prepared in methanol to achieve a response at the comparable level in HPLC-FLD. Working solutions were freshly prepared by diluting the mixed standard solution with HPLC-grade water to the required concentrations. All standards and working solutions were stored at 4 °C. All other reagents were of analytical grade.

### 2.3. Instrumentation

HPLC-FLD analysis of four PAHs was performed with a Dionex P680 HPLC pump (Germering, Germany) equipped with a Dionex Ultimate 3000 Fluorescence detector and the Chromeleon chromatography management software (Dionex, Germering, Germany). An Acentis Express reverse-phase C18 column (10 cm × 4.6 mm, particle size 5 μm, Supelco, Darmstadt, Germany) maintained at 25 °C was used for separation. Ultrasonic degassing was performed with an ultrasonic bath (Sarthak Scientific Services, Panchkula, India). The mobile phase consisted of water (solvent A) and acetonitrile (solvent B) and the flow rate was 1 mL/min. The isocratic elution mobile phase was set as follows: 15% A and 85% B with a total chromatographic run time of 10 min at $\lambda_{ex}$ = 260 nm and $\lambda_{em}$ = 420 nm using a fluorescent detector. The injection volume was 20 μL.

### 2.4. Water Sample Collection

The water samples selected for the investigation included two river water samples collected from the Chakki river (Pathankot, Punjab, India), two rain water samples and two bore-well drinking water samples collected from the Punjabi university campus (Patiala, Punjab, India), two metal factory wastewater samples were collected from a chemical factory (Patiala, India). Before the experiments were performed, all the water samples were filtered through Whatman filter paper and then through 0.22 μm micropore membranes and stored at 4 °C in a refrigerator.

### 2.5. Preparation of Sol–Gel C18 Coated FPSE Media

Due to the presence of starch and other finishing chemicals on the commercially available cellulose fabric used in apparel making, the substrate required a thorough cleaning. At the same time, a chemical treatment was applied to the substrate that maximizes the number of available hydroxyl groups which are required to effectively bind the sol–gel network via a condensation reaction. Preliminary cleaning of the substrate was accomplished by immersing a 100 cm$^2$ section of cotton cellulose fabric in deionized water for 30 min under constant sonication, followed by multiple rinsing steps with excess deionized (DI) water. For the surface chemical activation, the fabric was immersed in 1M NaOH solution for an hour under constant sonication. This step was followed by thorough rinsing with DI water and treatment with a 0.1M HCl solution for an hour to neutralize any residual NaOH which might still be on the fabric surface. Finally, the cleaned and activated cellulose substrate was dried and stored in an air-tight container until it was used for sol–gel C18 coating.

The sol solution used to create the sol–gel C18 coated FPSE media was prepared by sequentially mixing the sol–gel precursors methyl trimethoxysilane (MTMS); solvents methylene chloride and acetone; organically modified sol–gel precursor, octadecyl trimethoxysilane; sol–gel catalyst, trifluoroacetic acid (TFA) and water. In order to obtain a uniform sol solution for coating the fabric substrate, the molar ratio between methyl trimethoxysilane precursor:methylene chloride:acetone:trifluoroacetic acid:water was maintained at 1:2.33:1.94:0.5:0.20. The molar ratio between methyl trimethoxysilane and octadecyl trimethoxysilane was maintained at 1:0.38. To ensure that all the sol solution ingredients mixed homogeneously, the sol solution was vigorously vortexed for 3 min after adding each of the ingredients. The final sol solution was centrifuged for 5 min, followed by collection of the supernatant into a clean 3 oz. amber-colored glass reaction bottle. The solution was then sonicated for 10 min to remove trapped gaseous molecules from the sol solution.

The cleaned and pretreated cellulose fabric substrate was then gently immersed into the sol solution to initiate the substrate coating via a dip-coating process. The substrates were kept in the appropriate sol solutions for 2 h. During this surface coating period, a three-dimensional sol–gel network chemically bonded to the substrate was formed. At the end of the coating period, the sol solutions were expelled from the reaction bottle, the coated fabrics were dried in a desiccator and finally the sol–gel sorbent coated FPSE media was conditioned/aged in a home-made thermal conditioning device built inside a gas chromatography oven with continuous helium gas flow at 50 °C for 24 h. After conditioning/ageing, the sol–gel C18 coated FPSE media were cleaned sequentially with dichloromethane and methanol. Finally, the FPSE media were dried in the presence of continuous helium gas flow at 50 °C for 1 h. The FPSE media were then cut into 2.5 cm × 2.0 cm pieces, the typical application size for FPSE, and were stored in an airtight glass container for future use.

*2.6. Fabric Phase Sorptive Extraction Procedure*

Firstly, the sol–gel C18-coated FPSE media was conditioned in a mixture of 1 mL methanol and 1 mL acetonitrile for 5 min and then rinsed with 2 mL deionized water to remove any residual organic solvent. The FPSE procedure is shown in Figure 2.

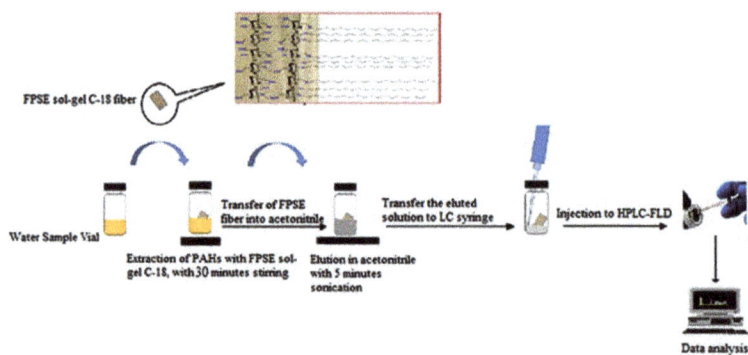

**Figure 2.** Chemical structure of sol–gel C18 coated fabric phase sorptive extraction (FPSE) media and the schematic representation of FPSE protocol.

The extraction process was performed quickly in a 20 mL glass vial containing 10 mL of the aqueous sample solution. A piece of sol–gel C18 coated FPSE media was directly immersed in the sample solution for 30 min under constant stirring at 1000 rpm at room temperature. After extraction from the sample solution, the PAHs were eluted from the sol–gel C18-coated FPSE media in 300 µL of acetonitrile with sonication for 5 min. Finally, 20 µL of this solution was injected directly into the HPLC-FLD system for analysis.

## 3. Results and Discussion

### 3.1. Selection of Fabric Phase Sorptive Extraction Sorbent Chemistry

Taking the nonpolar, hydrophobic characteristics of the selected PAHs into consideration (log $K_{ow}$ values between 4.45 and 5.16), it is obvious that a nonpolar sorbent would provide the best affinity towards the PAHs to selectively isolate them from complex environmental water sample matrix. As such, octadecyl trimethoxysilane was selected as the source of the C18 pendant group in the sol–gel silica network during the sol solution design. C18 has long been known as a nonpolar sorbent in solid phase extraction and in nonpolar stationary phase in reversed phase liquid chromatography. It is worth mentioning that the selectivity of commercially available C18 sorbent is substantially different from that of sol–gel C18 sorbent and often the latter demonstrates unique selectivity towards a wide range of analytes including polar, moderately polar and nonpolar analytes due the presence of residual surface silanol groups in the sol–gel C18 sorbent matrix.

Unlike the substrate used in C18 sorbent (silica particles), the substrate used in FPSE plays an active role in determining the overall polarity and selectivity of the FPSE media. A fabric substrate made of 100% cotton cellulose was selected as the support for sol–gel C18 coating because the hydrophilic nature of the substrate can cause water molecules to come close to the extraction device during extraction so that the requisite interactions between the C18 pendant groups and the PAHs can result in successful extraction of PAHs onto the FPSE media. The permeable structure of the fabric support also plays an important role as a pseudo-solid phase extraction disk and facilitates rapid analyte extraction due to the continuous diffusion of water through the FPSE media when under the influence of magnetic stirring during the extraction.

### 3.2. Characterization of Sol–Gel C18 Coated FPSE Media

#### 3.2.1. Scanning Electron Microscopy

Figure 3 represents the scanning electron micrographs (SEM) of (a) the uncoated Muslin cotton (100% cellulose) substrate at 100× magnification; (b) sol–gel C18 coated fabric phase sorptive extraction media at 100× magnification; and (c) sol–gel C18 coated fabric phase sorptive extraction media at 500× magnification.

**Figure 3.** Scanning electron microscopy images of (**a**) uncoated cotton (100%) cellulose fabric substrate at 100× magnification; (**b**) uniformly coated sol–gel C18 sorbent coating on fabric matrix at 100× magnification; (**c**) sol–gel C18 coated FPSE media at 500× magnification.

The extraction efficiency of FPSE primarily depends on the sponge-like porous architecture of the sol–gel sorbent coating (for faster analyte diffusion) and the permeability of the fabric substrate that mimics a solid phase extraction disk. The highly porous sorbent coating as well as the permeability and hydrophilicity of the cellulose fabric synergistically allow the aqueous sample matrix to flow through the FPSE medium, leading to rapid interaction between the sorbent and the analytes. Consequently, the analytes are adsorbed on the sorbent with high efficiency in a short period of time. As such, it is important to study the surface morphology of the FPSE media before and after the coating to ensure that the pores of the fabric substrate are well maintained even after the sol–gel coating.

The SEM images demonstrated that the microstructures and the pores of the cellulose substrate were well preserved even after the sol–gel C18 sorbent coating. This flow-through extraction mechanism is only exploited in solid phase extraction (SPE), and is totally absent in solid phase microextraction and related techniques (such as stir bar sorptive extraction, thin film microextraction, etc.) due to the impermeable nature of the substrate used in these microextraction techniques. The flow-through extraction system consequently helps to achieve faster extraction equilibrium. The enlarged image of the sol–gel C18-coated FPSE media demonstrates that C18 coatings are homogeneously distributed on the fabric substrate surface while maintaining the pores of the substrate.

### 3.2.2. Fourier-Transform Infrared Spectroscopy (FT-IR)

Figure 4 illustrates FT-IR spectra representing (a) uncoated Muslin cotton (100% cellulose) fabric; (b) octadecyl trimethoxysilane; (c) sol–gel C18 coated fabric phase sorptive extraction media.

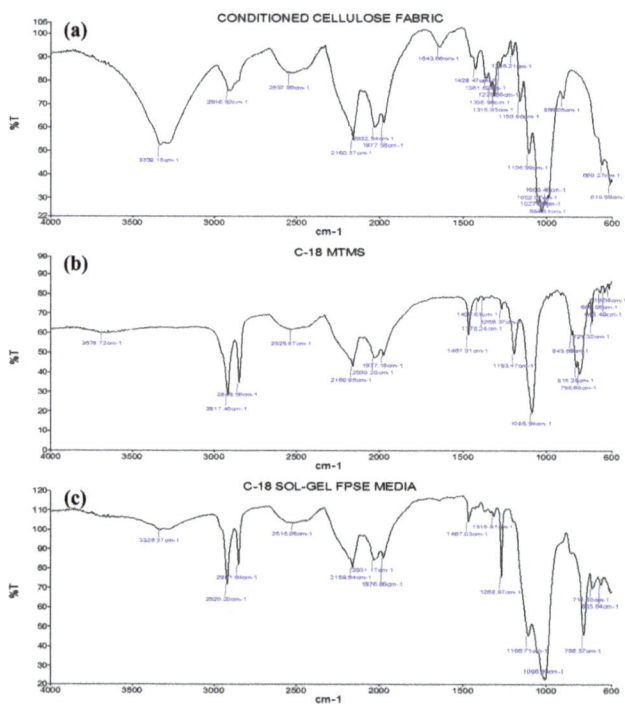

**Figure 4.** FT-IR spectra of (a) uncoated cellulose substrate; (b) octadecyl trimethoxysilane; and (c) sol–gel C18 coated FPSE medium.

The FT-IR spectra of the uncoated Muslin cotton 100% cellulose fabric demonstrates characteristic absorption bands at ~3330 cm$^{-1}$, 2916 cm$^{-1}$, 1315 cm$^{-1}$, and 1028 cm$^{-1}$ that correspond to O–H, C–H,

C–O vibration and C–H bending vibration, respectively [50]. The characteristic peaks of octadecyl trimethoxysilane appear at 2917 cm$^{-1}$, and 2849 cm$^{-1}$ which correspond to asymmetric and symmetric vibrations of –CH$_2$–, –CH$_2$– groups, respectively [51]. The strong peak at 1467 cm$^{-1}$ and 1085 cm$^{-1}$ are due to the vibration absorption of Si–O–C, and Si–O–Si, respectively. The strong peak at 815 cm$^{-1}$ is assigned to Si–C bonds and the peak at 795 cm$^{-1}$ is assigned to the vibration of (–CH$_2$–)$_n$ ($n \geq 4$) [52]. The characteristic peaks of sol–gel C18-coated FPSE media appeared at 2890 cm$^{-1}$ and 2851 cm$^{-1}$ which represent symmetric vibration of –CH$_2$– and asymmetric vibration of –CH$_3$, respectively. The same characteristic peaks are also seen in octadecyl trimethoxysilane spectra. In addition, the presence of ~1467 cm$^{-1}$, ~1268 cm$^{-1}$, ~1977 cm$^{-1}$ in both the sol–gel C18 coated FPSE media and octadecyl trimethoxysilane strongly suggests the successful integration of octadecyl moieties into the sol–gel network. The substantial reduction of the O–H stretching vibrations (at 3325 cm$^{-1}$) in sol–gel C18 coated FPSE media compared to uncoated cotton (100% cellulose) indicates the chemical integration of the sol–gel C18 network to the cellulose structure via condensation. Due to the chemical integration of sol–gel sorbent to the substrate surface, the resulting FPSE media offers a remarkably superior thermal, solvent and chemical stability than its commercial counterparts such as SPME, SBSE etc.

### 3.3. Optimization of the FPSE Procedure

In order to achieve accurate and sensitive chromatographic quantification of the trace PAHs in the water samples, the optimum conditions for using sol–gel C18 coated FPSE media were investigated. Several conditions affecting the extraction efficiency were optimized, including extraction time, sample volume, eluting solvent, elution time, volume of organic modifier, and salt concentration. Optimization experiments were performed using a standard aqueous solution of PAHs containing 0.10 µg/mL each of four PAHs to ensure a comparable level of response to each compound.

### 3.3.1. Optimization of Sample Volume

To obtain high enrichment factors and high recoveries for all PAHs, the initial sample volume should be as large as possible. Therefore, different volumes (5 mL, 10 mL, 15 mL, 20 mL) of an aqueous solution were investigated. It was found that the highest extraction efficiency was obtained with a sample volume of 10 mL, as shown in Figure 5a. As the sample volume was increased up to 15 mL, recovery increased and after 15 up to 20 mL no obvious change was observed. After 20 mL, recovery decreased up to 50 mL, inferring that the extraction efficiency was insufficient at volumes above 20 mL. Therefore, the initial sample volume was set at 15 mL for future FPSE protocols.

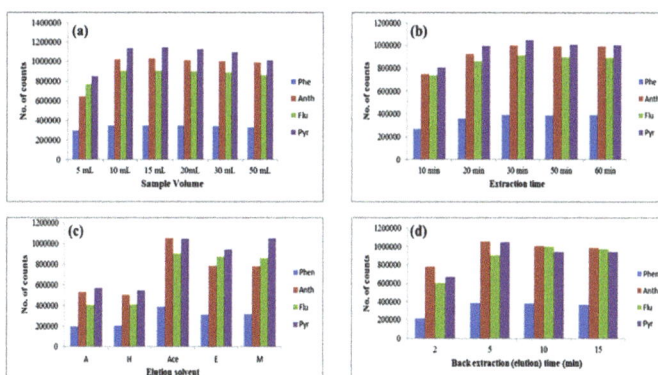

**Figure 5.** (**a**) Effect of sample volume; (**b**) Effect of extraction time; (**c**) Selection of the best elution solvent; (**d**) Optimization of back-extraction time. Extraction conditions: extraction time: 30 min; eluent solvent: acetonitrile; volume of elution solvent: 300 µL, desorption time: 5 min.

### 3.3.2. Optimization of Extraction Time

Another key factor affecting extraction efficiency is the extraction time and it so it was imperative that this factor be thoroughly investigated. The extraction time was set from 10 to 60 min. As shown in Figure 5b, recoveries of all four PAHs increased with extraction time as it was increased from 10 to 30 min, and then remained unchanged even when the time was increased up to 60 min, implying that the extraction equilibrium was achieved at about 30 min. Thirty minutes was therefore selected as the extraction time of FPSE protocol.

### 3.3.3. Optimization of Desorption Solvent and Time

As far as the FPSE protocol is concerned, PAH desorption from the sol–gel C18 coated FPSE media can significantly affect the sensitivity of PAH extraction. Therefore, choice of an appropriate elution solvent plays a key role in the process. Having considered the properties of PAHs, acetone (A), n-hexane (H), acetonitrile (Ace), ethanol (E) and methanol (M) were selected as potential elution solvents in this experiment. As shown in Figure 5c, acetone and n-hexane were poor eluents for PAHs. Methanol yielded the highest recovery for Pyr, whereas acetonitrile was preferable for recovery of all four PAHs. Therefore, acetonitrile was chosen as the elution solvent and was used for further studies. Furthermore, the elution efficiency also relies on the volume of the elution solvent. As shown in Supplementary Figure S1, for most of the PAHs, the recovery increased as the eluent volume increased from 100 µL to 300 µL, but remains nearly unchanged if the volume is further increased from 300 µL to 700 µL, after which dilution causes a decrease in recovery. To get maximum recovery with minimum solvent use, 300 µL of acetonitrile was selected for desorption.

In order to resolve any possible carry-over problems and to avoid any loss of PAHs, the sonication desorption time was further optimized. The process of desorption was carried out in an ultrasonic bath with desorption times of 1, 3, 5 and 10 min. The results shown in Figure 5d prove that the peak areas of PAHs increased as the desorption time increased from 1 min to 5 min, but remained unchanged as the desorption time was increased further. Therefore, 5 min was sufficient to achieve maximum desorption.

### 3.3.4. Effect of Salt Concentration and Organic Modifiers

Salt ions in the sample might also affect FPSE by competitive interaction between the salting-in and the salting-out effect. The salting-out effect causes the analyte to enhance its partition onto the sol–gel C18-coated FPSE media by decreasing its solubility in water, while the salting-in effect has the opposite effect. Thus, the effect of the addition of salt to the samples was investigated and is shown in Supplementary Figure S2. No obvious change was observed for the recoveries with KCl at 0–0.15 M, indicating that salt ion addition does not affect extraction efficiency. This is probably due to the low polarity of all four PAHs. Thus, no salt was added in the following experiments.

Addition of an organic modifier, such as methanol, might promote the extraction efficiency of FPSE for PAHs by preventing the FPSE C18 carbon chains from cross-linking and thus retaining the ability to contact the target analytes completely. It was found that methanol addition (0–3 mL) did not cause any changes in extraction efficiency as shown in Supplementary Figure S3. This was probably because of the fact that FPSE media coated with C18 hydrophobic long chains spreads out well in the form of a uniform film on both sides of cellulose fabric substrate, due to strong chemical bonding between the sol–gel sorbent and the fabric substrate. Therefore, no organic modifier is needed. This indicates that the sol–gel C18 coated FPSE media is stable in various solutions, and the extraction efficiency is independent of the salinity and the presence of an organic modifier.

### 3.4. Regeneration and Reusability of Sol–Gel C18 Coated FPSE Media

The durability of sol–gel C18 coated FPSE media was also investigated by extracting PAHs from a water sample 30 times using the same FPSE media. The FPSE media was regenerated with sonication in 10 mL acetonitrile for 15 min. The extraction efficiency of the sol–gel C18 coated FPSE media was almost unchanged after 30 extraction procedures the results of which are shown in Supplementary Figure S4. These results indicate that sol–gel C18 coated FPSE media can be repeatedly used for extraction.

### 3.5. Analytical Performance

All data were subjected to strict quality control procedures. The linearities of the chromatographic responses of the protocol were tested with calibration standards at seven concentration levels ranging from 0.01 to 10 ng/mL. Good linearities were observed for the four target PAHs, with all the correlation coefficients ($R^2$) above 0.99. Limits of quantification (LOQs) (signal-to-noise ratio = 10) and limits of detection (LODs) (signal-to-noise ratio = 3) of the four target compounds are shown in Table 1.

**Table 1.** Linear range, linearity curve, correlation coefficients, limits of detection (LODs) and limits of quantification (LOQs) for the determination of PAHs ($n$ = 5).

| Analyte | Linear Range (ng/mL) | Linearity Curve | $R^2$ | LOD (pg/mL) | LOQ (pg/mL) |
|---------|---------------------|-----------------|-------|-------------|-------------|
| Phen | 0.010–10 | $y = 34{,}591x + 772$ | 0.9997 | 1 | 3.33 |
| Anth | 0.010–10 | $y = 104{,}873x + 2147$ | 0.9997 | 0.1 | 0.33 |
| Flu | 0.010–10 | $y = 90{,}258x + 247$ | 0.9987 | 0.7 | 2.33 |
| Pyr | 0.010–10 | $y = 114{,}921x + 497$ | 0.9983 | 0.4 | 1.33 |

The accuracy, precision and recovery studies were performed on 15 mL deionized organic-free water spiked with (low, medium, and high concentration levels) 0.01, 0.5 and 5 ng of each PAH per mL (five replicates). Mean recoveries ranged from 88.1% to 90.5% for all the target compounds. The mixture of standards was analyzed five times within a day. The relative standard deviations (RSD) of concentrations for all the targets ranged from 1.1% to 4.1% (Table 2), demonstrating the high precision of the analytical protocol. In order to ensure the accuracy of the analysis, all samples were replicated five times, and the final concentration averages were used.

**Table 2.** Precision, accuracy and recovery of four selected PAHs in spiked deionized organic-free water at (low, medium, and high concentration levels) ($n$ = 5).

| Analyte | Phen | | | Anth | | | Flu | | | Pyr | | |
|---------|------|------|------|------|------|------|------|------|------|------|------|------|
| Precision (RSD%) | 0.01 | 0.5 | 5 | 0.01 | 0.5 | 5 | 0.01 | 0.5 | 5 | 0.01 | 0.5 | 5 |
| Intra-day | 5.6 | 4.7 | 3.6 | 1.9 | 2.8 | 2.2 | 2.1 | 2.8 | 1.8 | 1.9 | 2.1 | 1.1 |
| Inter-day | 4.9 | 5.01 | 4.1 | 2.2 | 2.3 | 2.5 | 1.8 | 2.9 | 1.9 | 3.1 | 2.7 | 2.0 |
| Accuracy (%) | | | | | | | | | | | | |
| Intra-day | 88 | 87 | 90 | 90 | 92 | 91 | 91.1 | 90.1 | 92.1 | 90.8 | 91.6 | 92.8 |
| Inter-day | 86 | 89 | 89.4 | 91 | 90 | 89.3 | 91 | 90 | 92 | 90 | 89 | 91 |
| Recovery (%) | 88.1 | 88 | 88.4 | 88.6 | 88 | 88.1 | 92 | 91 | 90.5 | 90.7 | 91 | 90.1 |
| (RSD%) | (2.9) | (4.1) | (3.8) | (3.1) | (3.2) | (2.4) | (2.2) | (2.8) | (1.8) | (3.2) | (4.2) | (1.4) |

### 3.6. Mathematical Model for Predicting Extraction Efficiency (Absolute Recovery, %)

A mathematical model was created for sol–gel C18 coated FPSE media that can be used as a predictive tool to assess the extraction efficiency (expressed as the absolute recovery) of analytes on sol–gel C18-coated FPSE media using their log $K_{ow}$ values. A carefully chosen test mixture was created to develop the mathematical model, consisting of 10 compounds representing different polarity and functionality (log $K_{ow}$ values ranging from 0.3 to 5.07). The selected test compounds included piperonal (PIP), phenol (PHE), furfuryl alcohol (FA), benzodioxole (BDO), naphthalene (NAP), 4-nitrotoluene (4NT), 9-anthracene methanol (9AM), 1,2,4,5-tetramethyl benzene (TMB), triclosan (TCL) and diethylstilbestrol (DESB). The extraction recovery value of each compound from an aqueous solution of the test mixture was determined and the values were plotted against their log $K_{ow}$ values to obtain a second order mathematical model: Extraction efficiency (absolute recovery, %) = $-2.274875$ + $20.816015 \times \log K_{ow} - 4.1478973 \times (\log K_{ow} - 2.737)^2$. This second order mathematical model can be used to predict the extraction efficiency of sol–gel C18 coated FPSE media for a given analyte using its $\log K_{ow}$ value. A graphical representation of the model is shown in Figure 6.

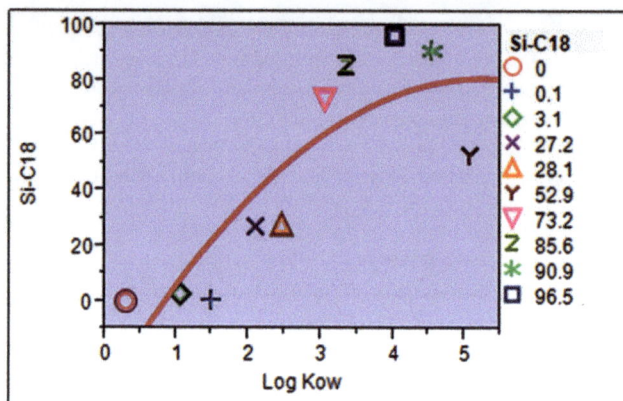

**Figure 6.** Correlation curve between absolute recovery and logarithmic values of octanol-coefficients of analytes with wide polarity range.

The predicted recovery values for four PAHs using the mathematical model and their actual recovery values are given in Table 3.

**Table 3.** Data demonstrating model predicted absolute recovery (%) and actual recovery (%) obtained by FPSE–HPLC–FLD method.

| Compound | Log $K_{ow}$ | Expected Recovery (%) | Actual Recovery (%) |
|----------|--------------|-----------------------|---------------------|
| Phen | 4.46 | 78.30 | 88.4 |
| Anth | 4.45 | 78.24 | 88.1 |
| Flu | 5.16 | 80.85 | 90.5 |
| Pyr | 4.88 | 80.32 | 90.1 |

For all the target PAHs, the actual extraction recovery values were found to be higher than the predicted values obtained from the model. This was attributed to the strong hydrophobic interactions between the PAHs and the sol–gel C18 sorbent, in addition to efficient trapping of the PAHs on the sol-gel C18 extraction media with high primary contact surface area, high loading of sol–gel nanocomposite sorbent (in the form of ultrathin film), as well as the permeability of the extraction device that mimics a solid phase extraction disk (characterized by exhaustive extraction).

### 3.7. Application to Real Water Samples

Subsequently, the FPSE-HPLC-FLD protocol developed was applied for PAH determination in real environmental water samples: bore-well water, river water, rain water and factory wastewater. Quintuplicate analyses were performed and the concentrations found for the target PAHs are summarized in Table 3. In the four types of environmental water samples, the target four PAHs were not detected in bore-well water and rain water. Phen and Flu were detected in two river water samples and all four target PAHs were found in each of the two metal-fabrication factory wastewater samples (Table 4).

**Table 4.** The four selected PAHs concentrations detected in real water samples (*n* = 5).

| Sample | Phen (ng/mL) | Anth (ng/mL) | Flu (ng/mL) | Pyr (ng/mL) |
|---|---|---|---|---|
| A [a] | n.d | n.d | n.d | n.d |
| B [a] | n.d | n.d | n.d | n.d |
| C [b] | 7.8 ± 2.6 | n.d | 5.6 ± 3.3 | n.d |
| D [b] | 8.8 ± 2.9 | n.d | 6.8 ± 3.4 | n.d |
| E [c] | n.d | n.d | n.d | n.d |
| F [c] | n.d | n.d | n.d | n.d |
| G [d] | 11 ± 3.1 | 7.8 ± 2.5 | 7.5 ± 3.8 | 5.8 ± 4.1 |
| H [d] | 4.1± 3.2 | 3.7 ± 3.1 | 7.7 ± 3.3 | 3.0 ± 2.8 |

[a]: bore-well water; [b]: river water; [c]: rain water; [d]: factory waste water; n.d: not detected.

The two river water samples, collected from two different locations in the Chakki River in Pathankot were analyzed. At one outlet point of metal factory wastewater, Phen, Anth, Flu, Pyr were detected at 11 ng/mL, 7.8 ng/mL, 7.5 ng/mL and 5.8 ng/mL, respectively, while at second outlet point, Phen, Anth, Flu, Pyr were detected at 4.1 ng/mL, 3.7 ng/mL, 7.7 ng/mL and 3.0 ng/mL, respectively, (shown in Table 3); suggesting that the PAHs came mainly from the wastewater. These results indicated that the method could be successfully applied to PAH analysis in real water samples. The typical chromatograms of a river water sample and a metal factory waste water sample are presented in Figure 7.

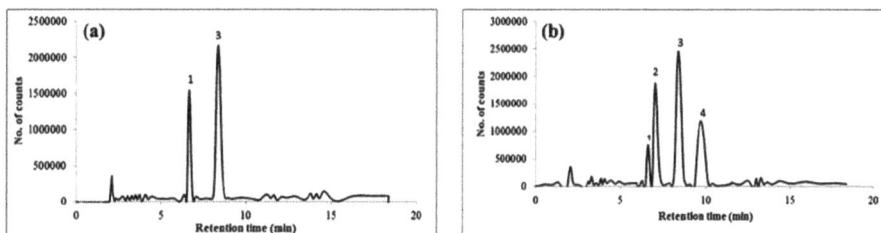

**Figure 7.** Chromatogram representing analysis of real environmental samples: (**a**) Chakki river water sample 2; (**b**) factory wastewater sample 2.

*3.8. Comparison of Sol–Gel C18 Coated FPSE Media with Other Sorbent Materials*

The performance of the new analytical protocol with the sol–gel C18-coated FPSE media as the adsorbent material was also compared with other materials reported in the literature (as shown in Table 5).

**Table 5.** Comparisons of the analytical performance of the method developed with reported methods in the literature.

| Method | Sorbent Material | Sorbent Preparation Time (Hours) | LOD (pg/mL) | References |
|---|---|---|---|---|
| MSPE-HPLC-FLD | TPA-functionalized MNPs | 37.5 | 0.04–3.75 | [53] |
| μ-SPE-GC-MS | Functionalized graphene sheet | 29.5 | 0.8–3.9 | [54] |
| SPE-HPLC-FLD | Cotton fiber | 0.5 | 0.1–2 | [55] |
| MIPs-SPE-GC-MS | Imprinted sol-gel adsorbent | 38 | 5.2–12.6 | [56] |
| FPSE-HPLC-FLD | Sol-gel C18 (Cellulose) | 24 | 0.1–1 | Current work |

Abbreviations: MIPs: molecularly imprinted polymers; MSPE: magnetic solid-phase extraction; TPA-functionalized MNPs: triphenylamine functionalized magnetic microspheres. HPLC-FLD: high performance liquid chromatography with fluorescence detector; GC-MS: gas chromatography mass spectrometry.

The time required in sorbent preparation and the LOD values were compared. The LOD values achieved in the present research are lower than those reported in the literature. Furthermore, sol–gel C18-coated FPSE media as a sorbent has obvious advantages: preparation of sorbents often requires several days to complete the synthesis process, involves dozens or hundreds of milliliters of organic solvents, is time consuming and may cause environment pollution, whereas sol–gel C18-coated FPSE media, an advanced inorganic–organic hybrid material with tunable porosity, selectivity, thermal and chemical stability, high reproducibility and solvent resistance, requires less organic solvents, is fast and is a green analytical endeavor. Therefore, the proposed FPSE-HPLC-FLD protocol is proven to be a green, convenient, efficient and reliable method for the pre-concentration of trace PAHs from water samples.

## 4. Conclusions

In summary, for the first time, sol–gel C18-coated FPSE media was directly and successfully applied as a selective adsorbent for PAH analysis in real environmental water samples. The LODs obtained for four target PAH compounds using the novel analytical protocol were 0.1–1 pg/mL. Compared to traditional SPE methods, the sol–gel C18-coated FPSE media is simple to prepare and regenerate for recurring usage, thereby meeting the need for rapid analysis. In addition, the costs of the preparation of the sol–gel C18-coated FPSE media are economical and organic solvent consumption is minimal. Sol–gel C18-coated FPSE media can also be regenerated and reused more than 30 times. The FPSE media is also degradable and therefore more environment friendly. Utilizing the unique attributes of FPSE, a novel, simple, efficient, fast, sensitive, green, economical and reliable FPSE-HPLC-FLD protocol is presented for trace level determination of four environmentally important PAHs. This protocol is suitable for many applications in health-related water contamination studies and offers new analytical capabilities for water quality assurance, particularly for routine monitoring of the presence of PAHs in water samples in order to ensure water quality, safety and consumer protection.

**Supplementary Materials:** The following are available online at http://www.mdpi.com/2297-8739/4/2/22/s1, Figure S1: Effect of eluent solvent volume: Extraction conditions: Sample volume, 15 mL; extraction time: 30 min; eluent solvent: acetonitrile; sonication desorption time 5 min. Figure S2: Effect of salt concentration: Extraction conditions: sample volume, 15 mL; extraction time: 30 min; eluent solvent, acetonitrile; volume of elution solvent, 300 μL desorption time, 5 min. Figure S3: Effect of organic modifier volume: Extraction conditions: sample volume, 15 mL; extraction time: 30 min; eluent solvent, acetonitrile; volume of elution solvent, 300 μL desorption time, 5 min. Figure S4: Regeneration and reusability of sol–gel C18 coated FPSE media. Extraction conditions: sample volume, 15 mL; extraction time: 30 min; eluent solvent: acetonitrile; volume of elution solvent: 300 μL desorption time, 5 min.

**Acknowledgments:** This study was financially supported by University Grants Commission (UGC), New Delhi, India.

**Author Contributions:** Shivender Singh Saini and Abuzar Kabir conceived and designed the experiments; Shivender Singh Saini and Abuzar Kabir performed the experiments; Shivender Singh Saini and Abuzar Kabir analyzed the data; Avasarala Lakshmi Jagannadha Rao, Ashok Kumar Malik and Kenneth G. Furton contributed reagents/materials/analysis tools; Shivender Singh Saini and Abuzar Kabir wrote the paper. Avasarala Lakshmi Jagannadha Rao, Ashok Kumar Malik and Kenneth G. Furton reviewed and edited the paper.

**Conflicts of Interest:** The authors declare no conflict of interest.

## References

1.  Agency for Toxic Substances and Drug Registry (ATSDR). *Toxicological Profile for Polycyclic Aromatic Hydrocarbons (PAHs) Update*; Agency for Toxic Substances and Drug Registry (ATSDR): Washington, DC, USA, 1995; Chapter 1, pp. 1–9.

2.  Agency for Toxic Substances and Drug Registry (ATSDR). *Toxicological Profile for Polycyclic Aromatic Hydrocarbons (PAHs) Update*; Agency for Toxic Substances and Drug Registry (ATSDR): Washington, DC, USA, 1995; Chapter 2, pp. 11–207.

3.  Bostrom, C.E.; Gerde, P.; Hanberg, A.; Jernstrom, B.; Johansson, C.; Kyrklund, T.; Rannug, A.; Tornqvist, M.; Victorin, K.; Westerholm, R. Cancer risk assessment, indicators, and guidelines for polycyclic aromatic hydrocarbons in the ambient air. *Environ. Health Perspect.* **2002**, *110*, 451–488. [CrossRef] [PubMed]

4. White, K.L. An overview of immunotoxicology and carcinogenic polycyclic aromatic hydrocarbons. *Environ. Carcinog. Rev.* **1986**, *4*, 163–202. [CrossRef]

5. Srogi, K. Monitoring of environmental exposure to polycyclic aromatic hydrocarbons: A review. *Environ. Chem. Lett.* **2007**, *5*, 169–195. [CrossRef]

6. Beyer, J.; Jonsson, G.; Porte, C.; Krah, M.M.; Ariese, F. Analytical methods for determining metabolites of polycyclic aromatic hydrocarbon (PAH) pollutants in fish bile: A review. *Environ. Toxicol. Pharmacol.* **2010**, *30*, 224–244. [CrossRef] [PubMed]

7. Van der Oost, R.; Beyer, J.; Vermeulen, N.P.E. Fish bioaccumulation and biomarkers in environmental risk assessment: A review. *Environ. Toxicol. Pharmacol.* **2003**, *13*, 57–149. [CrossRef]

8. Liu, L.Y.; Wang, J.Z.; Wei, G.L.; Guan, Y.F.; Wong, C.S.; Zeng, E.Y. Sediment records of polycyclic aromatic hydrocarbons (PAHs) in the continental shelf of China: Implications for evolving anthropogenic impacts. *Environ. Sci. Technol.* **2012**, *46*, 6497–6504. [CrossRef] [PubMed]

9. Baek, S.O.; Field, R.A.; Goldstone, M.E.; Kirk, P.W.; Lester, J.N.; Perry, R. A review of atmospheric polycyclic aromatic hydrocarbons: Sources, fate and behavior. *Water Air Soil Pollut.* **1991**, *60*, 279–300. [CrossRef]

10. Chen, H.Y.; Teng, Y.G.; Wang, J.S.; Song, L.T.; Zuo, R. Source apportionment of sediment PAHs in the Pearl River Delta region (China) using nonnegative matrix factorization analysis with effective weighted variance solution. *Sci. Total Environ.* **2013**, *444*, 401–408. [CrossRef] [PubMed]

11. Weinstein, J.E.; Crawford, K.D.; Garner, T.R.; Flemming, A.J. Screening level ecological and human health risk assessment of polycyclic aromatic hydrocarbons in stormwater detention pond sediments of Coastal South Carolina, USA. *J. Hazard. Mater.* **2010**, *178*, 906–916. [CrossRef] [PubMed]

12. Guo, J.F.; Guo, Q.Z.; Yan, G.P. Determination of polycyclic aromatic hydrocarbons in water samples by hollow fiber extraction coupled with GC-MS. *Anal. Methods* **2015**, *7*, 1071–1075. [CrossRef]

13. Shamsipur, M.; Hashemi, B. Extraction and determination of polycyclic aromatic hydrocarbons in water samples using stir bar sorptive extraction (SBSE) combined with dispersive liquid-liquid microextraction based on the solidification of floating organic drop (DLLME-SFO) followed by HPLC-UV. *RSC Adv.* **2015**, *5*, 20339–20345.

14. Zhou, Q.; Gao, Y. Determination of polycyclic aromatic hydrocarbons in water samples by temperature-controlled ionic liquid dispersive liquid-liquid microextraction combined with high performance liquid chromatography. *Anal. Methods* **2014**, *6*, 2553–2559. [CrossRef]

15. Manoli, E.; Samara, C. Polycyclic aromatic hydrocarbons in natural waters: Sources, occurrence and analysis. *TrAC Trends Anal. Chem.* **1999**, *18*, 417–428. [CrossRef]

16. Pánková, K. IACR Monographs on the Evaluation of the Carcinogenicity Risk of Chemical to Humans. *Biol. Plant.* **1986**, *28*, 354. [CrossRef]

17. Mahler, B.J.; Van Metre, P.C.; Bashara, T.J.; Wilson, J.T.; Johns, D.A. Parking lot sealcoat: An unrecognized source of urban polycyclic aromatic hydrocarbons. *Environ. Sci. Technol.* **2005**, *39*, 5560–5566. [CrossRef] [PubMed]

18. Dickhut, R.M.; Canuel, E.A.; Gustafson, K.E.; Liu, K.; Arzayus, K.M.; Walker, S.E.; Edgecombe, G.; Gaylor, M.O.; MacDonald, E.H. Automotive sources of carcinogenic polycyclic aromatic hydrocarbons associated with particulate matter in the Chesapeake Bay region. *Environ. Sci. Technol.* **2000**, *34*, 4635–4640. [CrossRef]

19. Sicre, M.A.; Marty, J.C.; Saliot, A.; Aparicio, X.; Grimalt, J.; Albaiges, J. Aliphatic and aromatic hydrocarbons in different sized aerosols over the Mediterranean Sea: Occurrence and origin. *Atmos. Environ.* **1967**, *21*, 2247–2259. [CrossRef]

20. Fermin, L.; Valley, C.; Alonso, E.J.; Olmo, M.D.; Vilchez, J.L. Determination of ultra-traces of anthracene in water samples by solid-phase spectrofluorometry. *Anal. Sci.* **1993**, *9*, 117–120.

21. Arbabi, M.; Nasseri, S.; Mesdaghinia, A.R.; Rezaie, S.; Naddafi, K.; Omrani, G.H.; Yunesian, M. Survey on physical, chemical and microbiological characteristics of PAH-contaminated soils in Iran. *Iran. J. Environ. Health Sci. Eng.* **2004**, *1*, 26–33.

22. Kanaly, R.; Harayama, S. Biodegradation of high-molecular-weight polycyclic aromatic hydrocarbons by bacteria. *J. Bacteriol.* **2000**, *182*, 2059–2067. [CrossRef] [PubMed]

23. Kweon, O.; Kim, S.J.; Jones, R.C.; Freeman, J.P.; Adjei, M.D.; Edmondson, R.D.; Cerniglia, C.E. A polyomic approach to elucidate the fluoranthene-degradative pathway in *Mycobacterium vanbaalenii* PYR-1. *J. Bacteriol.* **2007**, *189*, 4635–4647. [CrossRef] [PubMed]

24. Zhang, X.J.; Shi, Z.; Lyv, J.X.; He, X.; Englert, N.A.; Zhang, S.Y. Pyrene is a novel constitutive androstane receptor (CAR) activator and causes hepatotoxicity by CAR. *Toxicol. Sci.* **2015**, *147*, 436–445. [CrossRef] [PubMed]

25. Hollosi, L.; Wenzl, T. Development and optimisation of a dopant assisted liquid chromatographic-atmospheric pressure photo ionisation-tandem mass spectrometric method for the determination of 15 + 1 EU priority PAHs in edible oils. *J. Chromatogr. A* **2011**, *1218*, 23–31. [CrossRef] [PubMed]

26. Purcaro, G.; Moret, S.; Bucar-Miklavcic, M.; Conte, L.S. Ultra-high performance liquid chromatographic method for the determination of polycyclic aromatic hydrocarbons in a passive environmental sampler. *J. Sep. Sci.* **2012**, *35*, 922–928. [CrossRef] [PubMed]

27. Zhao, B.; Zhang, S.; Zhou, Y.; He, D.; Li, Y.; Ren, M.; Xu, Z.; Fang, J. Characterization and quantification of PAH atmospheric pollution from a large petrochemical complex in Guangzhou: GC–MS/MS analysis. *Microchem. J.* **2015**, *119*, 140–144. [CrossRef]

28. Mirivel, G.; Riffault, V.; Galloo, J.C. Simultaneous determination by ultra-performance liquid chromatography-atmospheric pressure chemical ionization time-of-flight mass spectrometry of nitrated and oxygenated PAHs found in air and soot particles. *Anal. Bioanal. Chem.* **2010**, *397*, 243–256. [CrossRef] [PubMed]

29. Fujiwara, F.; Guinez, M.; Cerutti, S.; Smichowski, P. UHPLC-(+)APCI-MS/MS determination of oxygenated and nitrated polycyclic aromatic hydrocarbons in airborne particulate matter and tree barks collected in Buenos Aires city. *Microchem. J.* **2014**, *116*, 118–124. [CrossRef]

30. Cruz-Vera, M.; Lucena, R.; Ardenas, S.C.; Valcarcel, M. One-step in-syringe ionic liquid-based dispersive liquid-liquid microextraction. *J. Chromatogr. A* **2009**, *1216*, 6459–6465. [CrossRef] [PubMed]

31. Wang, W.D.; Huang, Y.M.; Shu, W.Q.; Cao, J. Multiwalled carbon nanotubes as adsorbents of solid-phase extraction for determination of polycyclic aromatic hydrocarbons in environmental waters coupled with high-performance liquid chromatography. *J. Chromatogr. A* **2007**, *1173*, 27–36. [CrossRef] [PubMed]

32. Ma, J.; Xiao, R.; Li, J.; Yu, J.; Zhang, Y.; Chen, L. Determination of 16 polycyclic aromatic hydrocarbons in environmental water samples by solid-phase extraction using multi-walled carbon nanotubes as adsorbent coupled with gas chromatography-mass spectrometry. *J. Chromatogr. A* **2010**, *1217*, 5462–5469. [CrossRef] [PubMed]

33. Brum, D.M.; Cassella, R.J.; Pereira Netto, A.D. Multivariate optimization of a liquid-liquid extraction of the EPA-PAHs from natural contaminated waters prior to determination by liquid chromatography with fluorescence detection. *Talanta* **2008**, *74*, 1392–1399. [CrossRef] [PubMed]

34. Shamsipur, M.; Hassan, J. A novel miniaturized homogenous liquid-liquid solvent extraction-high performance liquid chromatographic-fluorescence method for determination of ultra traces of polycyclic aromatic hydrocarbons in sediment samples. *J. Chromatogr. A* **2010**, *1217*, 4877–4882. [CrossRef] [PubMed]

35. Elisabeth, R.; Rajasekhar, B. Optimization and validation of solid phase micro-extraction (SPME) method for analysis of polycyclic aromatic hydrocarbons in rainwater and stormwater. *Phys. Chem. Earth* **2009**, *34*, 857–865.

36. Bagheri, H.; Babanezhad, E.; Es-haghi, A. An aniline-based fiber coating for solid phase microextraction of polycyclic aromatic hydrocarbons from water followed by gas chromatography-mass spectrometry. *J. Chromatogr. A* **2007**, *1152*, 168–174. [CrossRef] [PubMed]

37. Wu, Y.L.; Xia, L.B.; Chen, R.; Hu, B. Headspace single drop microextraction combined with HPLC for the determination of trace polycyclic aromatic hydrocarbons in environmental samples. *Talanta* **2008**, *74*, 470–477. [CrossRef] [PubMed]

38. Yao, C.; Twu, P.; Anderson, J.L. Headspace single drop microextraction using micellar ionic liquid extraction solvents. *Chromatographia* **2010**, *72*, 393–402. [CrossRef]

39. Djozan, D.J.; Assadi, Y. Modified pencil lead as a new fiber for solid-phase microextraction. *Chromatographia* **2004**, *60*, 313–317. [CrossRef]

40. Djozan, D.; Assadi, Y.; Haddadi, S.H. Anodized aluminum wire as a solid-phase microextraction fiber. *Anal. Chem.* **2001**, *73*, 4054–4058.

41. Bai, D.S.; Li, J.H.; Chen, S.B.; Chen, B.H. A novel cloud-point extraction process for preconcentrating selected polycyclic aromatic hydrocarbons in aqueous solution. *Environ. Sci. Technol.* **2001**, *35*, 3936–3940. [CrossRef] [PubMed]

42. Heidari, H.; Razmi, H.; Jouyban, A. Preparation and characterization of ceramic/carbon coated $Fe_3O_4$ magnetic nanoparticle nanocomposite as a solid-phase microextraction adsorbent. *J. Chromatogr. A* **2012**, *1245*, 1–7. [CrossRef] [PubMed]
43. Yang, S.; Chen, C.; Yan, Z.; Cai, Q.; Yao, S. Evaluation of metal-organic framework 5 as a new SPE material for the determination of polycyclic aromatic hydrocarbons in environmental waters. *J. Sep. Sci.* **2013**, *36*, 1283–1290. [CrossRef] [PubMed]
44. Kabir, A.; Furton, K.G. Fabric Phase Sorptive Extractors. U.S. Patent 9,283,544, 15 March 2016.
45. Racamonde, I.; Rodit, R.; Quintana, J.B.; Sieira, J.B.; Kabir, A.; Furton, K.G.; Cela, R. Fabric phase sorptive extraction: A new sorptive microextraction technique for the determination of non-steroidal anti-inflammatory drugs from environmental water samples. *Anal. Chim. Acta* **2015**, *865*, 22–30. [CrossRef] [PubMed]
46. Rodríguez-Gomez, R.; Roldan-Pijuan, M.; Lucena, R.; Cárdenas, S.; Zafra-Gómez, A.; Ballesteros, O.; Valcárcel, M. Stir-membrane solid-liquid-liquid microextraction for the determination of parabens in human breast milk samples by ultra high performance liquid chromatography-tandem mass spectrometry. *J. Chromatogr. A* **2014**, *1354*, 26–33. [CrossRef] [PubMed]
47. Samanidou, V.; Kabir, A.; Galanopoulos, L.D.; Furton, K.G. Fast extraction of amphenicols residues from raw milk using novel fabric phase sorptive extraction followed by high-performance liquid chromatography-diode array detection. *Anal. Chim. Acta* **2015**, *855*, 41–50. [CrossRef] [PubMed]
48. Kayillo, S.; Dennis, G.R.; Shalliker, R.A. An assessment of the retention behaviour of polycyclic aromatic hydrocarbons on reversed phase stationary phases: Selectivity and retention on C18 and phenyl-type surfaces. *J. Chromatogr. A* **2006**, *1126*, 283–297. [CrossRef] [PubMed]
49. World Health Organization. *Guidelines for Drinking-Water Quality*, 2nd ed.; World Health Organization: Geneva, Switzerland, 1997.
50. Kumar, R.; Heena, G.; Malik, A.K.; Kabir, A.; Furton, K.G. Efficient analysis of selected estrogens using fabric phase sorptive extraction and high performance liquid chromatography-fluorescence detection. *J. Chromatogr. A* **2014**, *1359*, 16–25. [CrossRef] [PubMed]
51. Brambilla, R.; Pires, G.P.; dos Santos, J.H.Z.; Miranda, M.S.L.; Chornik, B. Octadecylsilane-modified silicas prepared by grafting and sol–gel methods. *J. Electron Spectrosc. Relat. Phenom.* **2007**, *156*, 413–420. [CrossRef]
52. Pan, H.; Wang, X.D.; Xiao, S.S.; Yu, L.G.; Zhang, Z.J. Preparation and characterization of $TiO_2$ nanoparticles surface-modified by octadecyltrimethoxysilane. *Indian J. Eng. Mater. Sci.* **2013**, *20*, 561–567.
53. Long, Y.; Chen, Y.; Yang, F.; Chen, C.; Pan, D.; Cai, Q.; Yao, S. Triphenylamine-functionalized magnetic microparticles as a new adsorbent coupled with high performance liquid chromatography for the analysis of trace polycyclic aromatic hydrocarbons in aqueous samples. *Analyst* **2012**, *137*, 2716–2722. [CrossRef] [PubMed]
54. Zhang, H.; Low, W.P.; Lee, H.K. Evaluation of sulfonated graphene sheets as sorbent for micro-solid-phase extraction combined with gas chromatography-mass spectrometry. *J. Chromatogr. A* **2012**, *1233*, 16–21. [CrossRef] [PubMed]
55. Song, X.; Li, J.; Xu, S.; Ying, R.; Ma, J.; Liao, C.; Liu, D.; Yu, J.; Chen, L. Determination of 16 polycyclic aromatic hydrocarbons in seawater using molecularly imprinted solid-phase extraction coupled with gas chromatography-mass spectrometry. *Talanta* **2012**, *99*, 75–82. [CrossRef] [PubMed]
56. Wang, J.; Liu, S.; Chen, C.; Zou, Y.; Hu, H.; Cai, Q.; Yao, S. Natural cotton fibers as adsorbent for solid phase extraction of polycyclic aromatic hydrocarbons in water samples. *Analyst* **2014**, *139*, 3593–3599. [CrossRef] [PubMed]

![separations logo] *separations*

MDPI

*Review*

# Recent Advances in Microextraction Techniques of Antipsychotics in Biological Fluids Prior to Liquid Chromatography Analysis

**Natalia Manousi [1,2], Georg Raber [1,*] and Ioannis Papadoyannis [2]**

[1]  Institute of Chemistry, Karl-Franzens-Universität Graz, 8010 Graz, Austria; nmanousi@chem.auth.gr
[2]  Laboratory of Analytical Chemistry, Department of Chemistry, Aristotle University of Thessaloniki, 54124 Thessaloniki, Greece; papadoya@chem.auth.gr
*  Correspondence: georg.raber@uni-graz.at; Tel.: +43-316-380-5305

Academic Editor: Juan F. García-Reyes
Received: 30 January 2017; Accepted: 5 May 2017; Published: 12 May 2017

**Abstract:** Antipsychotic drugs are a class of psychiatric medication worldwide used to treat psychotic symptoms principally in bipolar disorder, schizophrenia and other psycho-organic disorders. The traditional sample preparation techniques such as liquid-liquid extraction (LLE) or solid phase extraction (SPE), which were widely used, tend to have many drawbacks because they include complicated, time-consuming steps and they require large sample size as well large amounts of organic solvent. Therefore, due to the modern analytical requirements, such as miniaturization, automation and reduction of solvent volume and time, many microextraction procedures have been developed. In this review we aim to present an overview of those techniques which are used prior to liquid chromatography analyses both for forensic toxicology in different biological matrices as well as for therapeutic drug monitoring.

**Keywords:** antipsychotics; pharmaceuticals; microextraction; biofluids; HPLC; SPME; MEPS; LPME

## 1. Introduction

Antipsychotic drugs are a class of psychiatric medication primarily used to manage psychotic symptoms principally in schizophrenia, bipolar disorder and other psycho-organic disorders. Based on World Health Organization, sixty four compounds are classified as antipsychotics and for about 70% of these, analytical methods have been developed to determine them in human matrices [1].

Typical antipsychotics, known as first-generation antipsychotics were primarily discovered at 1950s and they tend to act on $D_2$ and $D_4$ receptors in the dopamine pathways of the brain. Those drugs are used much less frequently now because they show severe side-effects. For this reason second-generation antipsychotics, or atypical antipsychotics have been developed. Those drugs have a tendency to act less to those receptors, and therefore they show less side-effects than the primarily used typical antipsychotic drugs [2]. Figure 1 illustrates the chemical structures of common antipsychotics, which are currently used.

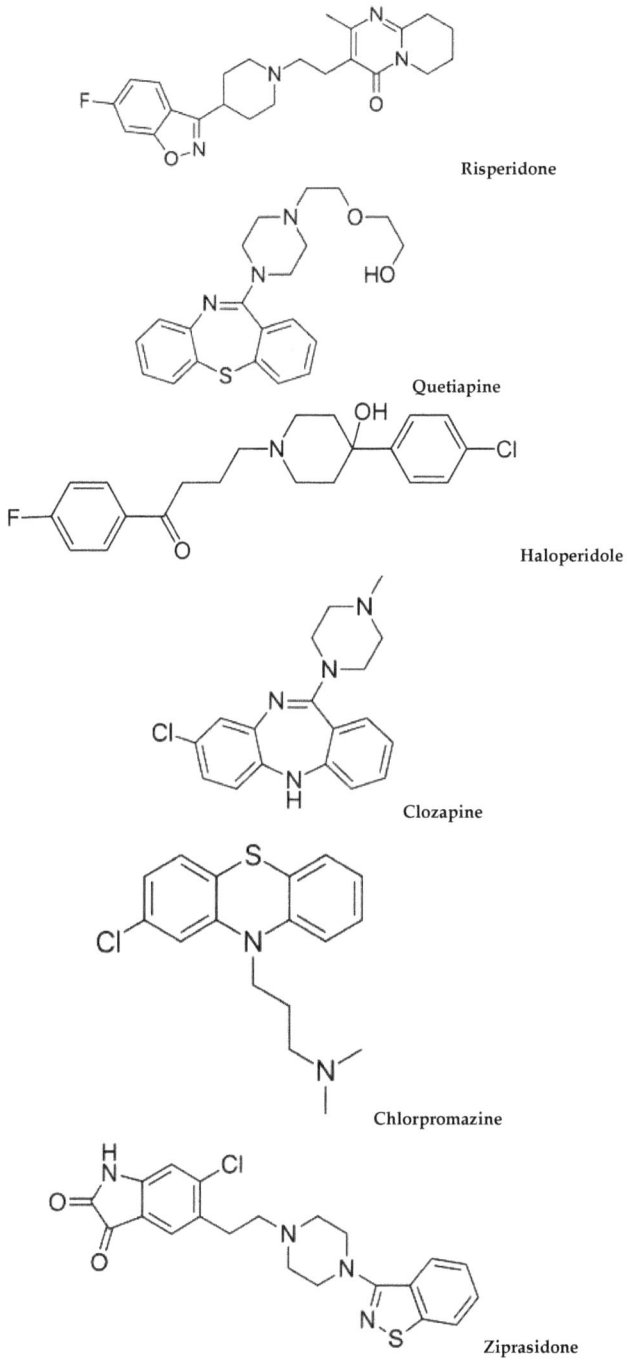

**Figure 1.** Chemical structures of common antipsychotic drugs.

Due to the wide use of these drugs worldwide, there is a great need of analytical methods in order to analyze biological samples. The quantitative determination of antipsychotics in human matrices is of great interest both for therapeutic drug monitoring and for forensic toxicology [1].

The modern trend in drug analysis is shifting from gas chromatography to liquid chromatography not only because of its good quantitative results, its high reproducibility, sensitivity and wide applicability, but also because most antipsychotic drugs are not volatile. For the moment ultrahigh performance liquid chromatography-tandem mass spectrometry UHPLC-MS/MS is the most preferred technique for the separation and analysis of antipsychotic drugs in biofluids.

Conventional matrices used for this purpose are serum, plasma and whole blood. However, other alternative matrices like oral fluid and urine; which are easily collected, keratinized matrices namely hair and nails; which are stable and capable of providing information for long periods of time, dry blood spots (DBS) and cerebrospinal fluid, are widely used [1].

Due to the high complexity of biological materials, which often contain proteins, salts, organic compounds with similar properties to the analytes and other endogenous compounds that may deteriorate the performance of separation, a sample preparation procedure is required. An ideal sample preparation technique should be fast and comprise the minimum number of working steps, should be easy to learn and easy to use, should be economical and environmental friendly and should be compatible with many analytical instruments [3].

The four major sample preparation techniques used for those matrices are liquid–liquid extraction (LLE), solid-phase extraction (SPE), protein precipitation (PP) and direct injection. However, these conventional techniques tend to have many fundamental drawbacks because they include complicated, time-consuming steps and they require large amounts of sample and organic solvents, while there are many difficulties in automation [1,3].

Therefore, there is a great need of developing novel, relatively simple, fast and solvent-free microextraction procedures which use smaller volumes of samples and solvents (microliter range or even smaller) and can be widely used to analyze these samples. To date, there is a big number of different microextraction techniques which are used for sample preparation of biological fluids and other biological matrices in order to enhance compatibility with modern analytical instrumentation, as well as to minimize the use of toxic chemicals and to decrease the size of biofluids or reagents' demand.

In this review we aim to present an overview of microextraction techniques which are used prior to liquid chromatography analysis in order to analyze biological fluids and to detect and quantify antipsychotics in conventional and alternative biological matrices.

## 2. Solid Phase Microextraction for the Determination of Antipsychotic Drugs in Biological Samples

Solid-phase microextraction (SPME) is an efficient solvent-free sample preparation method which was first introduced in the early 1990s by Pawliszyn and co-workers. It enables automation, miniaturization and high-throughput performance. This technique uses fibers and capillary tubes coated by stationary phases and it can be applied to samples in any state of matter gaseous, liquid and solid. The technique is based on partitioning of the analytes between the sample matrix and the extraction phase which is immobilized on a fused-silica SPME fiber coated with polymers, until the equilibrium is reached and subsequent thermal desorption of the extracts into a gas chromatograph, reconstitution in the mobile phase used for a separation with a liquid chromatograph, or direct injection to an HPLC injection port using suitable interface [4].

The type of the polymers which are used depends on the properties of the analyte. For drug analysis of biological matrices the most common coatings are the polydimethylsiloxane (PDMS) and polyacrylate (PA) while other polymers such as polypyrrole coatings, coatings based on restricted access materials, and those based on mixtures of biocompatible polymers with sorbents used for SPE have also been developed and used [5,6].

The most widely used technique is fiber-SPME. In this technique the analyte is directly extracted onto the coating of the fiber which is usually is inside a needle in a device with an assembly holder. For the procedure of SPME the sample is placed in a capped vial with a septum which is pierced by the needle of the device followed by the extension of the fiber either to the vapor above the sample (Head-Space SPME) for volatile analytes or directly to the sample (Direct Immersion-SPME) for the extraction of non-volatile analytes until equilibrium is reached. An alternative microextraction technique is in-tube SPME that uses a fused-silica capillary column. The extraction of the analytes takes place either onto the inner coating of the fiber or onto a sorbent bed. Compared to fiber SPME, in-tube SPME is more mechanically stable and can be used with on line coupling with HPLC or LC/MS instruments [7].

Various SPME methods have been developed for the analysis of antipsychotic drugs in biological matrices prior to liquid chromatography analysis.

Theodoridis et al. developed a method for the determination of a typical antipsychotic; haloperidol, together with other four drugs: quinine, naproxen, ciprofloxacin and paclitaxel in urine. Each analyte was studied independently. Haloperidol was determined using an Analyticals Erbasil Symmetry $C_{18}$ column, with a mixture of 0.05 M aqueous ammonium acetate and acetonitrile (35:65 $v/v$) as a mobile phase, while the detection was accomplished with a UV detector at 210 nm. For the SPME procedure, a PDMS 100 μm fiber was conditioned for 30 min in a GC injector operating at 250 °C. Then, 4 mL of a solution of each pharmaceutical in buffer (20 μg/mL in 0.9% NaCl, pH 9) was transferred in a glass vial containing a magnetic stirring bar, which was then capped and the sample was agitated at 700 rpm. Multiple SPME was also applied, but it did not provide yield enhancement for haloperidol, despite that this technique is supposed to be an excellent way to increase extraction yields. At the end of the extraction, the analyte was desorbed in 200 μL of methanol and an aliquot of 80 μL was injected to the HPLC. This method overcomes the problems that exist in conventional sample preparation techniques and can be applied to real urine samples [5].

In 2012, Bocato et al. published a method for the analysis of paliperidone after stereoselective fungal biotransformation of the atypical antipsychotic drug risperidone with SPME extraction prior to HPLC-MS/MS analysis. Paliperidone or 9-hydroxyrisperidone (9-RispOH) is characterized by the same pharmacologic activity of the parent drug risperidone. Another metabolite of risperidone, 7-hydroxyrisperidone, was also included in this study. The chromatographic separation was achieved using a Chiralcel OJ-H column with a mixture of methanol: ethanol (50:50, $v/v$) plus 0.2% triethylamine as a mobile phase at a flow rate of 0.8 mL·min$^{-1}$. Firstly, the SPME $C_{18}$ fiber probe 45 μm which was selected, was conditioned for 30 min with methanol and water (50:50, $v/v$). Extraction was performed by immersing the fiber in the fungal sample, the pH of which was controlled with a phosphate buffer (pH 7) and also 20% NaCl ($w/v$) was added. The addition of the electrolyte reduces its solubility of the organic analyte and increases its extraction yield. The adsorption lasted 30 min at room temperature with 600 rpm stirring speed following desorption in a 120 μL glass vial filled with the mobile phase. After 5 min, the fiber was withdrawn into the needle and an aliquot of 20 μL was injected into the HPLC–MS/MS system with no further treatment. To avoid carryover, after each desorption step the fiber was washed for 30 min with methanol. The obtained SPME recoveries were 28% for risperidone, 16% for 9-RispOH and 11% for 7-RispOH [6].

Kumazawa et al. have successfully developed an HPLC-MS/MS method to determine eleven phenothiazine derivatives (clospirazine, fluphenazine, perazine, thiethylperazine, thioridazine, flupentixol, thioproperazine, trifluoperazine, perphenazine, prochlorperazine and propericiazine) in human whole blood and urine using solid-phase microextraction (SPME) with a polyacrylate-coated fiber. The pH of the samples was adjusted to about 8 with KOH solution and the vial was sealed with a silicone-rubber septum cap. The syringe needle of the SPME device was passed through the septum and the polyacrylate fiber was pushed out from the needle and immersed directly in the sample solution in the vial at 40 °C and the extraction lasted 60 min at continuous stirring at 250 rpm. The fiber was then injected into the desorption chamber of the SPME-HPLC interface.

This was filled with distilled water containing 10 mM ammonium acetate plus 0.1% formic acid—100% acetonitrile (70:30, $v/v$). The desorption time was 10 min. Subsequently the entire contents of the desorption chamber were flushed directly on to the HPLC column by means of the mobile phase flow at 0.2 mL/min. This method was effectively applied to real samples after oral administration and it can be recommended for use in both therapeutic monitoring and clinical or forensic toxicology [8].

## 3. Microextraction by Packed Sorbent for the Determination of Antipsychotic Drugs in Biological Samples

Another microextraction technique developed in the last decade is microextraction by packed sorbent (MEPS). This novel technique is based on the same general principle of solid phase extraction (SPE), but with MEPS the packing is integrated directly into the syringe, in a very small barrel (BIN) which sets up the needle assembly of an HPLC syringe and not in a separate SPE cartridge. The sorbents that are used in MEPS are usually the same as conventional SPE columns. Most of the applied sorbents include silica-based sorbents ($C_2$, $C_8$ and $C_{18}$). When the biological sample passes through the solid support of the syringe, the analytes are adsorbed onto the sorbent which is packed in the BIN. Because MEPS and SPE build on the same principles there is the option of transferring a method from conventional SPE to MEPS relatively straight forward [9–12].

MEPS holds the high selectivity, the good sample purifying efficiency and extraction yields of SPE. Compared to traditional sample preparation techniques like LLE and SPE, MEPS procedure is faster, simpler, cheaper, more feasible, more environmental friendly, more user-friendly and uses both small amounts of biological sample (10 µL of plasma, urine or water) and large volumes (1000 µL). Compared to protein precipitation, this microextraction procedure is much more efficient. In general, MEPS can reduce sample volume and time necessary for the analysis. Moreover it can be fully automated and it can be connected to liquid chromatography (LC), gas chromatography (GC) or capillary electrochromatography (CEC) [9,13].

The most important factors in MEPS performance, which should be optimized before the sample analysis are conditioning, loading, washing and eluting solvents, sample flow rate, washing solution and the type and volume of the elution, which should be suitable for injection into LC or GC systems. Also, the volume of the sample should be optimized leading to the best equilibrium between a good analytical performance and a good extraction methodology. The optimum conditions will be contingent to the nature of the matrix being used and the retention capacity and specificity of the sorbent in order to obtain the highest recovery of the analytes [9].

To date, there are several MEPS methods that have been developed for the determination of antipsychotic drugs in biological matrices prior to liquid chromatography analysis.

In 2010 Saracino et al. aimed to develop an analytical method for the determination of an atypical antipschycotic drug; risperidone and its main active metabolite 9-hydroxyrisperidone in human plasma and saliva based on HPLC with coulometric detection and an innovative MEPS procedure. Those two analytes were also studied with a SPME procedure prior to HPLC-MS/MS analysis. The microextraction procedure was carried out using a BIN containing 4 mg of solid-phase material silica-$C_8$, after being activated with 100 µL of methanol for three times and conditioning with 100 µL of water for another three times, at a flow rate of 20 µL/s. For the extraction of the analytes, the samples were drawn up and down through the syringe 15 times (at a flow rate of 5 µL/s) without discarding and a washing step once with water (100 µL) and once with a mixture of water and methanol (95:5, $v/v$) took place, in order to remove biological interference from the samples. Then, the analytes were eluted with 250 µL of methanol and they were subsequently separated on a reversed phase $C_{18}$ column, using a mobile phase composed of acetonitrile (26%) and a pH 6.5 phosphate buffer (74%).After extraction, the sorbent was cleaned similarly to the activating and conditioning step in order to decrease memory effects and to condition for the next extraction. The same sorbent was used for about 50 extractions. The limit of quantitation for the two compounds was 0.5 ng/mL, while the limit of detection was 0.17 ng/mL. Extraction yields were higher than 90.1% and intra-day and inter-day precision results were good. As a result, this method was successfully applied to real saliva and blood samples from patients [13].

For the determination of the same analytes in plasma, urine and saliva, there is also a more recent MEPS-HPLC-UV method. The column which was used was a Chromsep $C_8$ reversed-phase (150 × 4.6 mm i.d., 5 µm) and the mobile phase consisted of a mixture of acetonitrile (27%, $v/v$) and a pH 3.0, 30 mM phosphate buffer containing 0.23% ($v/v$) triethylamine (73%, $v/v$) with a gradient elution program. The UV detector was set at 238 nm and diphenhydramine was used as the internal standard. The $C_8$ MEPS cartridges were activated and conditioned with 300 µL of methanol and then with 300 µL of water. For the loading step the samples were drawn into the syringe and discharged back 10 times. The cartridge was then washed with 200 µL of water and then with 200 µL of a water/methanol mixture. For the elution step, 500 µL of methanol were used. The eluate was dried and redissolved in mobile phase and 50 µL of the solution was injected into the HPLC system. For the MEPS procedure, extraction efficiencies were higher than 90%, while relative standard deviation (RSD) for precision was always lower than 7.9% for the two compounds. In the biological samples limits of quantification were lower than 4 ng/mL for risperidone and lower than 6 ng/mL for 9-hydroxyrisperidone. Finally, the developed method was successfully applied to the analysis of biological samples from patients and seems suitable for therapeutic drug monitoring [14].

In 2014, Mercolini et al. developed an HPLC method for the determination of a recent atypical antipsychotic; ziprasidone in plasma samples, using MEPS procedure. The analytes were separated on a RP $C_{18}$ column, with a mobile phase which was a mixture of acetonitrile (30%, $v/v$) and a pH 2.5, 50 mM phosphate buffer containing 0.2% ($v/v$) diethylamine (70%, $v/v$) that was delivered isocratically and the detection was performed at 320 nm. For the microextraction, the $C_2$ sorbent which was chosen was conditioned with 200 µL of methanol and equilibrated with 200 µL of water. The sample was loaded and discarded back 10 times. Washing of the sorbent took place with 100 µL of water and 100 µL of a water/methanol mixture (90/10, $v/v$). Finally, the elution was done by drawing and discharging 500 µL of methanol. The eluate was dried under vacuum, redissolved in 100 µL of mobile4 phase and injected in the HPLC-UV system. Extraction yields were higher than 90% while limit of quantitation was 1 ng/mL. The sensitivity and the selectivity of the method was also good. The developed method was compared to a SPE procedure, using $C_2$ cartridges and the results were satisfactory. As a result, this procedure was successfully applied to real plasma samples from patients who were using ziprasidone and can be used for therapeutic drug monitoring of patients undergoing treatment with ziprasidone [15].

In 2015, Souza et al. synthesized hybrid silica monoliths which were functionalized with aminopropyl- or cyanopropyl- groups by sol-gel process and used the mass selective stationary phase for MEPS to determine five antipsychotics, namely: olanzapine, quetiapine, clozapine, haloperidol and chlorpromazine) simultaneously with seven antidepressants, two anti-convulsants and two anxiolytics in plasma using UPLC-MS/MS. Due to the higher selectivity of the cyanopropyl hybrid silica for most of the drugs and its good mechanical strength, it was finally selected as the stationary MEPS phase. For the MEPS procedure, the stationary phase was conditioned with 4 × 200 µL of a methanol and acetonitrile mixture (50:50 $v/v$) and 4 × 200 µL of water. Then, 4 × 100 µL of plasma samples diluted with ammonium acetate solution (pH 10) was manually drawn. Then, the sorbent was washed with 150 µL of water and desorption took place using 100 µL of a 50:50 ($v/v$) mixture of methanol and acetonitrile. The extract was dried, and reconstituted with 50 µL of the mobile phase, which consisted of ammonium acetate solution 5 mmol/L (with 0.1% formic acid) and acetonitrile and then injected into a XSelects CSH $C_{18}$ (2.5 µm, 2.1 × 100 mm) column for analysis with liquid chromatography. The linearity of the method ranged from 0.05 to 1.00 ng/mL (limit of quantification) to 40–10,500 ng/mL. The absolute recoveries, the precision and the accuracy were good, so the developed method can be applied to the therapeutic drug monitoring of patients [16].

Clozapine and its metabolites were also determined in dried blood spots on filter paper with a HPLC method coupled with a coulometric detection, after being extracted with phosphate buffer and cleaned-up with MEPS procedure. The use of this matrix has many advantages because it eliminates the blood withdrawal, it has low cost and low biohazard risk and it is easy to use and to store. For the microextraction procedure, the sorbent which was 4 mg of solid phase silica-$C_8$ material, inserted into a syringe was activated using $3 \times 100$ μL of methanol and subsequently conditioned with $3 \times 100$ μL of water. For the clean-up $10 \times 150$ μL of the extract from DBS was drawn up and down, followed by a washing step first with 100 μL of water and second with a mixture of water and methanol (95:5, *v/v*). The elution step took place using 150 μL of the mobile phase and the liquid was injected into the HPLC system. For the HPLC analysis a reversed phase $C_{18}$ column was used with a mobile phase composed of methanol, acetonitrile and phosphate buffer. All MEPS steps namely: activation, loading, washing and elution were carried out in manual mode. The extraction yields were higher than 90%, the method validation gave satisfactory results for accuracy, precision, sensitivity and selectivity. The developed method was successfully applied to real samples obtained from patients. Therefore, this developed method is suitable for therapeutic drug monitoring for patients undergoing treatment with clozapine [17].

Hendrickx et al. developed a capillary UHPLC-UV method in combination with MEPS as a sample clean-up procedure, in order to determine chlorpromazine, olanzapine and their flavin-containing monooxygenase mediated N-oxides in rat brain microdialysates. The analysis was carried out with an Acclaim Pepmap RP $C_{18}$ capillary column. For the MEPS procedure 4 mg of a mixed solid phase M1 (80% $C_8$, 20% SCX) cartridge was selected. Firstly, the sorbent was activated using first 100 μL of a solution consisting of 5% ammonia in 80% methanol (*v/v*) and second 100 μL of methanol and then conditioned with a 1:3 mixture (*v/v*) of Ringer's solution and phosphate buffer (pH 2.5). For the adsorption $3 \times 50$ μL microdialysate sample diluted with the same buffer was drawn through the syringe and ejected. Then the sorbent was washed first with 100 μL 5% acetic acid (*v/v*) and second with 100 μL of a mixture of methanol and water (10:90, *v/v*). The elution step took place with 50 μL of a solution containing 5% ammonia in 80% methanol (*v/v*) and the extracts were diluted with 150 μL of mobile phase which consisted of 10 mM ammonium acetate with 0.05% triethylamine, adjusted to pH 3.00 using formic acid and acetonitrile and injected into the UHPLC system. For the examined analytes MEPS recoveries were higher than 92% and intra- and inter-day variabilities were below 15%. The applicability of the method was checked by analyzing real samples from patients, thus proving that it can be used for therapeutic drug monitoring [18].

## 4. Liquid Phase Microextraction for the Determination of Antipsychotic Drugs in Biological Samples

Liquid Phase Microextraction (LPME) is a miniaturized form of liquid-liquid extraction, which was firstly introduced at 1990s, when Dasgupta [19] and Jeannot and Cantwell [20] suggested almost at the same time the use of extraction solvents in the low microliter range. It is considered as a simple, rapid and cheap sample preparation technique, which requires only several microliters of organic solvents in contrast to traditional LLE, which requires several hundred of milliliters.

Based on hydrodynamic features, this technique can be classified into static LPME and dynamic LPME. In the static LPME, a solvent is used as an extractant and it is suspended in the sample. As a result transference of the target compounds to the extractant is carried out. On the other hand, in the dynamic mode, the exractant solvent forms a microfilm inside of an extraction unit, such as a microsyringe and the mass transfer of the analytes takes place between the sample and the microfilm [21].

The main forms of LPME are (1) single drop microextraction; the oldest form of LPME, which is less frequently used today compared to more recently developed techniques, because it was based on a droplet of solvent hanging at a needle of a syringe and it was not considered very robust, (2) Dispersive LPME (DLLME) and (3) Hollow fiber LPME (HF-LPME). Hollow fiber LPME is a recently developed technique, which is based on immobilized organic solvents inside pores of hollow fibers. This technique partly consists of (1) a donor phase, which is the aqueous sample containing the target compounds,

(2) the porous fiber with the organic solvent trapped inside and (3) a receptor phase inside the hollow fiber lumen.

Prior to analysis, the organic solvent is immobilized in the fiber's pores by dipping the hollow fiber in a vial containing the solvent in order to form a layer. Next, the lumen is filled with the acceptor phase, which could be an organic, an acidic or a basic solution. For the analysis, the fiber is inserted in the sample, which is the donor phase containing the analytes and extraction takes place into the immobilized organic solvent [22].

Depending on the acceptor phase that is used HF-LPME can be classified into (1) two phase HF-LPME, in which the organic solvent, which is immobilized in the hollow fiber and the receptor phase, which is inside the lumen of the fiber are the same solvent. In this case the solution that is used as a receptor phase can be directly injected to the gas chromatography system, whereas, for liquid chromatography and capillary electroapothesis, evaporation of the solvent and reconstitution in an aqueous solution are mandatory, so that the sample is compatible with the analytical apparatus.

Another form of HF-LPME is the three-phase HF-LPME, in which the acceptor phase is an acidic or basic aqueous solution. In this case, extraction of the analytes takes place from the aqueous sample primarily into the immobilized organic solvent. After that, back extraction from the organic solvent takes place in the final receptor solution, which is the aqueous solution placed into the lumen of the hollow fiber. This extraction mode is limited to basic or acidic analytes that can be ionized [22].

Hollow fiber-LPME can also be classified as static HF-LPME, which includes magnetic stirring of the solution and dynamic HF-LPME, in which small volumes of the sample are repeatedly pulled in and out of the fiber in order to increase the extraction speed [21].

During the development of an ideal HF-LPME process, many parameters should be optimized in order to achieve the best results. These parameters are the material of the fiber, the type of organic solvent, the pH of the sample and the acceptor phase, the volume of sample and of solvent, the time, the temperature, the ionic strength and the stirring speed [22].

Dispersive LLPME is a recent novel approach of liquid-phase microextraction, introduced by Assadi and their co-workers in 2006 [23]. This technique is based on a ternary solvent system consisting of an extraction solvent, a disperser solvent and an aqueous sample. A mixture consisting of the organic and the disperser solvent is rapidly and vigorously injected in the aqueous sample, which contains the target analytes. For this purpose a syringe is used [24]. As a result, a cloudy solution is formed, which is supposed to be stable for a specific time. As a next step, phase separation takes place by gently shaking and centrifuging the mixture. If the density of the organic solvent is higher than this of water, the solvent goes to the bottom of the tube and it can be removed by a microsyringe, after discarding the aqueous solution. The crucial parameters in this procedure are the type and the volume of extraction and disperser solvents, the extraction time after the formation of the cloudy solution, the pH of the sample and its ionic strength. As for the extraction solvent, it should be miscible with the disperser solvent and it should be able to extract the target analytes. Moreover, its high density and low solubility in water assist the centrifugation step. As for the disperser solvent, it has to be soluble in the organic solvent and miscible in water in order to enable the organic solvent to be dispersed in the sample and to form a cloudy solution. The most common disperser solutions are acetone, methanol and acetonitrile [22].

## 4.1. DLLME for the Determination of Antipsychotic Drugs in Biological Samples

Several LPME methods have been developed for the determination of antipsychotic drugs in biological matrices prior to liquid chromatography analysis using either HF-LPME or DLLME.

Cruz-Vera et al. published an almost solvent-less HPLC-UV method for the determination of seven phenothiazine derivatives in urine using the dynamic liquid-phase microextraction (dLPME) procedure. The whole process took place under dynamic conditions in an automatic flow system. The extraction unit was consisted of a syringe pump and a 1 mL syringe connected to a Pasteur pipette. For the microextraction, 100 µL of a mixture of ionic liquid 1-butyl-3-methyl imidazolium hexafluorophosphate and acetonitrile (50:50, $v/v$) were picked up in the pipette, which was then inserted into a vial containing the sample, the pH of which was primarily fixed at 8. A volume of 10 mL was drawn with a flow rate of 0.5 mL·min$^{-1}$. When the extraction was completed, 50 μL of the ionic liquid were drawn out at a flow rate of 0.05 mL·min$^{-1}$ and recovered in a vial containing 50 μL of acetonitrile. Finally, 20 μL of the mixtures were injected into tandem LiChrosorb C$_8$ (4.6 mm × 150 mm)–LiChrosorb C$_{18}$ (4.6 mm × 150 mm) cartridge columns and determined with a mobile phase consisting of acetonitrile/water/acetic acid/trimethylamine 40/40/20/2 ($v/v/v/v$). A new pipette was used for each extraction so there is no carry-over effect. The recovery values was between 72% and 98%, the limits of detection were between 21 ng/mL and 60 ng/mL and the repeatability expressed as RSD varied between 2.2% and 3.9% and the method was successfully validated [22].

Xiong et al. developed a HPLC-UV method for the separation and quantitative determination of three psychotropic drugs (amitryptiline, clomipramine and thioridazine) in urine, using DLLME as a sample preparation technique. For the microextraction, 5 mL of the sample, the pH of which was adjusted to 10 using NaOH, was placed in a test tube, in which 0.50 mL of acetonitrile (as a disperser solvent) containing 20 μL of carbon tetrachloride (as an extraction solvent) were rapidly and vigorously injected, in order to form a cloudy solution. Accordingly, the mixture was shaken and then the separation of the phases took place by centrifugation at 4000 rpm for 3 min. The extraction time was fixed at 3 min. After slowly discarding the aqueous solution, the resulting droplet and the lipidic solid were dissolved in 200 μL acetonitrile, filtered and injected into the HPLC system for analysis with a C$_8$column and a mixture of ammonium acetate (0.03 mol/L, pH 5.5)–acetonitrile (60:40, $v/v$) as a mobile phase. The absolute recoveries were between 96% and 101%, the limit of detection was 3 ng/mL, while the limit of quantification was 10 ng/mL. The developed method was efficiently applied to urine samples obtained from patients to estimate and personalize the drug dose [24].

In 2011, Chen et al. developed a DLLME-HPLC-UV method for the determination of two antipsychotic drugs; clozapine and chlorpromazine in urine. For the DLLME procedure 10 mL of the sample was placed in a test tube after adjusting the pH to 10 with NaOH and 200 μL of ethanol (as a disperser solvent) containing 40 μL CCl$_4$ (as an extraction solvent) was fast and vigorously injected, in order to form a cloudy solution, which was then shaken and centrifuged for 2 min at 4000 rpm to achieve phase separation. After that, the precipitate was dissolved by 0.5 mL methanol after careful removal of the supernatant solution, the extract was filtered and injected into the HPLC. For the separation, a Symmetry® C$_{18}$ column packed with 5.0 μm particle size of dimethyloctylsilyl bounded amorphous silica was used with a mixture of CH$_3$COONH$_4$ (0.03 g/mL, pH 5.5)-CH$_3$CN (60:40, $v/v$) as a mobile phase. With these conditions, the limits of detection were lower than 6 ng/mL, and the limits of quantification lower than 39 ng/mL. The absolute extraction efficiencies were higher than 97%. The method was successfully applied to the analysis of real samples obtained from patients [25].

In 2011, Zhang et al. developed a DLLME-HPLC-UV method for the determination of tetrahydropalmatine and tetrahydroberberine in rat urine; two active components in *Rhizoma corydalis*, which possess strong antipsychotic actions. For the DLLME procedure 1.00 mL of the sample solution was placed in a glass tube and 100 mL of 1 mol/L of NaOH solution were added. The mixture was vortexed and 100 μL of methanol (dispersive solvent) containing 37 μL of chloroform (extraction solvent) was injected rapidly and vigorously in order to form a cloudy solution which was then centrifuged for 3 min at 4000 rpm. After discarding the upper layer solution, the sedimented phase was evaporated to dryness and the dry residue was reconstituted in 50 μL of a acetonitrile-0.1% phosphoric acid solution which was used as a mobile phase and a 20 μL aliquot was injected into a Ultimate XB-C$_{18}$ (150 × 4.6 mm i.d., packed with 5 mm particles) column. The extraction recoveries with this technique were higher than 69% and this method was efficiently applied in real urine samples [26].

In 2015, Fisichella et al. developed a DLLME method for the determination of many different classes of drugs including main drugs of abuse (cocaine and metabolites, amphetamines and analogues, LSD, ketamine, opiates, methadone and fentanyl and analogues), Z-compounds and 44 benzodiazepines and antipsychotic drugs in blood samples followed by analysis with liquid UHLC-MS/MS. For the microextraction procedure, 100 µL of chloroform (extraction solvent) and 250 µL of methanol (disperser solvent) were rapidly and vigorously injected into the blood which was primarily deproteinized with 500 µL of methanol and its pH was fixed at 9 with the use of 0.2 g of NaCl and 100 µL of saturated carbonate buffer. A cloudy solution was formed, which was then shaken for 1 min using an ultrasonic water bath and centrifuged at 4000 rpm for 5 min. About 50 µL of the organic phase was evaporated to dryness and reconstituted in 100 µL of mobile phase. Then, 10 µL was then injected into a superficially porous Kinetex Biphenyl column (2.6 µm, 100 × 2.1 mm) with UHPLC-MS/MS instrument. The mobile phase consisted of $H_2O$ with 0.1% HCOOH MeOH with 0.1% HCOOH and a gradient elution program was chosen. The limits of detection were lower than 2 ng/mL and limits of quantification were lower than 10 ng/mL, with satisfactory accuracy and precision. The developed method was subsequently applied to the analysis of 50 blood samples from forensic cases [27].

### 4.2. HF-LPME for the Determination of Antipsychotic Drugs in Biological Samples

For trace amounts of chlorpromazine in biological fluids, a hollow fiber liquid phase microextraction HF-LPME-HPLC-UV method was also developed. The drug was extracted from 11 mL of sample into an organic phase which was n-dodecane trapped in the pores of the fiber followed by the back-extraction into a receiving aqueous solution consisting of 0.01 M phosphate buffer (pH 2.0), located inside the lumen of the hollow fiber. For the extraction 11 mL of the aqueous sample solution was placed into a glass vial with a stirring bar and the vials were put on a magnetic stirrer. The stirring speed was 1000 rpm. Then, 20 µL of the receiving phase were injected into the polypropylene fiber, which was placed into the organic solution for 5 s and then into water for 5 s to remove the extra organic solution from its surface. After that, the fiber was bent and placed into the sample for 60 min and at the end of the extraction the fiber was removed, the receiving phase was withdrawn into the syringe and 10 µL of the receiving phase was injected into the HPLC. The whole procedure was carried out in absence of salt. The detection limit for chlorpromazine was 0.5 µg/L and intra-day and inter-day assay (RSD %) were lower than 10.3%. The method was successfully applied to drug level monitoring in biological fluids (urine and serum) of patients and gave satisfactory results [28].

## 5. Novel LPME and SPME Techniques for the Determination of Antipsychotic Drugs in Biological Samples

Except for the conventional DLLME process, which was described above, many DLLME variations are widely used nowadays. Firstly, in-situ DLLME is also gaining attention, while temperature assisted-DLLME, UV-assisted-DLLME, Microwave-assisted-DLLME and vortex-assisted DLLME are becoming more and more popular as observed in the literature. Additionally, DLLME variations which use deep eutectic solvents (DES's), surfactants, or ionic liquis (ILs) in order to form emulsions are also reported through scientific papers. Some of these processes are described below.

These procedures tend also to avoid the main problems existed in conventional sample preparation techniques such as LLE, SPE and PP which have been widely used to extract antipsychotic drugs from biological fluids.

In 2013 Fisher et al. developed an HPLC-MS/MS method for the determination of amisulpride, aripiprazole, dehydroaripiprazole, clozapine, norclozapine, olanzapine, quetiapine, risperidone, 9-hydroxyrisperidone, and sulpiride in small volumes of plasma or serum and they also investigated its ability to be applied to haemolysed whole blood as well as to oral fluid. The extraction took place in glass test tubes, where 200 µL of the sample and 100 µL of tris solution (2 mol/L pH 10.6) was added. For the extraction, 100 µL of butyl acetate:butanol (9 + 1, *v*/*v*) was added and the mixture was firstly vortexed for 30 s and then centrifuged (13,600× *g*, 4 min). After taking the upper layer an aliquot of

200 μL was injected into a Waters Spherisorb S5SCX sulfopropyl-modified silica column for HPLC analysis, using 50 mmol/L methanolic ammonium acetate, (pH 6.0) as a mobile phase. The limits of quantification varied between were 1–5 μg/L depending on the analyte. Recoveries varied between 16% and 107%and the reproducibility of the method was good. Thus, the developed method can be applied in the therapeutic drug monitoring of patients who undergo treatment with the examined drugs [29].

Also, in 2013, Ebrahimzadeh et al. developed a high-performance liquid chromatography coupled to photodiode array detector (HPLC-DAD) method to preconcentrate and determine the antipsychotic drug (haloperidol) in biological samples using ultrasound-assisted emulsification microextraction. The use of ultrasonic radiation helps to speed up different procedures such as homogenization, mass transfer and emulsion formation. In this microextraction procedure, a small volume of a solvent which is not miscible with water is injected into an aqueous sample solution and an emulsion is formed using sonication without any need for a dispersive solvent. When the microextraction is over the mixture is centrifuged and the two different phases are separated. In this case, 30 μL of 1-undecanol was injected into a glass-centrifuge tube containing 4 mL of the sample solution after fixing its pH to 10 (NaCl 4% $w/v$). The mixture was put in an ultrasonic water bath for 20 min at 25 °C in order to form an emulsion, which was then centrifuged at 4000 rpm for 5 min. The droplets of the extraction solvent floated at the top of the tubes and they were solidified after being cooled in an ice bath. Finally, they were removed and they were melt again at room temperature and analyzed with HPLC. An ODS-H $C_{18}$ (250 × 4.6 mm i.d., 5 μm) column was used, using a mixture of methanol and monobasic potassium phosphate solution (0.02 mol/L, pH 4) (60:40, $v/v$) as a mobile phase. The extraction yields were higher than 90% and the limits of quantification varied between 4 and 8 μg/L. As a result, the method can be successfully applied in real plasma and urine samples obtained from patients who undergo treatment with haloperidol [30].

In 2015, Zare et al. developed an ionic-liquid-based surfactant-emulsified microextraction procedure accelerated by ultrasound radiation followed by HPLC analysis for the determination of antidepressant and antipsychotic drugs; doxepine and perphenazine in urine. This procedure is based on the replacement of less green solvents with ionic liquids, which pose some unique properties such as tunability of their viscosity and surface tension and are widely used in green chemistry. For the microextraction of the drugs, 4 mg of the lipophilic SDS surfactant was poured into 6 mL of the sample, followed by the addition of 50 μL of the ionic liquid 1-hexyl-3-methylimidazolium hexafluorophosphate, which was used as an extracting solvent. The mixture was vortexed for 10 min and sonicated for another 10 min and a cloudy solution was formed. After centrifugation for 5 min at 5000 rpm the aqueous phase was removed and the phase containing the analytes was dissolved in methanol and injected into the HPLC system. For the analysis a Zorbax SB-$C_8$ (250 mm × 4.6 mm, 5 μm) column was used with a mixture of acetate buffer (pH 4)/acetonitrile (70:30, $v/v$) as a mobile phase. For the examined drugs extraction recoveries were between 89% and 98%. The performance of the developed method was compared with two other methods, one including dispersive liquid–liquid microextraction and one including ultrasound-assisted surfactant-based emulsification microextraction. The novel microextraction procedure showed some significant advantages and can be used for preconcentration, separation and determination of above mentioned drugs in real urine samples [31].

In 2016 Li et al., prepared novel magnetic octadecylsilane (ODS)—polyacrylonitrile (PAN) thin-films for microextraction of the antipsychotic drugs; quetiapine and clozapine in plasma and urine samples followed by HPLC-UV analysis. Thin-film microextraction (TFME) is a novel form of solid-phase microextraction for preconcentration and clean-up of analytes in biofluids. Thus, those films are supposed to be superior to conventional SPME because of their high extraction rate, their sensitivity and their low extraction time. These parameters can be optimized by using novel film coatings. In this work, the thin firms were made magnetic by adding superparamagnetic $SiO_2@Fe_3O_4$ nanoparticles to the films. For the drug microextraction procedure, the films were preconditioned with methanol and water and then they were added into the sample, the pH of which was adjusted to 9.5 by adding a 0.1 mol/L NaOH solution. The mixture was mechanically shaken for 50 min and then the film was removed by a strong magnet and cleaned with 3 mL of water. For the desorption of the analytes the film was again mechanically shaken in 1 mL of methanol for 5 min and the obtained solution was evaporated and redissolved in 100 μL of methanol and injected into a $C_{18}$ reversed-phase column for HPLC-UV analysis. The repeatability of the method was good, the films were reusable up to 15 times, extraction recoveries were higher than 99% and detection limits were lower than 0.015 μg/mL. The method was successfully applied to real plasma and urine samples and it can be used for therapeutic drug monitoring of patients [32].

All of the developed methods that are mentioned in the following paragraphs are summarized in Table 1.

**Table 1.** Microextraction techniques used for determination of antipsychotic drugs in biological fluids (HPLC: high performance liquid chromatography; UPLC: Ultra performance liquid chromatography UV: ultraviolet visible; MS: mass spectrometry; SPME: solid-phase microextraction; MEPS: microextraction by packed sorbent; LPME: liquid phase microextraction; DLLME: dispersive liquid/liquid microextraction).

| Analyte | Microextraction | Determination | Matrix | Recovery | Reference |
|---|---|---|---|---|---|
| Haloperidol, quinine, naproxen, ciprofloxacin and paclitaxel | SPME | HPLC-UV | Urine | 85–95% | [5] |
| Risperidone and its biotransformation products | SPME | HPLC-MS/MS | Liquid culture medium | 11–28% | [6] |
| Clospirazine, fluphenazine, perazine, thiethylperazine, thioridazine, flupentixol, thioproperazine, trifluoperazine, perphenazine, prochlorperazine, propericiazine | SPME | HPLC-MS/MS | Fungi pool Whole blood, urine | 0.0002–39.8% | [8] |
| Risperidone and 9-hydroxyrisperidone | MEPS | HPLC-Coulometric Detection | Plasma, saliva | >90.1% | [13] |
| Risperidone and 9-hydroxyrisperidone | MEPS | HPLC-UV | Plasma, urine, saliva | >90.0% | [14] |
| Ziprasidone | MEPS | HPLC-UV | Plasma | >90.0% | [15] |
| Olanzapine, quetiapine, clozapine, haloperidol and chlorpromazine in combination with seven antidepressants, two anticonvulsants and two anxiolytics | MEPS | UPLC-MS/MS | Plasma | Not mentioned | [16] |
| Clozapine and its metabolites | MEPS | HPLC-Coulometric detection | DBS | >90% | [17] |
| Chlorpromazine, olanzapine and their FMO mediated N-oxides in rat brain | MEPS | Capillary UHPLC-UV | Rat Brain Microdialysates | 92–98% | [18] |
| Seven phenothiazine derivatives | dynamic LPME | HPLC-UV | Urine | 72–98% | [21] |
| Amitriptiline, clomipramine and thioridazine | DLLME | HPLC-UV | Urine | 96–101% | [24] |
| Clozapine and chlorpromazine | DLLME | HPLC-UV | Urine | >97% | [25] |
| Tetrahydropalmatine and tetrahydroberberine | DLLME | HPLC-UV | Rat urine | >69% | [26] |
| Drugs of abuse (cocaine and metabolites, amphetamines and analogues, LSD, ketamine, opiates, methadone and fentanyl and analogues), Z-compounds and 44 benzodiazepines and antipsychotic drugs | DLLME | UHPLC-MS/MS | Plasma | Not mentioned | [27] |
| Chlorpromazine | HFLPME | HPLC-UV | Urine, serum | >70% | [28] |
| Amisulpride, aripiprazole, dehydroaripiprazole, clozapine, norclozapine, olanzapine, quetiapine, risperidone, 9-hydroxyrisperidone, and sulpiride | Liquid extraction with small volumes | HPLC-MS/MS | Plasma, serum, oral fluid and hemolysed whole blood | 16–107% | [29] |
| Haloperidol | Ultrasound-assisted emulsification microextraction. | HPLC-DAD | Urine, plasma | >90% | [30] |

Table 1. *Cont.*

| Analyte | Microextraction | Determination | Matrix | Recovery | Reference |
|---|---|---|---|---|---|
| Doxepine and perphenazine | Ionic-liquid-based surfactant-emulsified microextraction procedure accelerated by ultrasound radiation | HPLC-UV | Urine | 89–98% | [31] |
| Quetiapine and clozapine | Microextraction with magnetic ODS-PAN thin-films | HPLC-UV | Urine, plasma | 99–110% | [32] |

## 6. Conclusions

Even in cases where simple biological matrices are involved, sample pretreatment cannot be avoided. Based on their useful benefits, microextraction techniques in the extraction and pre-concentration of various antipsychotic drugs in different biological matrices are growing. With the use of novel procedures, the main disadvantages of traditional sample preparation techniques such as LLE, SPE and protein precipitation can be overcome. At the same time, microextraction techniques are compatible with green chemistry, which is nowadays a trend in analytical chemistry. It also agrees with simplification and miniaturization which are trends gaining more and more interest day by day. Thus, more innovative methods can be developed, for the determination of a greater variety of typical and atypical antipsychotics in different biological samples. Nevertheless, a lot of progress is expected to be made with the use of new sorbents and/or new solvents which would make the whole sample preparation process simpler, faster, more economical, more efficient and more environmental friendly.

**Acknowledgments:** Natalia Manousi was receiving scholarship by Onassis Foundation during preparation of this review.

**Conflicts of Interest:** The authors declare no conflict of interest.

## References

1. Patteet, L.; Cappelle, D.; Maudens, E.K.; Crunelle, L.C.; Sabbe, B.; Neels, H. Advances in detection of antipsychotics in biological matrices. *Clin. Chim. Acta* **2015**, *441*, 11–22. [CrossRef] [PubMed]
2. Saar, E.; Gerostamoulos, D.; Drummer, O.H.; Beyer, J. Identification and quantification of 30 antipsychotics in blood using LC-MS/MS. *J. Mass Spectrom.* **2010**, *45*, 915–925. [CrossRef] [PubMed]
3. Kataoka, H. New trends in sample preparation for clinical and pharmaceutical analysis. *Trends Anal. Chem.* **2003**, *22*, 232–244. [CrossRef]
4. Arthur, L.C.; Pawliszyn, J. Solid phase microextraction with thermal desorption using fused silica optical fibers. *Anal. Chem.* **1990**, *62*, 2145–2148. [CrossRef]
5. Theodoridis, G.; Lontou, M.; Michopoulos, F.; Sucha, M.; Gondova, T. Study of multiple solid-phase microextraction combined off-line with high performance liquid chromatography: Application in the analysis of pharmaceuticals in urine. *Anal. Chim. Acta* **2004**, *516*, 197–204. [CrossRef]
6. Bocato, Z.M.; Simões, A.R.; Calixto, A.L.; de Gaitani, C.M.; Pupo, T.M.; de Oliveira, R.A.M. Solid phase microextraction and LC-MS/MS for the determination of paliperidone after stereoselective fungal biotransformation of risperidone. *Anal. Chim. Acta* **2012**, *742*, 80–89. [CrossRef] [PubMed]
7. Kataoka, H.; Ishizaki, A.; Saito, K. Recent progress in solid-phase microextraction and its pharmaceutical and biomedical applications. *Anal. Methods* **2016**, *8*, 5773–5788. [CrossRef]
8. Kumazawa, T.; Seno, H.; Watanabe-Suzuki, K.; Hattori, H.; Ishii, A.; Sato, K.; Suzuki, O. Determination of phenothiazines in human body fluids by solid-phase microextraction and liquid chromatography/tandem mass spectrometry. *J. Mass Spectrom.* **2000**, *35*, 1091–1099. [CrossRef]
9. Sarafraz-Yazdi, A.; Amiri, A. Liquid-phase microextraction. *Trends Anal. Chem.* **2010**, *20*, 1–14. [CrossRef]

10.  D'Angelo, V.; Tessari, F.; Bellagamba, G.; De Luca, E.; Cifelli, R.; Celia, C.; Primavera, R.; Di Francesco, M.; Paolino, D.; Di Marzio, L.; et al. Microextraction by packed sorbent and HPLC-PDA quantification of multiple anti-inflammatory drugs and fluoroquinolones in human plasma and urine. *J. Enzyme Inhib. Med. Chem.* **2016**, *31*, 110–116. [CrossRef] [PubMed]

11.  Locatelli, M.; Ciavarella, M.T.; Paolino, D.; Celia, C.; Fiscarelli, E.; Ricciotti, G.; Pompilio, A.; Di Bonaventura, G.; Grande, R. Determination of ciprofloxacin and levofloxacin in human sputum collected from cystic fibrosis patients using microextraction by packed sorbent-high performance liquid chromatography photodiode array detector. *J. Chromatogr. A* **2015**, *1419*, 58–66. [CrossRef] [PubMed]

12.  Locatelli, M.; Ferrone, V.; Cifelli, R.; Barbacane, R.C.; Carlucci, G. Microextraction by packed sorbent and high performance liquid chromatography determination of seven non-steroidal anti-inflammatory drugs in human plasma and urine. *J. Chromatogr. A* **2014**, *1367*, 1–8. [CrossRef] [PubMed]

13.  Saracino, A.M.; de Palma, A.; Boncompagni, G.; Raggi, A.M. Analysis of risperidone and its metabolite in plasma and saliva by LC with coulometric detection and a novel MEPS procedure. *Talanta* **2010**, *81*, 1547–1553. [CrossRef] [PubMed]

14.  Mandrioli, R.; Mercolini, L.; Lateana, D.; Boncompagni, G.; Raggi, A.M. Analysis of risperidone and 9-hydroxyrisperidone in human plasma, urine and saliva by MEPS-LC-UV. *J. Chromatogr. B* **2011**, *879*, 167–173. [CrossRef] [PubMed]

15.  Mercolini, L.; Protti, M.; Fulgenzi, G.; Mandrioli, R.; Ghedini, N.; Conca, A.; Raggi, A.M. A fast and feasible microextraction by packed sorbent (MEPS) procedure for HPLC analysis of the atypical antipsychotic ziprasidone in human plasma. *J. Pharm. Biomed. Anal.* **2014**, *88*, 467–471. [CrossRef] [PubMed]

16.  Souza, D.I.; Domingues, S.D.; Queiroz, E.C.M. Hybrid silica monolith for microextraction by packed sorbent to determine drugs from plasma samples by liquid chromatography–tandem mass spectrometry. *Talanta* **2015**, *140*, 166–175. [CrossRef] [PubMed]

17.  Saracino, A.M.; Lazzara, G.; Prugnoli, B.; Raggi, A.M. Rapid assays of clozapine and its metabolites in dried blood spots by liquid chromatography and microextraction by packed sorbent procedure. *J. Chromatogr. A* **2011**, *1218*, 2153–2159. [CrossRef] [PubMed]

18.  Hendrickx, S.; Uğur, Y.D.; Yilmaz, T.I.; Şener, E.; Schepdael, V.A.; Adams, E.; Broeckhoven, K.; Cabooter, D. A sensitive capillary LC-UV method for the simultaneous analysis of olanzapine, chlorpromazine and their FMO-mediated N-oxidation products in brain microdialysates. *Talanta* **2017**, *162*, 268–277. [CrossRef] [PubMed]

19.  Liu, H.; Dasgupta, P.K. Analytical chemistry in a drop. Solvent extraction in a microdrop. *Anal. Chem.* **1996**, *68*, 1817–1821. [CrossRef] [PubMed]

20.  Jeannot, M.A.; Cantwell, F. Solvent microextraction into a single drop. *Anal. Chem.* **1996**, *68*, 2236–2240. [CrossRef] [PubMed]

21.  Sharifi, V.; Abbasi, A.; Nosrati, A. Application of hollow fiber liquid phase microextraction and dispersive liquid liquid microextraction techniques in analytical toxicology. *J. Food Drug Anal.* **2016**, *24*, 264–276. [CrossRef]

22.  Cruz-Vera, M.; Lucena, R.; Cárdenas, S.; Valcárcel, M. Determination of phenothiazine derivatives in human urine by using ionic liquid-based dynamic liquid-phase microextraction coupled with liquid chromatography. *J. Chromatogr. B* **2009**, *877*, 37–42. [CrossRef] [PubMed]

23.  Rezaee, M.; Assadi, Y.; Milani Hosseini, M.-R.; Aghaee, E.; Ahmadi, F.; Berijani, S. Determination of organic compounds in water using dispersive liquid–liquid microextraction. *J. Chromatogr. A* **2006**, *1116*, 1–9. [CrossRef] [PubMed]

24.  Xiong, C.; Ruan, J.; Cai, Y.; Tang, Y. Extraction and determination of some psychotropic drugs in urine samples using dispersive liquid–liquid microextraction followed by high-performance liquid chromatography. *J. Pharm. Biomed. Anal.* **2009**, *49*, 572–578. [CrossRef] [PubMed]

25.  Chen, J.; Xiong, C.; Ruan, J.; Su, Z. Dispersive liquid-liquid microextraction combined with high-performance liquid chromatography for the determination of clozapine and chlorpromazine in urine. *Food Chem.* **2011**, *31*, 277–284. [CrossRef] [PubMed]

26.  Zhang, M.; Le, J.; Wen, J.; Chai, Y.; Fan, G.; Hong, Z. Simultaneous determination of tetrahydropalmatine and tetrahydroberberine in rat urine using dispersive liquid–liquid microextraction coupled with high-performance liquid chromatography. *J. Sep. Sci.* **2011**, *34*, 3279–3286. [CrossRef] [PubMed]

27. Fisichella, M.; Odoardi, S.; Strano-Rossi, S. High-throughput dispersive liquid/liquid microextraction (DLLME)method for the rapid determination of drugs of abuse, benzodiazepines and other psychotropic medications in blood samples by liquid chromatography–tandem mass spectrometry (LC-MS/MS) and application to forensic cases. *Microchem. J.* **2015**, *123*, 33–41.

28. Sobhi, R.H.; Yamini, Y.; Abadi, R.H.B. Extraction and determination of trace amounts of chlorpromazine in biological fluids using hollow fiber liquid phase microextraction followed by high-performance liquid chromatography. *J. Pharm. Biomed. Anal.* **2007**, *45*, 769–774. [CrossRef] [PubMed]

29. Fisher, S.D.; Partridge, J.S.; Handley, S.A.; Couchman, L.; Morgan, E.P.; Flanagan, J.R. LC-MS/MS of some atypical antipsychotics in human plasma, serum, oral fluid and haemolysed whole blood. *Forensic Sci. Int.* **2013**, *229*, 145–150. [CrossRef] [PubMed]

30. Ebrahimzadeh, H.; Dehghani, Z.; Asgharinezhad, A.A.; Shekari, N.; Molaei, K. Determination of haloperidol in biological samples with the aid of ultrasound-assisted emulsification microextraction followed by HPLC-DAD. *J. Sep. Sci.* **2013**, *36*, 1597–1603. [CrossRef] [PubMed]

31. Zare, F.; Ghaedi, M.; Daneshfar, A. Ionic-liquid-based surfactant-emulsified microextraction procedure accelerated by ultrasound radiation followed by high-performance liquid chromatography for the simultaneous determination of antidepressant and antipsychotic drugs. *J. Sep. Sci.* **2015**, *38*, 844–851. [CrossRef] [PubMed]

32. Li, D.; Zou, J.; Cai, P.; Xiong, C.; Ruan, J. Preparation of magnetic ODS-PAN thin-films for microextraction of quetiapine and clozapine in plasma and urine samples followed by HPLC-UV detection. *J. Pharm. Biomed. Anal.* **2016**, *125*, 319–328. [CrossRef] [PubMed]

*separations*

MDPI

*Review*

# On the Extraction of Antibiotics from Shrimps Prior to Chromatographic Analysis

**Victoria Samanidou \*, Dimitrios Bitas, Stamatia Charitonos and Ioannis Papadoyannis**

Laboratory of Analytical Chemistry, University of Thessaloniki, GR 54124 Thessaloniki, Greece; dimitrisbgr@gmail.com (D.B.); stamatia_chrt@windowslive.com (S.C.); papadoya@chem.auth.gr (I.P.)
* Correspondence: samanidu@chem.auth.gr; Tel.: +30-2310-997698; Fax: +30-2310-997719

Academic Editor: Frank L. Dorman
Received: 2 February 2016; Accepted: 26 February 2016; Published: 4 March 2016

**Abstract:** The widespread use of antibiotics in veterinary practice and aquaculture has led to the increase of antimicrobial resistance in food-borne pathogens that may be transferred to humans. Global concern is reflected in the regulations from different agencies that have set maximum permitted residue limits on antibiotics in different food matrices of animal origin. Sensitive and selective methods are required to monitor residue levels in aquaculture species for routine regulatory analysis. Since sample preparation is the most important step, several extraction methods have been developed. In this review, we aim to summarize the trends in extraction of several antibiotics classes from shrimps and give a comparison of performance characteristics in the different approaches.

**Keywords:** sample preparation; extraction; aquaculture; shrimps; chromatography; antibiotics

## 1. Introduction

According to FAO (CWP Handbook of Fishery Statistical Standards, Section J: AQUACULTURE), "aquaculture is the farming of aquatic organisms: fish, mollusks, crustaceans, aquatic plants, crocodiles, alligators, turtles, and amphibians. Farming implies some form of intervention in the rearing process to enhance production, such as regular stocking, feeding, protection from predators, *etc.*" [1].

Since 1960, aquaculture practice and production has increased as a result of the improved conditions in the aquaculture facilities. Such improvements include better water quality, infection control, high nutrition feeds and improved aquatic species, through newly developed hybridization techniques, particular species breeding and the use of molecular genetics [2]. According to FAO 2005, in the time span from 1990 to 2005, aquaculture production each year has tripled from 16.8 million tons to 52.9 million tons. By 2015, it was also predicted that aquaculture would constitute 39% of the seafood production in weight worldwide, dramatically increasing from 4% in 1970 and 28% in 2000. Eleven of the fifteen elite aquaculture producing countries are located in Asia, with 94% of the total worldwide production, while China on its own has 71% of the total production [3].

Shrimp aquaculture is one of the most important aquacultures and makes a considerable contribution to the national economies, both in developed and developing countries. According to the "Global Study of Shrimp Fisheries" from FAO, the biggest domestic product percentage of shrimp farming belongs to Madagascar (1%), excluding the traditional shrimp fishing. The gross domestic values for other developing countries range between $2.72 million–$558 million US. Shrimp is the most profitable exported product in Cambodia, Indonesia, Kuwait, Madagascar, Mexico, Nigeria, and Trinidad and Tobago, and to a lesser extent Australia and Norway. Shrimp consumption, on the other hand, is high in most developed countries, such as Australia and Norway, with the United States presenting the highest consumption and, as a result, being the greatest shrimp market worldwide [4].

The increased aquaculture practice has resulted in increased levels of infections among the species. Usually the farming is done in cages, where high populations are confined to a limited space, and infection outbreaks are common despite good hygiene levels. Bacteria, parasites, viruses and fungi can infect the confined animals, with bacteria being the main source of infections [5].

Antibiotics are used in aquaculture in order to control the infection outbreaks. They are natural, semisynthetic or synthetic compounds and their antibacterial effect resides on their ability to eliminate the bacteria or hinder their growth. Antibiotics used for human disease treatment, such as penicillins, macrolides, sulfonamides, tetracyclines and quinolones/fluoroquinolones, are often used in aquaculture. Specifically, oxytetracycline, florfenicol, sarafloxacin, enrofloxacin, chlortetracycline, ciprofloxacin, norfloxacin, oxolinic acid, perfloxacin, sulfamethazine, gentamicin, and tiamulin are commonly used in aquaculture infections. Besides the use of antibiotics as bacterial infection treatment, sulfonamides, β-lactams and macrolides can be used as growth-promoting or infection-preventing agents. They are used in sub-therapeutic doses in animal feed or veterinary drugs [6–9].

The extensive use of antibiotics, however, may lead to residues in edible animal tissues and cause allergic or toxic effects to sensitive groups or the development of persistent microorganisms. It poses a risk to human health through the migration of antibiotics from aquaculture products to the human organism. As a result, authorities in many countries have published regulations on the antibiotic usage and residues in aquaculture and aquaculture products to minimize the risk to human health associated with consumption of their residue [9].

These regulations are strict in Europe, North America and Japan, where only few antibiotics are approved and maximum residue levels (MRLs) are introduced. However, the majority of aquaculture production and export takes place in countries where few or no regulations exist [6,9].

To comply with the EU regulation, state laboratories have to put into practice methods for both screening and confirming the presence in seafood.

Until every aquaculture country complies with regulations, controls are essential when importing aquaculture products. Sensitive analytical methods have been developed in order to control the product compliance to the regulations and ensure that the residue levels are lower than the MRLs. Sample preparation is the most important step during the development and the application of such analytical methods.

A significant number of multi-residue or single analytical methods have been reported in the literature for the determination of antibiotics in shrimps.

In general, the most common sample preparation techniques are solid phase extraction (SPE), using appropriate columns for each class examined, and solid-liquid extraction (SLE). However, liquid-liquid extraction (LLE) has been also used in some cases. In addition to this, recently developed materials, such as molecular imprinted polymers, have also been applied in some studies. The distribution of sample preparation techniques for the extraction of each class of antibiotics from shrimps is illustrated in the pie charts of Figure 1.

## Quinolones

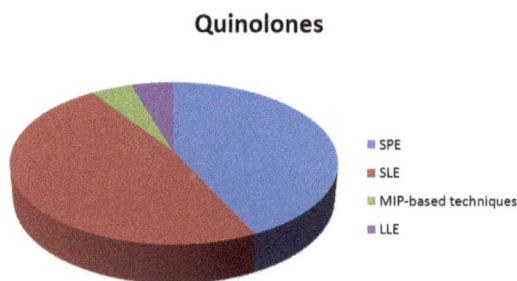

**Figure 1.** *Cont.*

## Tetracyclines

- SPE
- SLE (including ASE and PLE)

## Amphenicols

- SPE
- SLE
- IAC
- MISPE
- MSPD
- LLE
- SFE

## Sulfonamides

- SLE
- SPE
- MIP-based techniques
- QuEChERS
- SLE + LLE

## Nitrofurans

- SPE (after derivatization)
- SLE (prior to derivatization)

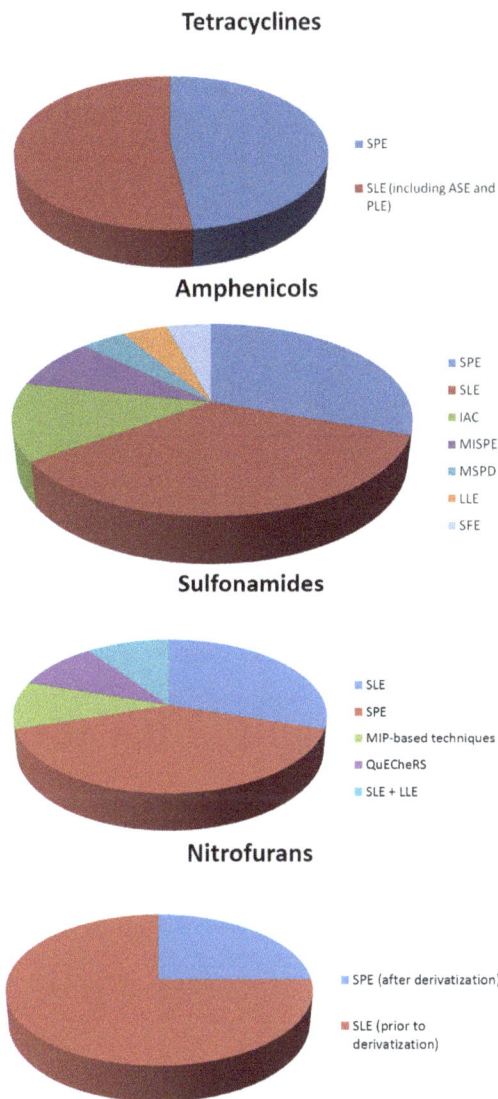

**Figure 1.** Sample preparation techniques used in the extraction of antibiotics from shrimps.

In this review, emphasis is put on extraction methods with regard to the isolation and purification steps. Results of published methods are summarized in the text and presented comparatively in tables.

## 2. Antibiotics

The most effective and useful antibacterial agents inhibit or prevent the development of the cell wall, the protein synthesis or the DNA replication and transcription. Less effective and clinically useful are those agents that act on the cell membrane or inhibit a metabolic path of the cell. Penicillins, cephalosporins and β-lactams inhibit the cell synthesis, chloramphenicol, tetracyclins and macrolides inhibit the protein synthesis, and quinolones, nitrofurans and sulfonamides inhibit the DNA synthesis [10].

Quinolones are synthetic antibiotics with a broad-spectrum antibacterial effect. This antibiotic group includes plain quinolones, such as oxolinic acid and nalidixic acid, and fluorinated quinolones, known as fluoroquinolones, such as ciprofloxacin, flumequine and sarafloxacin [5].

Quinolones have a dual heterocyclic aromatic ring structure as shown in Figure 2, with the first ring having a nitrogen atom at position 1, a carboxyl group at position 3 and a carbonyl group at position 4, and the second ring having a carbon atom at position 8. Fluoroquinolones result from the addition of a fluorine atom at position 6 of the second ring. Substitution at position 1 and 7 results in new enhanced fluoroquinolones [11–13].

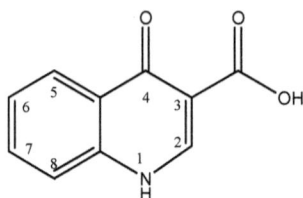

**Figure 2.** General chemical structure of quinolones.

The maximum residue limit in muscle tissue according to the Commission Regulation (EU) No. 37/2010 for danofloxacin, enrofloxacin-ciprofloxacin and oxolinic acid is 100 µg/kg [14].

Tetracyclines are broad-spectrum antibiotics, and their group includes tetracycline, oxytetracycline, chlortetracycline, demeclocycline, lymecycline, doxycycline, minocycline and tigecycline [15].

Tetracyclines were discovered in 1945 and were the first broad-spectrum antibiotics. The first generation of tetracyclines includes chlortetracycline and tetracycline, which were introduced for clinical use in 1948 and 1953, respectively [16,17]. Tetracycline antibiotics have a linearly arranged naphthalene ring structure (Figure 3), with a nitrogen-containing functional group region (2N region) and an oxygen-containing functional group region (C3-C4 region) [16].

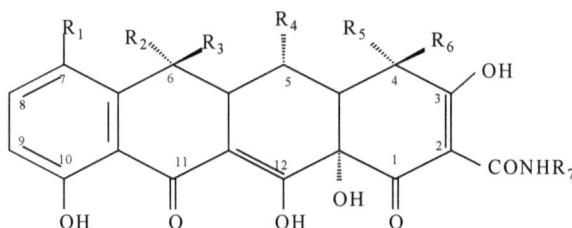

| Compound | | $R_1$ | $R_2$ | $R_3$ | $R_4$ | $R_5$ | $R_6$ | $R_7$ |
|---|---|---|---|---|---|---|---|---|
| TC | Tetracycline | H | $CH_3$ | OH | H | $N(CH_3)_2$ | H | H |
| OTC | Oxytetracycline | H | $CH_3$ | OH | OH | $N(CH_3)_2$ | H | H |
| CTC | Chlortetracycline | Cl | $CH_3$ | OH | H | $N(CH_3)_2$ | H | H |

**Figure 3.** General chemical structure of tetracyclines.

The maximum residue limit in muscle tissue according to the Commission Regulation (EU) No. 37/2010 [14] for chlortetracycline, oxytetracycline and tetracycline is 100 µg/kg, while only oxytetracycline hydrochloride and oxytetracycline dihydrate are approved for use in aquaculture from the U.S. Food and Drug Administration (FDA) [18].

Amphenicols are a broad-spectrum antibiotic group that includes chloramphenicol and its metabolites, thiamphenicol and florfenicol. Florfenicol also has its own metabolite, florfenicol amine [5].

Chloramphenicol is the oldest and the most known member of this antibiotic group. It was originally isolated from cultures of Streptomyces venezuelae and was first used for clinical purposes in 1947. It is effective against many bacteria strains, but its toxicity and unwanted effects limited its use over the years [19,20].

The structure of chloramphenicol is shown in Figure 4.

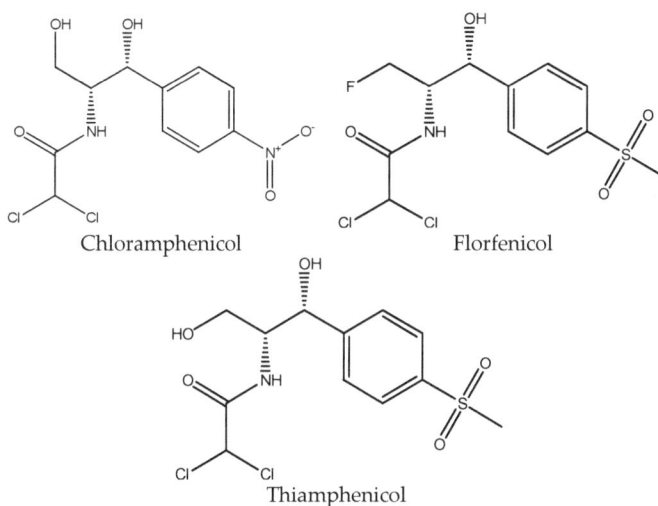

**Figure 4.** Chemical structure of chloramphenicol, florfenicol and thiamphenicol.

The maximum residue limit in muscle tissue according to the Commission Regulation (EU) No. 37/2010 [14] for florfenicol and florfenicol amine is 100 μg/kg; for thiamphenicol, it is 50 μg/kg, and chloramphenicol is completely prohibited. Florfenicol is only approved for use in aquaculture from the U.S. Food and Drug Administration (FDA) [18].

Macrolides are a category of semi-synthetic medium-spectrum with a macrolyclic lactone nucleus of 14–16 atoms to which different sugars are attached, forming the different types of the macrolide antibiotics. The category's most common antibiotic is erythromycin with a cladinose at C3 and desosamine at C5 (Figure 5).

Macrolides were discovered in natural products in 1950. Especially erythromycin was discovered in 1952, and it is still the most widely used macrolide drug in medicine, while at the end of the 1980s, two more semisynthetic derivatives of erythromycin were discovered.

The antibacterial activity of macrolides is due to their binding to the subunit 50S in the bacterial ribosome; as a result, it prevents the bacterial protein synthesis [21].

The MRL set by the Committee for veterinary medicinal products is 200 μg/kg in muscles, liver and kidneys of animal origin, 40 μg/kg in milk, and 150 μg/kg in eggs for the macrolide drugs [14].

Sulfonamides are derivatives of para-aminobenzenesulfonamide and their structure is similar to the structure of para-aminobenzoic acid (PABA), a molecule which takes part in the biosynthesis of dihydrofolic and folic acids by microorganisms (Figure 6). The basic structure of their molecule consists of an unsubstituted amine (–NH$_2$) on a benzene ring at C4 position and a sulfonamide group para to the amine (Figure 5). Sulfonamides are separated into four groups: (1) short—or medium acting sulfonamides; (2) long-acting sulfonamides; (3) topical sulfonamides and (4) sulfonamide derivatives for inflammatory bowel disease [22,23].

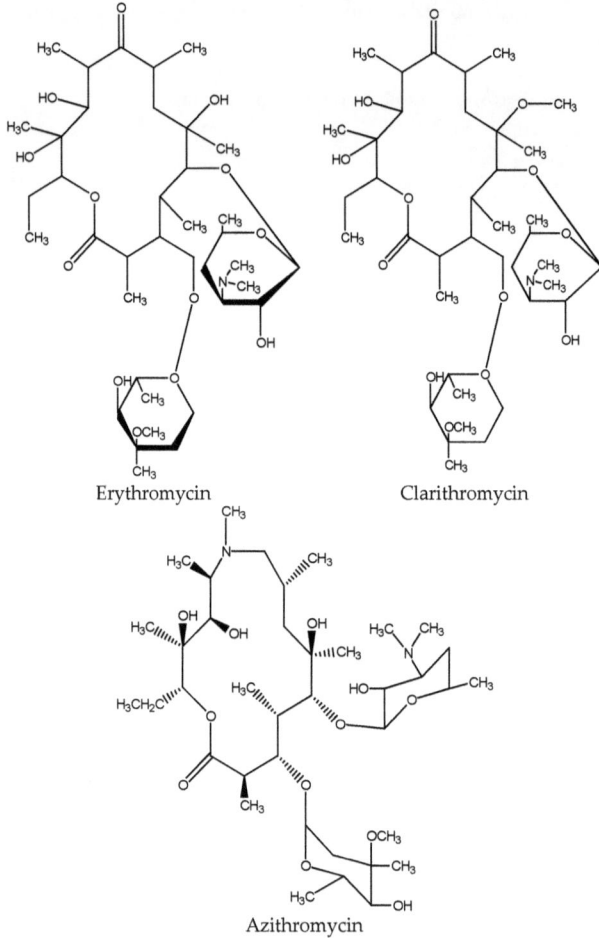

**Figure 5.** Chemical structure of common macrolides.

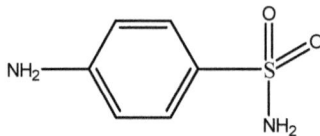

**Figure 6.** General chemical structure of sulfonamides.

The MRL set by the Committee for veterinary medicinal products is 100 µg/kg for the parent drug or the residues of sulfonamides in milk, fish and other seafood [14].

Most common nitrofurans are furazolidone, furaltadone, nitrofurazone and nitrofurantoin and their metabolites, 3-amino-2-oxazolidinone (AOZ), 3-amino-5-morpholinomethyl-2-oxazolidinone (AMOZ), semicarbazide (SEM) and 1-aminohydantoin (AHD), respectively. Due to the binding nitrofurans form, it is not easy to determine the parent nitrofuran, but it is possible to determine its metabolite in tissue samples. The chemical structure of nitrofurans is shown in Figure 7 [24].

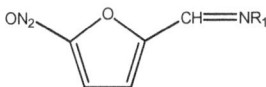

**Figure 7.** Chemical structure of Nitrofurans.

Nitrofurans are used as broad-spectrum antibiotics in veterinary practice, as a treatment to gastrointestinal infections [25] or against *Salmonella* sp., *Mycoplasma* sp. and some protozoa [26]. Since 1993, they have been banned in most of the countries in the world, but they are still used in some others. No MRL is set by the Committee for veterinary medicinal because nitrofurans and their metabolites are banned in EU [27].

## 3. Trends in the Extraction of Antibiotics from Shrimps

Shrimp tissue contains high amounts of protein. It also contains unsaturated fatty acids, such as the necessary eicosapentaenoic and docosahexaenoic acids, and minerals, such as calcium. The tissue composition depends on the feed given to the shrimps [28].

As mentioned above, a significant number of multi-residue or single analytical methods have been reported in the literature for the determination of antibiotics in shrimps.

To begin with, quinolones are mostly determined in shrimps after using SPE or SLE as the sample preparation technique. Furthermore, LLE and MIP-based techniques are equally applied in some of the studies. The same phenomenon appears in the determination of the class of tetracyclines, where SPE and SLE, including accelerated solvent extraction (ASE) and pressurized liquid extraction (PLE), are almost equally and most frequently used in the analysis of shrimps.

The class of amphenicols is determined by using a wide variety of sample preparation techniques. SLE is once again the primary preferable technique, followed by SPE. A different approach of the extraction is achieved with the use of immunoaffinity columns (IAC) for the determination of amphenicols in shrimps. Molecular imprinted—SPE (MISPE) are also used in a smaller number of studies. In addition to these, there were some cases where matrix solid phase dispersion (MSPD), LLE and supercritical fluid extraction (SFE) is performed in shrimp samples.

For the class of sulfonamides, SPE is the sample preparation of choice, followed by SLE. In addition, there are some studies in which a combination of SLE and LLE is used for the extraction of sulfonamide drugs from shrimp samples. Furthermore, MISPE and QuEChERS are applied for the determination of sulfonamides.

The determination of nitrofurans is achieved by determining the derivatives of the drugs. Derivatization takes place before or after the sample preparation. Derivatization is preformed after sample preparation using SLE, or prior to SPE.

In the following paragraphs, analytical methodologies for the extraction of antibiotics from shrimp tissue are presented and classified according to the category of antibiotics.

### 3.1. Extraction of Quinolones

Enrofloxacin and ciprofloxacin were extracted using 10 mL of acetonitrile. The extract was evaporated to dryness at 37 °C, and the residue was re-dissolved with an ammonium acetate buffer to a final volume of 2 mL. A SPE cleanup step was applied with a SDB-RPS cartridge (polyStyrene Divinylbenzene-Reverse Phase sorbent) preconditioned twice with 1 mL of ethanol, 1 mL of water and 1 mL of the ammonium acetate buffer, sequentially. Target compounds were eluted with 4 mL methanol and ammonium hydroxide solution 1 M (75:25, $v/v$). The eluates were evaporated to dryness at 37 °C, and the residue was re-dissolved in 300 μL of formic acid solution (pH = 2.5). The extraction procedure yielded recoveries between 94.0%–106.0%, 97.0%–103.0% for ENR and CIP, respectively. Analysis was carried out by an LC-MS/MS system, separation was achieved by a Polaris C18A 3 μm (150 × 2.0 mm) with a Chromsep guard column SS (10 × 2.0 mm), and the mobile phase consisted of an acetonitrile and formic acid solution (pH = 2.5) delivered in gradient conditions. The LOD was 4 μg/kg and 3 μg/kg for ENR and CIP, respectively [29].

Ciprofloxacin, danofloxacin, enrofloxacin and sarafloxacin were extracted from shrimp samples using 16 mL of acidic acetonitrile and the addition of dichloromethane (to a final volume of 25 mL). A SPE cleanup step with a Strata C18 E was preconditioned with 2 mL of acetonitrile. The antibiotics were eluted from the SPE cartridges with 2 × 2 mL of acetonitrile. The eluates were evaporated to dryness under a nitrogen stream at 45 °C, and the residue was re-dissolved in 200 μL of acetonitrile and 800 μL of deionized water. The extraction procedure yielded recoveries between 63.0%–117.0%, 71.0%–87.0%, 72.0%–92.0%, 95.0%–125.0% for CIP, DAN, ENR and SAR, respectively. Analysis was carried out by a UPLC-MS system, separation was achieved by a HSS T3 $C_{18}$ column (1.8 mm, 2.1 × 50 mm) (Waters, Milford, MA, USA), and the mobile consisted of 4 mM $NH_4OH$/50 mM formic acid buffer in either 10% MeCN or 90% MeCN (gradient elution). The LOD values were 0.13, 0.14, 0.19, 0.14 ng/g for for CIP, DAN, ENR and SAR, respectively. This method allows a single analyst to prepare 25 samples each day [30].

Ofloxacin, norfloxacin, ciprofloxacin and lomefloxacin were extracted from the spiked samples with 30 mL of a 1% acetic acid ethanol solution. A cleanup step using SPE was applied. SPE cartridge was washed with 10 mL of methanol, water, methanol in order, and the quinolones were eluted with 10 mL of 25% ammonia methanol. The eluate was evaporated to dryness under a nitrogen stream at 35 °C, and the residues were re-dissolved in mobile phase. The extraction procedure for a peeled prawn sample without shell yielded recoveries between 88.3%–99.8%, 95.9%–109.4%, 91.2%–107.0%, 88.9%–103.4% for OFL, NOR, CIP and LOME, respectively. Analysis was carried out by a HPLC system coupled with a chemiluminescence detector, and separation was achieved by a XDB-$C_8$, 150 mm × 4.6 mm i.d., 5 μm column. The LOD was 0.43, 0.36, 0.40 and 2.4 ng/mL for OFL, NOR, CIP and LOME, respectively. This study established a novel HPLC chemiluminescence detection method for quinolone determination, which was based on the Ce(IV)–Ru(bpy)$_3$$^{2+}$–$HNO_3$ system [31].

Enrofloxacin and ciprofloxacin were extracted from the spiked samples with 5 mL of methanol:acetic acid (98:2, *v*/*v*), the extracts were evaporated (to a final volume of 2 mL) under a nitrogen stream at 50 °C, and the residue was re-dissolved in 10 mL water:acetic acid (98:2, *v*/*v*). The SPE cleanup step involved a Sep-Pak $C_{18}$ (500 mg, 6 mL) cartridge preconditioned and equilibrated with 6 mL of methanol and 6 mL of Milli-Q water sequentially. The quinolones were eluted with 6 mL of methanol:(1 M) phosphoric acid (9:1, *v*/*v*) and 4 mL of methanol, the eluate was evaporated to dryness under nitrogen stream at 50 °C, and the dry residue was re-dissolved in 1 mL of Tris buffer solution (pH 9.1). The extraction yielded 88.43%, 80.41% average recoveries for ENR and CIP, respectively. Analysis was carried out by a HPLC system coupled with a fluorescence detector, separation was achieved by a PLRP-S column (5 μm, 4.6 × 150 mm) with a RP18-E guard column (5 μm, 4 × 40 mm) (Polymer Laboratories Inc., Church Stretton, UK), and the mobile phase consisted of orthophosphoric acid, acetonitrile and tetrahydrofuran (gradient elution). The LOD was 0.015, 0.025 μg/g for ENR and CIP, respectively [32].

Nine fluoroquinolones and 3 acidic quinolones were extracted from the spiked samples with 20 mL of AcCN/MeOH (1:1 *v*/*v*), a SPE cleanup step with a $Fe^{3+}$ immunoaffinity cartridge followed, and the quinolones were eluted with 0.5 mL of a McIlvaine-EDTA-NaCl buffer. The extraction procedure yielded inter-day recoveries between 73.7%–89.7% for the fluoroquinolones and 75.7%–87.6% for the acidic quinolones. Analysis was carried out by a HPLC system coupled with a spectrofluorometric detector, separation was achieved by an Atlantis $dC_{18}$ IS column (4.6 × 20 mm, 3 mm), and the mobile phase consisted of (15:85:0.1 *v*/*v*) MeOH-water-formic acid for the fluoroquinolones and (35:65:0.1 *v*/*v*) MeOH-water-formic acid for the acidic quinolones (isocratic elution). The LOQ ranged between 1.5–50.0 mg/kg for the fluoroquinolones and 1.5–3.0 mg/kg for the acidic quinolones [33].

Flumequine, nalidixic acid and oxolonic acid were extracted from the spiked samples with 5 mL of acetonitrile. Two mL of 0.1 mol/L an ammonia solution and 2 mL *n*-hexane were added to the extracts in order to remove the colored and fatty components. The extracts were evaporated under a nitrogen stream at 45 °C and 6 mL of hydrochloric acid 0.1 mol/L, and 6 mL of ethyl acetate were added. The ethyl acetate extract was evaporated to dryness at 40 °C, and the residue was re-dissolved in 300 μL of methanol. The extraction procedure yielded recoveries between 73.3%–84.5%, 80.4%–90.4%, 79.2%–88.3% for OXO, NAL and FLU, respectively. Analysis was carried out by a HPLC system coupled with a fluorescence detector, separation was achieved by a $C_{18}$-Nucleosil HD column (4 mm × 250 mm, 5 μm), and the mobile phase consisted of 0.01 mol/L oxalic acid (pH 2.3) and acetonitrile (65/35, $v/v$) (isocratic elution). The $CC_\beta$ was 610.9 μg/kg, 13 μg/kg, and 117.3 μg/kg for FLU, NAL and OXO, respectively. This method gave good results concerning the complexity of the matrix and allows evaluation of the shrimp samples being compliant to the current European legislation [34].

Oxolinic acid, flumequine and nalidix acid were extracted from the spiked samples with 12 mL of ethyl acetate (re-extracted with another 12 mL) and the addition of 2 g anhydrous sodium sulfate, the extract was evaporated to dryness, and the residue was re-dissolved in 2 mL of a 0.2% formic acid aqueous solution. The extraction procedure yielded 92.6%, 79.3%, 79.8% average recoveries for OXO, FLU and NAL, respectively. Analysis was carried out by an LC system coupled with a fluorescence detector, separation was achieved by an Agilent Zorbax Eclipse XDB $C_8$ column (4.6 mm × 150 mm, 5 μm), and the mobile phase consisted of 60% oxalic acid (0.01 M), 30% acetonitrile and 10% methanol ($v/v/v$) (isocratic elution). MDL was 3, 2.7 and 2.3 ng/g for OXO, NAL and FLU, respectively. The simple extraction scheme provided LC-MS compound confirmation with increased sample throughout, over previews GC-MS methods and selectivity for the above antibiotics [35].

Flumequine, oxolinic acid and nalidix acid were extracted from the spiked samples with 5 mL of 1% acetic acid, 10 mL of acetonitrile and the addition of 2 g sodium chloride. The extract were evaporated to dryness under nitrogen stream at 55 °C and re-dissolved in 2.5 mL of reconstitution solution containing 40 ng/mL of piromidic and 100 mg/mL of EDTA in acetonitrile/water (1:1, $v/v$). The extraction procedure yielded recoveries between 79.0%–88.0%, 91.0%–95.0%, 100.0%–101.0% for FLU, OXO and NAL, respectively. The analysis was carried out by a LDTD source coupled to a triple quadrupole mass spectrometer. The MDL was 1.7, 2.6, 4.4 ng/g for FLU, OXO and NAL, respectively. This method was found to meet many of the drug residue analysis requirements in shrimp tissue samples, using a single solvent extraction step, resulting in decreased sample analysis and increased sample throughput [36].

OXO and FLU were extracted by mixing the spiked samples with 400 μL of supramolecular solvent. The extraction procedure yielded recoveries between 100%–102%, 100%–101.4% for FLU and OXO, respectively. Analysis was carried out by an LC system coupled with a fluorescence detector, separation was achieved by a Kromasil $C_{18}$ column (5 μm, 150 mm × 4.6 mm), and the mobile phase consisted of 55% oxalic acid (0.01 M) and 45% acetonitrile/methanol (75:25, $v/v$) delivered isocratically. The $CC_\beta$ was 109 g/kg, and 622 g/kg for OXO and FLU, respectively. This method proved to be reliable, fast and low-cost. It demonstrates high extraction efficiency regardless of the matrix composition, and a simple one-step analyte extraction with neither cleanup nor evaporation was needed [37].

Shrimp samples were mixed with trichloroacetic acid aqueous solution (15%, $w/v$), and the resulting extracts were spiked with ciprofloxacin. Yeast@MIPs or yeast@NIPs were dispersed in the extracts, collected, and washed with a 10-mL methanol-acetic acid solution (59:1 $v/v$). The resulting extracts were dried under nitrogen stream at 298 K, and the residues were re-dissolved in 0.4 mL methanol. The extraction procedure yielded a 86.4% recovery. Analysis was carried out by HPLC system coupled with ultraviolet detector, separation was achieved by a $C_{18}$ (150 × 4.6 mm$^2$) column, and the mobile phase consisted of methanol-water (24:76, $v/v$) (isocratic elution). The surface imprinted yeast@MIPs developed for this paper exhibited high adsorption capacity, high selectivity, rapid binding ability for CIP, and could be used at least five times without losing their adsorption capacity. Moreover, they were successfully used in real sample analysis for CIP in shrimps yielding good recoveries [38].

An overview of the extraction methodologies for the determination of quinolones in shrimps is presented in Table 1.

*3.2. Extraction of Tetracyclines*

Tetracycline, oxytetracycline and chlorotetracycline were extracted from the spiked samples with 10 mL of succinic acid and an addition of 1–1.5 g sodium chloride and tissue disruptor. A SPE cleanup step with OASIS hydrophobic-lipophilic-balanced (HLB) SPE columns (6 mL, 200 mg, Waters Corp, Milford, MA, USA) were conditioned with 4 mL of methanol, water and succinic acid sequentially. The tetracyclines were eluted with 2 mL of methanol, the eluates were evaporated to dryness under nitrogen stream at 60 °C, and the residues were re-dissolved in 2 mL of 0.1% formic acid. The extraction procedure yielded 82.9%, 93.2%, 76.8% average recoveries for TC, OTC, and CTC, respectively. Analysis was carried out by an LC-MS/MS system, separation was achieved by a MacMod HydroBond PS $C_8$, 100 mm × 2.1 mm, column, and the mobile phase consisted of 75%, 0.1% formic acid, 18% acetonitrile and 7% methanol (isocratic elution). The average LOQ was 50 ng/g for all analytes. The developed method is ideal for routine analysis, avoids the use of complex buffers and provides a simple and fast extraction procedure [39].

Oxytetracycline, tetracycline, chlorotetracycline and doxycycline were extracted from shrimp samples with 12.5 mL of $Na_2$EDTA-McIlvaine buffer at pH 4. A SPE cleanup step with the C-18E cartridge (500 mg, 6 mL) (Phenomenex, Torrance, CA, USA) was activated with 10 mL of methanol and 10 mL of Milli-Q water sequentially. The tetracyclines were eluted with 10 mL of methanol, the solvent was removed under room temperature, and the residues were passed through 0.45 μm PTFE filter. The extraction procedure yielded recoveries between 83.3%–96.5%, 88.4%–96.9%, 86.0%–93.3%, 90.6%–102.0% for OTC, TC, CTC and DC, respectively. Analysis was carried out by an LC system coupled with an electrochemical detector with a nickel-implanted boron-doped diamond thin film electrode (Ni-DIA), separation was achieved by a ODS-3 Inertsil $C_{18}$ (5 μM 4.6 mm × 250 mm) column, and the mobile phase consisted of 0.01 M phosphate buffer (pH 2.5)-acetonitrile (80:20, $v/v$) delivered isocratically. The LOD ranged between 0.1–0.5 g/mL for all analytes. This paper demonstrates the first use of Ni-DIA electrodes for the electroanalysis of tetracyclines, with excellent performance for the oxidative detection of tetracyclines, exhibiting well-defined voltammograms, high sensitivity and significant advantages over the BDD and glassy carbon electrode [40].

Oxytetracycline, tetracycline, chlorotetracycline and doxycycline were extracted from the shrimp samples with HPLC grade methanol. The extract was evaporated to dryness, and the residue was re-dissolved in mobile phase. The extraction procedure yielded recoveries between 91.0%–98.0%, 81.0%–99.0%, 84.0%–101.0%, 80.0%–85.0% for TC, OTC, CTC and DC, respectively. Analysis was carried out by an LC-MS/MS system, separation was achieved by a reverse phase Zorbax Eclipse Plus $C_{18}$ (5 μm particle size, 4.6 × 100 mm) column, and the mobile phase consisted of 0.1% formic acid in water and 0.1% formic acid in methanol under gradient elution. The LOD was 11, 12, 20, 23 ng/g for TC, OTC, CTC and DC, respectively [41].

Seven tetracyclines were extracted from the spiked samples with a Dionex accelerated solvent extractor 200 (Dionex, Sunnyvale, CA, USA), which provides the use of solvents at temperatures up to 80 °C and pressures up to 85 bar, and methanol and 1 mmol/L trichloroacetic acid at pH 4.0 as solvents. The spiked samples were mixed with 5 g of $Na_2$EDTA-washed sand and packed in an extraction cell at pH 4.0. The extraction procedure yielded 75.6%–103.5% average recovery for all analytes. Analysis was carried out by a HPLC system coupled with a dual λ absorbance detector, separation was achieved by a ZORBAX SB-$C_{18}$ (150 mm × 4.6 mm I.D., 5 μm) (Agilent Technology, Santa Clara, CA, USA) column, and the mobile phase consisted of methanol, acetonitrile and 0.01 M oxalic acid (gradient elution). The $CC_\beta$ ranged between 7.8–108.1 μg/kg. This method provided fast sample extraction with pressurized liquid extraction, compared to conventional liquid-liquid extractions, with reduced solvent use [42].

An overview of the extraction methodologies for the determination of tetracyclines in shrimps is presented in Table 2.

**Table 1.** Overview of extraction methodologies for the determination of quinolones in shrimps.

| Reference | Analytes | Sample | Sample Preparation | Analytical Technique | LOD-LOQ, CCα-CCβ | Recovery (%) |
|---|---|---|---|---|---|---|
| [29] | ENR and CIP | fish and prawn | Homogenized prawn tissue (1 g), spiked (100 μL IS, 3 μg/mL), extraction (10 mL ACN, vortex (1 min), shaker (15 min), centrifuged (10 min, 3700 g). Supernatant evaporated (dryness, 37 °C), add ammonium acetate buffer, vortex (15 s), sonicated (15 min). Purification (SPE cartridges, SDB-RPS), conditioned (successively 2 × 1 mL MeOH, 2 × 1 mL water and 2 × 1 mL ammonium acetate buffer), sample loaded, cartridge dried (centrifugation, 5 min, 3700 g), eluted (4 mL elution solution). Evaporation to dryness, 37 °C, reconstituted (300 μL formic acid), vortex (15 s), filtration (0.2 μm). | LC-MS/MS | LOD (μg/kg): ENR: 4, CIP: 3<br><br>LOQ (μg/kg): ENR: 14, CIP: 10<br><br>CCα (μg/kg): ENR and CIP: 111 | ENR: 94.0–106.0, CIP: 97.0–103.0 |
| [30] | CIP, DAN, ENR and SAR | salmon, shrimp and tilapia | Tissue (4 g), mixed with acidic ACN (16 mL), CH₂Cl₂ added (to 25 mL), rotated (10 min), centrifuged (10 min, 2000 rpm), supernatant (10 mL) removed, evaporated (45 °C, N₂, to 2 mL). SPE column preconditioned (2 mL ACN), samples rinsed (2 × 2 mL ACN), passed through column, add 1 mL ACN. Eluent collected, evaporated (dryness, 45 °C, N₂), residue reconstituted (vortexing, 200 μL ACN, deionized water 800 μL), hexane (1 mL) added, vortexed, centrifuged (1000 rpm, 5 min), aqueous layer filtered (0.2 μm syringe filter). | UPLC-MS/MS | LOD (ng/g): CIP: 0.13, DAN: 0.14, ENR: 0.19, SAR: 0.14<br><br>LOQ (ng/g): CIP: 0.4, DAN: 0.43, ENR: 0.56, SAR: 0.41 | CIP: 63.0–117.0, DAN: 71.0–87.0, ENR: 72.0–92.0, SAR: 95.0–125.0 |
| [31] | OFL, NOR, CIP and LOME | prawn | Sample (5 g) dissolved (30 mL, 1% acetic acid ethanol solution), homogenized, centrifuged (4500 rpm, 5 min). Sample poured into SPE cartridge, adsorbed quinolones washed (10 mL MeOH, water, MeOH in order, 2.0 mL/min), quinolones absorbed eluted (10 mL, 25% ammonia MeOH, 1.5 mL/min). Eluent evaporated (N₂, 35 °C), dissolved (mobile phase), filtered (0.45 μm filter). | HPLC-CL | LOD (ng/mL): OFL: 0.43, NOR: 0.36, CIP: 0.40, LOME: 2.4<br><br>LOQ (ng/mL): - | OFL: 90.3–101.4, NOR: 89.5–107.8, NOR: 88.0–107.2, LOME: 94.4–106.0 (prawn sample with shell)<br>OFL: 88.3–99.8, NOR: 95.9–109.4, NOR: 91.2–107.0, LOME: 88.9–103.4 (peeled prawn sample without shell) |
| [32] | ENR and CIP | prawns | Prawn muscle (1 g) extracted (5 mL MeOH: CH₃COOH, 98:2 v/v), vortexed (2 min), sonication (10 min), centrifuged (1968 g, 10 min), pellet tissue extracted, evaporated (to 2 mL, 50 °C, N₂), filtration, water: CH₃COOH (10 mL, 98:2 v/v) added. Sample applied to SPE cartridge (C₁₈, 500 mg, 6 mL), washed (6 mL water: phosphoric acid 1 M, 3:2 v/v), cartridges preconditioned and equilibrated (6 mL MeOH and 6 mL Milli-Q water), eluted (6 mL MeOH: phosphoric acid 1 M, 9:1 v/v and 4 mL MeOH). Eluate evaporated (dryness, N₂, 50 °C), residue re-dissolved (1 mL Tris buffer, pH 9.1), filtered (syringe filter 0.45 μm, nylon). | HPLC-FLD | LOD (μg/g): ENR: 0.015, CIP: 0.025<br><br>LOQ (μg/g): - | ENR: 88.43 (ave.), CIP: 80.41 (ave.) |

**Table 1.** *Cont.*

| Reference | Analytes | Sample | Sample Preparation | Analytical Technique | LOD-LOQ, $CC_\alpha$-$CC_\beta$ | Recovery (%) |
|---|---|---|---|---|---|---|
| [33] | 9 FQ: MARB, OFL, NOR, CIP, ENR, DAN, ORB, DIF and SAR and 3 AQ OXO, NAL and FLU | muscle (cattle, swine and chicken), liver (chicken), raw fish (shrimp and salmon), egg (chicken), and processed food (ham, sausage and fish sausage) | Sample (2 g), ACN/MeOH (20 mL, 1:1) added, homogenized (20 s), centrifuged (2500 g, 5 min), supernatant diluted (×2 with MeOH). Diluted extract (1 mL) added to the $Fe^{3+}$ IMAC cartridge, allowed to pass, cartridge washed (1 mL MeOH and water), eluted (0.5 mL McIlvaine-EDTA-NaCl buffer). | LC-FLD | LOQ (mg/kg): DAN: 0.8, SAR: 6.5, ORB, DIF, OXO, FLU: 1.5, NOR, OFL, CIP, ENR: 2.5, NAL: 3, MARB: 50 | MARB: 88.2–89.5, OFL: 76.6–87.5, NOR: 85.4–86.4, CIP: 86.1–91.8, ENR: 88.2–90.0, DAN: 98.7–103.5, ORB: 87.3–88.7, DIF: 86.8–92.3, SAR: 81.1–81.8, OXO: 81.6–84.2, NAL: 87.7–94.9, FLU: 83.1–85.3 |
| [34] | FLU, NAL and OXO | shrimps | Samples spiked (100 μL IS), extraction (5 mL ACN), homogenized, centrifuged (10 min, 4000× g, 20 °C), supernatants transferred, re-extraction. ACN extracts combined, ammonia solution (2 mL, 0.1 mol/L) and n-hexane (2 mL) added, vortexed, centrifuged, n-hexane supernatant removed, procedure subsequently repeated (without ammonia). ACN extract evaporated (water-bath, 45 °C, vacuum), HCl (6 mL, 0.1 mol/L) and ethyl acetate (6 mL) added, vortexed, centrifuged, extraction twice replicated. Acetate supernatants combined, water-bath, 45 °C, vacuum, evaporation to dryness, 40 °C, N2), dissolved (300 μL MeOH, ultrasonic bath). | HPLC-FLD | LOD (μg/kg): FLU: -, NAL: 6.9, OXO: -; $CC_\alpha$ (μg/kg) FLU: 599.7, NAL: 10.3, OXO: 110.1; $CC_\beta$ (μg/kg) FLU: 610.9, NAL: 13, OXO: 117.3 | OXO: 73.3–84.5, NAL: 80.4–90.4, FLU: 79.2–88.3 |
| [35] | OXO, FLU and NAL | shrimp | Sample (2.0 g), fortification, vortexed (10 s), equilibrated (15 min), ethyl acetate (12 mL) and anhydrous sodium sulfate (2 g) added, shaken (1 min), centrifuged (1500 rpm, 10 min, 4 °C), supernatant transferred, evaporated (50–55 °C, N2), sample re-extracted (additional 12 mL ethyl acetate, as above). Supernatant added to original, evaporated (dryness or until oily residue), residue re-dissolved (2 mL, 0.2% formic acid), vortex (30 s), hexane (2 mL) added, mixed, centrifuged (5 min, 2600 rpm, 4 °C), hexane layer discarded (aspiration), aqueous liquid filtered (0.45 μm glass microfiber syringe filter). | LC-FLD, LC-MS/MS | LC-FL: MDL (ng/g): OXO: 3, FLU: 2.7, NAL: 2.3; LOQ (ng/g): OXO: 9, FLU: 8.1, NAL: 6.9 | OXO: 88.0–97.7 (ave. 92.6), FLU: 77.1–80.7 (ave. 79.3), NAL: 78.6–81.7 (ave. 79.8) |
| [36] | FLU, OXO and NAL | catfish, shrimp, and salmon | Sample (2.5 g) fortified, equilibrated (15 min), vortex-mixed (30 s), 5 mL 1% acetic acid), ACN (10 mL) and NaCl (2 g) added, shaken (5 min, 2500 rpm), centrifuged (10,000 rpm, 5 min, 4 °C), organic layer, transferred, evaporated (55 °C, dryness, N2, 5 psi first 5 min, then 30–35 min at 12–15 psi). Reconstitution solution (2.5 mL) added to dried extracts, sonicated (1 min), vortex-mixed (30 s), centrifuged (10 min, 17,250 g, 4 °C), supernatants passed through filters (0.2 μm nylon syringe). Aliquots of each sample (3 mL) spotted into individual wells (96-well LazWell microtiter plate), evaporate (dryness, room temperature). | LDTD-MS/MS | MDL (ng/g): FLU: 1.7, OXO: 2.6, NAL: 4.4; LOQ (ng/g): FLU: 7.9, OXO: 17.3, NAL: 7.0 | FLU: 79.0–88.0 (ave.), OXO: 91.0–95.0 (ave.), NAL: 100.0–101.0 (ave.) |

**Table 1.** *Cont.*

| Reference | Analytes | Sample | Sample Preparation | Analytical Technique | LOD-LOQ, $CC_\alpha$-$CC_\beta$ | Recovery (%) |
|---|---|---|---|---|---|---|
| [37] | FLU and OXO | fish and shellfish (salmon, sea trout, sea bass, gilt-head bream, megrim and prawns) | Sample (200 mg) and supramolecular solvent (400 µL) mixed, micro PTFE-coated bar introduced, vortex-shaken (2500 rpm, 15 min), thermostated (15 °C), centrifuged (15,000 rpm, 15 min). | LC-FLD | $CC_\alpha$ (g/kg): OXO: 104, FLU: 611<br>$CC_\beta$ (g/kg): OXO: 109, FLU: 622 (salmon) | FLU: 100.0–102.0, OXO: 100.0–101.4 |
| [38] | CIP | shrimp | Sample (5 g) homogenated, dispersed in trichloroacetic acid aqueous solution (15%, $w/v$), stirred (2 h, 398 K), centrifugation and filtration, extraction solution collected and spiked. yeast@MIP's or yeast@NIP's (5 mg) dispersed in spiked samples (10 mL), incubated (6 h, 298 K). yeast@MIP's or yeast@NIP's collected (centrifugal filtration), washed (10 mL MeOH-acetic acid solution, 59:1 $v/v$), extracts dried (N2, 298 K), residues redissolved (0.4 mL MeOH). | HPLC | - | 86.4 |
| [43] | MQCA and QCA | porcine, chicken (muscles and livers), fish and shrimp (muscles) | Samples deproteinated (5% metaphosphoric acid in 10% MeOH), LLE, cleanup (solid phase extraction, mixed mode anion-exchange columns). | LC-MS/MS | LOQ (µg/kg): MQCA and QCA: 0.1 | MQCA and QCA: 62.4–118.0 |
| [44] | DES-CIP, NOR, CIP, DAN, ENR, ORB, SAR, and DIF | shrimp | Shrimp tissue extracted (ammoniacal ACN), extract defatted, evaporated, dissolution (basic phosphate buffer). | LC-FLD-MS | LOD (ng/g): -<br>LOQ (ng/g): 0.1–1 | 75.0–92.0 |
| [45] | 8 FQ: (NOR, OFL, DAN, CIP, DES-CIP, ENR, SAR and DIF) and 4 AQ (OXO, FLU, NAL | salmon, trout, and shrimp | Drugs extracted with mixture of ethanol and 1% acetic acid, diluted (aqueous HCl), defatted (hexane), cation-exchange solid phase extraction. | LC-MS/MS | LOD (ng/g):<br>QNs: 0.1, FQNs: 0.4 | 57.0–96.0 |
| [46] | CIP, DAN, DIF, ENR, FLU, MARB, NAL, NOR, OFL, ORB, OXO and SAR | muscle, liver, chicken eggs, milk, prawn and rainbow trout | Sample extracted (ACN–water, 95:5), 1/5 of filtered extract diluted (water, keep ACN ratio at ca. 60%), passed through C18 mini-column, eluate evaporated (dryness), residues dissolved (MeOH–water, 30:70, $v/v$) | HPLC-FLD | LOD (µg/g): -<br>LOQ (µg/g): 0.005 | CIP: 75.7, DAN: 96.5, DIF: 77.9, ENR: 97.3, FLU: 75.7, MARB: 80.7, NAL: 71.0, NOR: 74.4, OFL: 96.1, ORB: 74.3, OXO: 70.2, SAR: 72.3 |
| [47] | 6 FQ | fish, shrimp and crab | Sample extraction (acid ACN), defatted (n-hexane, water removed (Na2SO4). | UPLC-MS/MS | LOD (µg/kg): 0.1 (all)<br>LOQ (µg/kg): 0.2 (all) | 76.9–95.9 |

**Table 1.** *Cont.*

| Reference | Analytes | Sample | Sample Preparation | Analytical Technique | LOD-LOQ, $CC_\alpha$-$CC_\beta$ | Recovery (%) |
|---|---|---|---|---|---|---|
| [48] | 12 FQ (MARB, NOR, ENR, CIP, DES-CIP, LOME, DAN, SAR, DIF, OFL, ORB and ENO) and 6 AQ (OXO, NAL, FLU, CIN, piromidic acid and pipemidic acid) | milk, chicken, pork, fish and shrimp | Extraction with ACN-1% HCOOH, diluted (10% ACN), defatted (hexane). | LC-MS/MS | $CC_\alpha$ (ng/g): 0.18–0.68  $CC_\beta$ (ng/g): 0.24–0.96 (all) | - |
| [49] | MARB, NOR, CIP, LOME, DAN, ENR, SAR, DIF, OXO, NAL and FLU | chicken, pork, fish and shrimp | Extraction with 0.3% metaphosphoric acid and ACN (1:1, $v/v$), cleanup (HLB cartridge). | HPLC-FLD | LOD (ng/g): -  LOQ (ng/g): 5.0–28.0 | 71.7–105.3 |
| [50] | MARB, CIP, NOR, LOME, DAN, ENR, SAR, DIF, OXO and FLU | swine, chicken, and shrimp tissues | Samples (≤2.0 g) and small volume of organic reagent (≤4.6 mL) of a nonchlorinated solvent. | HPLC-FLD | LOQ (ng/g): -  LOD (ng/g): 0.3–1 (all) | 72.8–106.8 |

**Table 2.** Overview of extraction methodologies for the determination of tetracyclines in shrimps.

| Reference | Analytes | Sample | Sample Preparation | Analytical Technique | LOD-LOQ, CC$_\alpha$-CC$_\beta$ | Recovery (%) |
|---|---|---|---|---|---|---|
| [39] | TC, OTC, and CTC | shrimp and whole milk | Shrimp extracted (1–1.5 g sodium chloride, 10 mL succinic acid, tissue disruptor, blending 20–30 s), tissue disruptor rinsed (2 × 3 mL succinic acid), centrifuged (5 min, 4000 rpm, 4 °C), supernatant decanted (25 mL depth filter, into centrifuge tube with 1–1.5 g alumina and 5 mL, 0.01 M oxalic acid), re-extracted (10 mL succinic acid), blending (tissue disruptor), centrifuged (5 min, 4000 rpm, 4 °C), supernatant decanted (depth filter), filter washed (4 mL water), extracts shaken (10 s), centrifuged (5 min, 4000 rpm, 4 °C) (HLB) SPE columns (6 mL, 200 mg) conditioned (sequentially 4 mL MeOH, 4 mL water and 4 mL succinic acid, 4 mL succinic acid above column bed), extracts applied (flow rate 45 drops/min), column washed (4 mL water), dried (5 min, full vacuum), eluted (2.0 mL MeOH), extracts evaporated (dryness, 60 °C, N$_2$, 15 min), ACN (1 mL) added, residue dissolved (2.0 mL, 0.1% HCOOH), vortexed. | LC-UV, LC-MS/MS | LOD (ng/g): - LOQ (ng/g): 50 (all, ave.) | OTC:82.9 (ave.), TC: 93.2 (ave), CTC: 76.8 (ave) |
| [40] | OTC, TC, CTC and DC | shrimp | Sample (2.50 g), Na$_2$EDTA-McIlvaine buffer (12.5 mL, pH 4) added, blended (30 s), shaken (10 min), centrifuged (30 min, 3500 rps). Supernatant loaded into SPE cartridge, activated (10 mL MeOH and 10 mL Milli-Q water), washed (10 mL Milli-Q water), eluted (10 mL MeOH), solvent removed at room temp., residues filtered (0.45 µm PTFE filter). | Ni-DIA electrode, HPLC/Ni-implanted electrode, HPLC/Ni-DIA electrode | Ni-DIA electrode: LOD (g/mL):OTC: 0.1, TC: 0.1, CTC: 0.5, DC: 0.5, LOQ (g/ml): - | Ni-implanted electrode: OTC: 84.8–102.5, TC:85.9–97.0, CTC: 91.48–97.9, DC: 88.4–103.7 Ni-DIA electrode OTC: 83.3–96.5, TC: 88.4–96.9, CTC: 86.0–93.3, DC: 90.6–102.0 |
| [41] | TC, OTC, CTC and DC | prawns | Sample, homogenized, HPLC grade MeOH added, centrifuged (15 min, 3000 rpm), supernatant evaporated (dryness), dissolved (mobile phase, 0.1% formic acid in MeOH), filtered (0.22 µm membrane filter). | LC-MS/MS | LOD (ng/g): TC: 11, OTC: 12, CTC: 20, DC: 13 LOQ (ng/g): TC: 19, OTC: 20, CTC: 20, DC: 20 | TC: 91.0–98.0, OTC: 81.0–99.0, CTC: 84.0–101.0, DC: 80.0–85.0 |
| [42] | OTC, TC, CTC, MNC, MTC, DMC and DC | egg, fish and shrimp | Extraction with Dionex accelerated solvent extractor 200 using MeOH and 1 mmol/L TCA at 80 °C/85 bar/pH 4.0. Sample (5 g) mixed (5 g, Na$_2$EDTA-washed sand), packed in extraction cell (circular glass microfiber filters, 1.98 cm, above and below the packing). Resulting extracts diluted, evaporated (dryness, 40 °C, N$_2$), residue dissolved (1 mL mobile phase), vortexed, filtered (0.22 µm nylon Millipore chromatographic filter). | HPLC | CC$_\alpha$ (µg/kg): MNC: 5.5, OTC: 101.8, TC: 102.2, DMC: 6.5, CTC: 106.8, MTC: 8.8, DC: 13.0 CC$_\beta$ (µg/kg): MNC:7.8, OTC:104.2, TC: 104.4, DMC:8.6, CTC:108.1, MTC: 105, DC: 15.3 | OTC: 80.5–101.8, TC: 81.5–85.4, CTC: 78.7–85.7, DC: 83.4–89.1, DMC: 82.1–85.3, MNC: 80.4–99.7, MTC: 81.4–86.2 |

*3.3. Extraction of Amphenicols*

Chloramphenicol was extracted from shrimps using 4 mL of a phosphate extraction solution. An aliquot of 4.5 mL of ethyl acetate was added in the supernatants, the organic layer was evaporated to dryness under mild nitrogen stream at 45 °C, and the residue was re-dissolved in 300 μL of methanol/water (50:50 $v/v$). The extraction procedure yielded recoveries between 98.3%–100.0%. Analysis was carried out by an LC-MS/MS system, separation was achieved by a $C_{18}$ reverse-phase analytical column (100 mm × 2.1 mm, 4 μm) connected to a $C_{18}$ pre-column (1 cm × 4 mm, 4 μm), and the mobile phase consisted of water and methanol (gradient elution). The LOD was 0.03 ng/g. The extraction procedure used in this study proved to be simple and fast, with no cleanup step needed and there was no matrix interference [51].

Chloramphenicol was extracted from shrimps with 7 mL of ethyl acetate and the addition of 1 g sodium sulfate anhydrous. The extracts were evaporated to dryness under nitrogen stream at 40 °C, and the residues were re-dissolved in 1 mL of acetonitrile. Analysis was carried out by an LC-MS/MS system, separation was achieved by a X-Terra (2.1 × 150 mm, 3.5 μm) column, and the mobile phase consisted of methanol and water containing 0.1% $NH_4OH$ delivered using a gradient elution program. The $CC_\beta$ was 0.04 μg/kg [52].

Chloramphenicol was extracted from the spiked samples with 5 mL of ethyl acetate, and after some evaporate/re-dissolve steps the final ethyl acetate extract was evaporated to dryness under nitrogen stream at 45 °C. The residue was re-dissolved in 0.5 mL of hexane:$CCl_4$ (1:1 $v/v$) and mixed with 0.7 mL of HPLC-grade acetonitrile-water (1:1 $v/v$), and the upper clear phase was used for the analysis. The extraction procedure yielded recoveries between 95.88%–96.96%. Analysis was carried out by an LC-ESI-MS/MS system, separation was achieved by a $C_{18}$ reverse phase Unisil HPLC column (150 × 4 mm i.d., 5 μm) (Gasukuro Kogyo, Inc., Tokyo, Japan), and the mobile phase consisted of water 90% plus acetonitrile 10% and water 10% plus acetonitrile 90% (isocratic elution). The $CC_\beta$ was 0.098 mg/kg [53].

Chloramphenicol was extracted from the spiked samples with 5 mL of acetonitrile. An aliquot of 5 mL of chloroform was added to the extracts; after vortexing and centrifugation, the chloroform layer was discarded. The acetonitrile extracts were evaporated to dryness under nitrogen stream, and the residues were re-dissolved in 1 mL of mobile phase. The extraction procedure yielded recoveries between 85.5%–115.6%. Analysis was carried out by an LC-MS/MS system, separation was achieved by a Luna 5 μm $C_{18}$ (150 × 4.6 mm) column, and the mobile phase consisted of water and acetonitrile delivered under gradient conditions. The LOD was 0.02 mg/kg. This method performed a simple and rapid liquid–liquid extraction without using any other cleanup step such as SPE [54].

Chloramphenicol was extracted from the spiked samples with 20 mL of acetonitrile/4% NaCl in water solution (1:1, $v/v$), the extract was de-fatted with 2 × 10 mL of *n*-hexane, and 7 mL of water-saturated ethyl acetate was added to the remainder. The ethyl acetate extracts were evaporated to dryness and re-dissolved in 3 mL of water/acetonitrile (95:5, $v/v$). A SPE cleanup step with a $C_{18}$ (500 mg/3 mL) cartridge preconditioned with 10 mL of methanol and 10 mL of water/acetonitrile (95:5, $v/v$) followed, and the CAP was eluted with 3 mL of water/acetonitrile (45:55, $v/v$). 4 mL of water-saturated ethyl acetate was added to the eluate, the extracts were evaporated to dryness under nitrogen stream, and the residue was re-dissolved in 1 mL of acetone/toluene (20:80, $v/v$). A second SPE step with a Silica cartridge preconditioned with 6 mL of acetone/toluene (20:80, $v/v$) followed. CAP was eluted with 6 mL of acetone/toluene (70:30, $v/v$). The eluent was evaporated to dryness, and 50 μL of derivatization mixture N,O-bis(trimethylsilyl)acetamide (BSA)/*n*-heptane (1:1, $v/v$) was added to the residue. The extraction procedure yielded 95.0% average recovery. Analysis was carried out by a GC-MS system, and separation was achieved by a 30 m × 0.2, 5 μm I.D. column, 0.25 mm ZB5 column. The $CC_\beta$ was 0.087 mg/kg [55].

Chloramphenicol and Chloramphenicol glucuronide were extracted from the spiked samples with 7 mL of acetonitrile and a SPE cleanup step with a Chem-Elut cartridge followed. Analytes were eluted twice with 15 mL and 15 mL of dichloromethane, the eluate was evaporated to dryness under a nitrogen stream at 45–50 °C, and the residue was re-suspended with 5 mL of hexane:ethyl acetate (50:50 *v:v*). A second SPE cleanup step followed with a Bond Elut-NH$_2$ cartridge. Analytes were eluted with 3 mL of ethyl acetate:methanol (50:50 *v:v*), the eluate was evaporated to dryness under nitrogen stream at 45–50 °C, and the residue was re-dissolved in 300 μL of HPLC grade water. Analysis was carried out by an LC-MS/MS system, separation was achieved by a Nucleodur 5 μm C-18 (EC), 125 × 2.0 mm column, and the mobile phase consisted of 55% 10 mM ammonium acetate and 45% methanol delivered isocratically. The CC$_\beta$ was 0.17 μg/kg [56].

Chloramphenicol was extracted from spiked samples with 40 mL of a 0.05-mol/L phosphate buffer (pH 7.0), and 3 mL of 15% trichloroacetic acid in water were added to the extracts in order to precipitate the proteins. The extract was loaded into MISPE cartridges, the eluates were evaporated to dryness under nitrogen stream, and the residue was re-dissolved in mobile phase. The extraction procedure yielded recoveries between 84.9%–89.0%. Analysis was carried out by a HPLC system coupled with a UV detector, separation was achieved by a Beckman C$_{18}$ column cartridge (4.6 mm × 250 mm, 5 μm), and the mobile phase consisted of methanol and water (40:60, *v/v*) delivered isocratically. The developed MISPE method provides simple cleanup and preconcentration of CAP with high efficiency, which can increase the sensitivity of conventional chromatographic methods. Additionally, MISPE can be used for enrichment, purification and determination of trace CAP from complex food matrices [57].

Chloramphenicol, thiamphenicol, florfenicol and florfenicol amine spiked samples were blended with 2 g of C$_{18}$ material (dispersion adsorbent). The mixture was transferred to a glass column with degreased cotton packed at the bottom and at the top of the sample mixture, and the column was tightly compressed. The analytes were eluted with 5 mL of ethyl acetate-ACN-25% ammonium hydroxide (10/88/2, *v/v/v*), the eluate was dried under nitrogen stream at 50 °C, and the residue was reconstituted with 1 mL of 5% MeOH in a 0.1% formic acid-5 mmol/L ammonium acetate solution. The extraction procedure yielded recoveries between 84.0%–98.8%. Analysis was carried out by an LC-MS/MS system, separation was achieved by a Hypersil C$_{18}$ column (150 mm × 2.1 μm, 5 μm), and the mobile phase consisted of 0.1% formic acid solvent (including 5 mmol/L ammonium acetate) and methanol delivered using gradient elution. The CC$_\beta$ was 0.05, 0.11, 0.13, 0.04 μg/kg for florfenicol, florfenicol amine, thiamphenicol and CAP, respectively [58].

Chloramphenicol, florfenicol and thiamphenicol shrimp samples were mixed with 0.5 g of sea sand and 1 g of anhydrous sodium sulfate using a glass mortar. The dry mixture was placed into a SFE chamber that was closed and attached to the SFE system. The extract was evaporated to dryness under nitrogen stream, and the residue was re-dissolved with ethyl acetate for GC-MS analysis. The extraction procedure yielded 92.0%, 87.0%, 85.0% recovery for chloramphenicol, florfenicol and thiamphenicol, respectively. The analytes were collected by *in situ* silylation with Sylon BFT, analysis was carried out by a GC-MS system, and separation was achieved by a TR-5MS (30 m × 0.25 mm i.d. × 0.25 μm film thickness (Thermo Electron, Waltham, MA, USA). The LOD was 8.7, 15.8, 17.4 pg/g for chloramphenicol, florfenicol and thiamphenicol, respectively [59].

Chloramphenicol was extracted from the shrimp samples with ethyl acetate, the extract was evaporated and the dry residue was re-dissolved in 15 mL of salting out solution. In order to remove the fatty components, n-hexane was also added. A second extraction with ethyl acetate followed, the extract was evaporated, and the dry residue was re-dissolved in 5 mL of ACN-water (10:90, *v/v*). A cleanup step with sol-gel filter column on-line coupled to an immunoaffinity column containing anti-CAP antibodies followed, and CAP was eluted with 10 mL of ACN-water (40:60, *v/v*). For better analyte concentration, the eluate was extracted twice with 3 mL of ethyl acetate and the addition of 2 g sodium sulfate. The extract was evaporated to dryness under nitrogen stream, and the residue was re-dissolved in 1 mL of ACN-water (10:90, *v/v*). The extraction procedure yielded 68.0% recovery. Analysis was carried out by a HPLC system coupled with a UV detector, separation was achieved by a Spherisorb S ODS1, 250 × 4.6 mm I.D., 5 μm, column, and the mobile phase consisted of 10 mM sodium acetate pH 5.4 and ACN using gradient elution. The LOD was LOD 1.8 ng/g. The immunoaffinity columns developed in this study could efficiently remove the shrimp matrix interferences and could be repeatedly used without a decrease in their cleanup efficiency [60].

Ultrasound-assisted matrix solid phase dispersion (MSPD) was applied for the extraction of Chloramphenicol, thiamphenicol and florfenicol. Preconditioned SPE sorbent was mixed with the spiked sample, the mixture was replaced in the SPE cartridge, and the antibiotics were eluted with 1 mL of acetonitrile and then with 1 mL of methanol. The extract was evaporated to dryness under nitrogen stream at 40 °C, and the residue was re-dissolved in 400 µL of lamotrigine aqueous solution 10 ng/mL. The extraction procedure yielded recoveries between 81.3%–114.5%, 72.0%–103.3%, 89.1%–120.6% for thiamphenicol, florfenicol and CAP, respectively. Analysis was carried out by a HPLC system coupled with a diode array detector, separation was achieved by a LiChroCART-LiChrospher® 100 RP-18 (5 µm, 250 × 4 mm) (Merck, Darmstadt, Germany) column, and the mobile phase consisted of ammonium acetate (0.05M) and ACN (70:30 $v/v$, isocratic elution). The LOQ was 20 µg/kg for all analytes [61].

An overview of the extraction methodologies for the determination of amphenicols in shrimps is presented in Table 3.

### 3.4. Extraction of Sulfonamides

Sulfadiazine, sulfamerazine, sulfameter, sulfamethazine, sulfamethoxazole and sulfadimethoxine were extracted using MIP-based SPE and determined with HLPC-UV. Extraction was performed on the shrimp samples with an aquatic solution of acetic acid followed by vortexing, sonication and centrifugation. The supernatant was loaded into the MISPE/NISPE cartridges. The cartridges were washed with acetonitrile in water and elution took place with MeOH/acetic acid, and the eluate was evaporated to dryness under nitrogen flow. The residue was dissolved in acetonitrile in water, filtered and analyzed. Solvent A in the mobile phase was acetic acid/water and solvent B was acetonitrile. Gradient analysis was performed at a flow rate of 1 mL· min$^{-1}$, and the UV detector was set at 270 nm. The recoveries obtained ranged from 85.5% to 106.1%, while LOD was among 8.4 to 10.9 µg/kg, and LOQ was among 22.4 to 27.7 µg/kg [62].

Fourteen sulfonamides were determined in shrimp samples. Extraction was performed with acetonitrile followed by a sonication step and a centrifugation step. The supernatant was kept, and the same process of extraction was followed once more. C$_{18}$ powder was added into the 2 supernatants after being put together, the solution was homogenized and centrifuged. The supernatant of this step was evaporated to dryness under nitrogen flow. The residue was dissolved with potassium dihydrogen phosphate and filtered before HLPC injection. The mobile phase consisted of potassium dihydrogen (A) and methanol (B). A Capcellpak C$_{18}$ column was used, flow rate was set at 1 mL· min$^{-1}$, and the detection was performed at 270 nm. The recoveries ranged from 51.8% to 89.7%. The LOD for SAs was among 3 and 6 µg/kg, and the LOQ among 9 and 18 µg/kg [63].

Fourteen sulfonamides, sulfanilamine, sulfadiazine, sulfathiazole, sulfapyridine, sulfamerazine, sulfamethazine, sulfamethizole, sulfamethoxypyridazine, sulfachloropyridazine, sulfadimethoxine and sulfadoxine, sulfamethoxazole, sulfamethoxine, and sulfaquinoxaline residues were determined in shrimps. The samples were homogenized and a mixture of acetic acid-methanol-acetonitrile and acetonitrile were added to them. After centrifugation, the supernatant was transferred to a separatory funnel with DI water and DEG. Methylene chloride was added to the funnel, and separation was performed; the bottom layer was collected and evaporated to a final volume of 2–3 mL with a rotary evaporator. The solution was applied to a SCX SPE cartridge, and the sulfonamides were eluted with acetone-ammonium acetate mix. The eluate was evaporated to a final volume of 2 mL, prior to LC analysis. The analytical column used for the separation was a Symmetry C$_{18}$, 3.5 µm, 150 mm × 4.6 mm I.D. The mobile phase consisted of aqueous acetic acid-methanol-acetonitrile (A) and acetonitrile (B). Post-column derivatization took place with fluorescamine solution. The fluorescent detection was performed at 400 nm excitation wavelength and 495 nm emission wavelength. The mean recoveries for the spiked shrimp samples at three levels ranged from 67.3% to 90.5% [64].

Seven sulfonamides (sulfaguanidine, sulfadiazine, sulfamethazine, sulfamonomethixine, sulfamethoxazole, sulfadimethoxine and sulfaquinoxaline) were determined in shrimp samples with a monolithic column coupled with boron-doped diamond amperometric detection. Extraction was performed to the samples with Na$_2$EDTA-McIlvaine's buffer solution, and the mixture was sonicated and centrifuged. The supernatant was applied to an Oasis SPE cartridge and elution was performed with methanol. The eluate was evaporated under nitrogen flow, and the residue was dissolved with the mobile phase and filtered, prior to HPLC-EC analysis. Separation was achieved with the use of a monolithic column. The mobile phase consisted of phosphate buffer, acetonitrile and ethanol. The LOD value was found between 1.2 ng/mL and 3.4 ng/mL, and the LOQ value was found between 4.1 ng/mL and 11.3 ng/mL [65].

Eight sulfonamides (sulfachloropyridazine, sulfadiazine, sulfamerazine, sulfamethazine, sulfamethizole, sulfmethoxazole, sulfanilamine and sulfathiazole) were determined in shrimp samples using LC-MS/MS analysis. The shrimp samples were spiked, the extraction of sulfonamides was achieved with acetonitrile in acidic conditions, and EDTA was added to the extract, prior to SPE step. An Oasis HLB SPE cartridge was loaded with the extract, the drugs were eluted with methanol, and then the solution was evaporated to a final volume of 4 mL. The analysis was carried out using a triple quadrupole LC-MS/MS with positive electrospray ionization and MRM mode. A reverse-phased C$_{18}$ column was used for the separation of the drugs, and the mobile phase consisted of HPLC-grade water with formic acid and ammonium formate (A) and a mixture 1:1 (*v:v*) of methanol-acetonitrile [66].

Eight sulfonamides (sulfadiazine, sulfamerazine, sulfaguanidine, sulfisoxazole, sufladimethoxine, sulfamonomethoxine, sulfadoxine and sulfamethoxazole) were determined in shrimp samples with UPLC analysis using a graphene/polyaniline modified screen-printed carbon electrode. The samples were extracted with a Na$_2$EDTA-Mckkvaine buffer solution. The mixture was vortexed, sonicated and centrifuged. The supernatant was applied to a Microcolumn Vertipak$^{TM}$ for the SPE procedure, the sulfonamides were eluted with methanol, and the extract was filtered before UPLC injection. The mobile phase consisted of phosphate solution: acetonitrile: ethanol, and the potential used was +1.4 V *vs.* Ag/AgCl. The LOD was found between 1.162 ng/mL and 2.900 ng/mL, and the LOQ was found between 3.336 ng/mL and 20.425 ng/mL [67]

An overview of the extraction methodologies for the determination of sulfonamides in shrimps is presented in Table 4.

### 3.5. Extraction of Macrolides

Nine macrolides (erythromycin, tylosin, josamycin, spiromicyn, neospiromycin, tlmicosin, gamithromycin, tildipirosin and oleandomycin) were determined in shrimp samples with LC-MS/MS analysis. Extraction was performed to homogenized samples using bearing balls, water and acetonitrile. The mix was centrifuged, the supernatant was kept to another tube, and extraction was performed again to the sample with acetonitrile and phosphate buffer. The complete supernatant was centrifuged and then applied to a Bond-Elut C$_{18}$ SPE cartridge. Elution was achieved with methanolic ammonium acetate, and hexane was added to remove the fat. Hexane was removed by aspiration, and the solution was evaporated under nitrogen flow to a final volume of less than 1 mL. Methanolic ammonium acetate and methanol was added to the solution to a final volume of 1 mL, and the mix was centrifuged. A quantity of it was filtered and remained overnight before LC-MS/MS analysis. The analytical column used for the separation of the macrolides was a Kinetex 2.6 μm XB-C$_{18}$ 2.1 mm I.D. × 100 mm and a SecurityGuard Ultra C$_{18}$ 2.1 mm guard. The mobile phase consisted of an aqueous solution of formic acid and acetonitrile, and the detection was performed with a triple quadrupole MS/MS combined with an electrospray ionization source in the positive mode. The LOD was 0.5 μg/kg, and recoveries were found between 47% and 99% [68].

Six macrolides (erythromycin, elandomycin, azithromycin, clarithromycin, tilmicocin and roxithromycin) were determined in shrimp samples using HPLC-UV analysis, after sample preparation with magnetic MIP-based SPE. The samples were spiked and remained overnight. NaOH and acetonitrile were added the next day, and the mix was vortexed and centrifuged. MSPE was performed to the supernatant with MMIPs, which were magnetically separated from the solution. Elution of the macrolides from the MMIPs was achieved with a mixture of methanol and $KH_2PO_4$. The eluate was evaporated to dryness under nitrogen flow, and the residue was dissolved with a mixture of acetonitrile and $KH_2PO_4$. The separation of the macrolides was achieved with a SunFireTM $C_{18}$ column (250 mm × 4.6 mm I.D., 5 μm, Waters). The mobile phase consisted of acetonitrile (A) and $KH_2PO_4$ (B), and the detection was performed at 210 nm. The LOD was found between 0.015 μg/kg and 0.2 μg/kg, and the LOQ was found between 0.075 μg/kg and 0.5 μg/kg [69].

### 3.6. Extraction of Nitrofurans

Furazolidone, furaltadone, nitrofurazone and nitrofurantoin were determined in shrimp samples with UPLC-MS analysis. The four drugs were extracted from the samples by using HCl and were derivatized with 2-nitrobenzaldehyde (NBA). After remaining overnight, the samples were neutralized and centrifuged, and the supernatant was applied to preconditioned Oasis HLB cartridges. Elution was performed with ethyl acetate, and the eluate was evaporated to dryness under nitrogen flow, reconstituted in the mobile phase and filtered. The mobile phase was methanol and an aqueous solution of ammonium formate. The LOD was between 0.5–0.8 mg/kg, and LOQ was 1 mg/kg [70].

**Table 3.** Overview of extraction methodologies for the determination of amphenicols in shrimps.

| References | Analytes | Sample | Sample Preparation | Analytical Technique | LOD-LOQ, $CC_\alpha$-$CC_\beta$ | Recovery (%) |
|---|---|---|---|---|---|---|
| [51] | CAP | fish, shrimp, poultry, eggs, bovine and swine | Samples homogenized, IS and phosphate extraction solution (4 mL) added, ultrasonic bath (15 min), centrifugation (3000 $g$, 10 min).Supernatant transferred, 4.5 mL ethyl acetate added, vortexed (1 min), centrifuged (3000 $g$, 10 min). Organic layer transferred, evaporated (45 °C, $N_2$). Residue re-suspended (300 µL MeOH water, 50:50 $v/v$), vortexed (20 s). | LC-ESI-MS/MS | LOQ (ng/g): 0.03, LOQ (ng/g): 0.1 | 98.3–100.0 |
| [52] | CAP | honey, shrimp, and poultry meat | Sample (1 g), spiked (50 µL IS, 10 ng/mL), sodium sulfate anhydrous (1 g) and ethyl acetate (7 mL) added, shaken (20 min), centrifuged (15 min, 2000 rpm). Supernatant transferred, ethyl acetate (7 mL) added to sediment, both supernatants combined, evaporated (dryness, $N_2$, 40 °C).ACN (1 mL) added on residues, evaporated ($N_2$, 40 °C), re-suspended (500 µL MeOH/$H_2O$, 10:90), vortexing (15 s), filtered | LC-ESI-MS/MS | $CC_\alpha$ (µg/kg): 0.03, $CC_\beta$ (µg/kg): 0.04 | - |
| [53] | CAP | shrimp | Homogenized shrimp meat (3.0 g), water (2 mL) added, spiked (50-CAP IS, final concentration 0.5 mg/kg), ethyl acetate (5 mL), shaken (10 min, 100 rpm), vortexed (30 s), centrifugation (3000 $g$, 10 min). Extract evaporated (dryness, $N_2$, water bath, 45 °C). Residue dissolved (1 mL, 1 M ammonium acetate and 4 mL petroleum ether), vortexed (60 s), centrifuged (5 min, 3000 $g$), upper phase discarded, isooctane (2 mL) added, vortexed (30 s), centrifuged (3000 $g$, 10 min), upper phase discarded, ethyl acetate (3 mL), vortexing (60 s), centrifugation (3000 $g$, 5 min). Organic layer collected, reduced (dryness, $N_2$, water bath, 45 °C. Residue dissolved (0.5 mL hexane:$CCl_4$, 1:1 $v/v$), mixed (0.7 mL HPLC-grade CAN-water, 1:1 $v/v$), vortexed (30 s), centrifuged (4000 $g$, 6 min), upper phase filtered (0.22 µm PVDF syringe filter). | LC-ESI-MS-MS | $CC_\beta$ (mg/kg): 0.057, $CC_\beta$ (mg/kg): 0.098 | 95.88–96.96 |
| [54] | CAP | honey, fish and prawns | Homogenized tissue (1.0 g) weighed, spiked (50 µL $d^5$-CAP IS, 6 ng/mL), vortexed (30 s), allowed (20 min), ACN added (5 mL), vortexed (15 s), shaken (20 min, 180 rpm), centrifuged (5 min, 4000 rpm). Supernatant transferred; chloroform added (5 mL), vortexed (15–20 s), agitation and centrifugation, centrifugation (5 min, 2000 rpm), chloroform layer discarded. ACN phase evaporated (dryness, $N_2$, water bath at 40–45 °C). Residue re-constituted (1 mL of mobile phase, water–ACN, 90:10 $v/v$). | LC-ESI-MS/MS | $CC_\alpha$ (mg/kg): 0.04, $CC_\beta$ (mg/kg): 0.06, LOD (mg/kg): 0.02, LOQ (mg/kg): 0.06 | 85.5–115.6 (all) |

**Table 3.** *Cont.*

| References | Analytes | Sample | Sample Preparation | Analytical Technique | LOD-LOQ, CC$_\alpha$-CC$_\beta$ | Recovery (%) |
|---|---|---|---|---|---|---|
| [55] | CAP | shrimp, crayfish and prawns | Freeze-dried samples (2 g), reconstituted (water, original sample 10 g, IS ($d^5$-CAP) added, extraction (ACN/4% NaCl, 1:1 $v/v$, 20 mL), homogenized, centrifuged (4000 rpm, 15 min). Supernatant separated, de-fatted (2 × 10 mL $n$-hexane), ethyl acetate (7 mL) added, vortexed, supernatant transferred, extraction repeated. Combined supernatants evaporated (dryness), re-dissolved (3 mL of water/ACN, 95:5 $v/v$). Re-dissolved sample applied to C$_{18}$ cartridge (500 mg, Separtis). Cartridge preconditioned (10 mL MeOH and 10 mL water/ACN, 95:5 $v/v$), sample eluted (3 mL water/ACN, 45:55 $v/v$), ethyl acetate (4 mL) added to eluted sample, vortexed, extraction repeated. Combined extracts evaporated (dryness, N$_2$), residue re-dissolved (1 mL of acetone/toluene, 20:80 $v/v$), applied onto a Silica cartridge (1 g). Cartridge preconditioned (6 mL of acetone/toluene, 20:80 $v/v$), washed (2 × 3 mL acetone/toluene 20:80 $v/v$), eluted (6 mL acetone/toluene, 70:30 $v/v$), extract evaporated (dryness), derivatization mixture (50 µL, BSA/$n$-heptane, 1:1 $v/v$) added, react (45 min, 60 °C). | GC/NCI/MS | CC$_\alpha$ (mg/kg): 0.074, CC$_\beta$ (mg/kg): 0.087 | 95.0 (all) |
| [56] | CAP and CAP glucuronide | honey and prawns | Sample (3 g) homogenized, 7 mL ACN added, centrifugation (10 min, 3900 G, 4 °C). Supernatant applied to cartridge (10 mL Chem-Elut, 5 min), elution (15 mL and 10 mL CH$_2$Cl$_2$) evaporation (dryness, 45–50 °C), residue re-suspended (5 mL hexane:ethyl acetate, 50:50 $v:v$), extract loaded (pre-conditioned SPE cartridge, 500 mg, 3 cc). SPE cartridge washed (3 mL ethyl acetate), eluting (3 mL ethyl acetate:MeOH, 50:50 $v/v$), evaporation (dryness, N$_2$, 45–50 °C), dissolved (300 µL HPLC grade water). | Biacore Q biosensor, LC–MS/MS | Biosensor: CC$_\alpha$ (µg/kg): 0.04, CC$_\beta$ (µg/kg): 0.17. LC-MS/MS: CC$_\alpha$ (µg/kg): 0.09, CC$_\beta$ (µg/kg): 0.17 | - |
| [57] | CAP | milk and shrimp | Samples (10 g), spiked, placed statically (15 min), phosphate buffer (40 mL, 0.05 mol/L, pH 7.0) added, vortexed (2 min), sonicated (15 min), centrifuged (1.4 × 10$^3$ g, 10 min). Supernatant transferred, to precipitate proteins TCA in water (3 mL, 15%) added, vortexed (2 min), centrifugation (1.4 × 10$^3$ g, 10 min), filtered (microfilters, 0.45 µm). Eluent samples from MISPE cartridges evaporated (dryness, N$_2$), re-dissolved (mobile phase). | HPLC-UV | - | 84.9–89.0 |

**Table 3.** *Cont.*

| References | Analytes | Sample | Sample Preparation | Analytical Technique | LOD-LOQ, $CC_\alpha$-$CC_\beta$ | Recovery (%) |
|---|---|---|---|---|---|---|
| [58] | CAP, THI, FFC and FFC amine | shrimp and fish | Sample (1 g) placed into a mortar, blended (2 g, $C_{18}$ material-dispersion adsorbent, 5 min) with a pestle, homogeneous mixture transferred in glass column (300 × 15 mm i.d.) with degreased cotton packed at the bottom/top of the sample, column tightly compressed. Extraction solvent mixture ethyl acetate-ACN-25% $NH_4OH$ (10/88/2, v/v/v, 5 mL) used for elution, eluate dried (N2, 50 °C), residue reconstituted (1.0 mL 5% MeOH in 0.1% formic acid 5 mmol/L $CH_3COONH_4$). | LC–MS/MS | $CC_\alpha$ (µg/kg): FFC: 0.01, FFC amine: 0.05, THI: 0.07, CAP: 0.01 ; $CC_\beta$ (µg/kg): FFC: 0.05, FFC amine: 0.11, THI: 0.13, CAP: 0.04 | 84.0–98.8 |
| [59] | CAP, FFC and THI | shrimp | Sample (0.5 g), 500 µL working standard solution added, homogenized, dehydrated (0.5 g sea sand and 1 g anhydrous $Na_2SO_4$) in a glass mortar. Dry mixture placed into the SFE chamber, modifier introduced, chamber closed and attached to SFE system. Extracted substances collected *in situ* (silylation in a glass tube), tube filled with solvent containing derivatization reagent (20 mL, Sylon BFT), placed in a column oven. After extraction, solvent evaporated (dryness, $N_2$), residue resolved (200 µL ethyl acetate). | NCI–GC/MS | LOD (pg/g): CAP: 8.7, FFC: 15.8 and THI:17.4 | CAP: 92.0, FFC: 87.0, THI: 85.0 |
| [60] | CAP | shrimp | Sample (10 g) homogenized, extracted (×2, ethyl acetate), acetate extracts evaporated, residue dissolved (15 mL salting out solution), fatty components removed (n-hexane), extraction (ethyl acetate), extract evaporated, residue dissolved (5 mL ACN-water, 10:90 v/v). Solution pumped through sol-gel filter column (flow-rate 0.5 mL/min) on-line coupled to immunoaffinity column (containing 0.67 mg anti-CAP antibodies). Flushing (10 mL ACN-water, 10:90 v/v, 0.5 mL/min), filter column removed, IAC washed (10 mL ACN-water, 10:90 v/v), eluting CAP (10 mL ACN-water, 40:60 v/v, 1.0 mL/min). Eluate extracted (2 × 3 mL ethyl acetate), combined extracts dried (2 g $Na_2SO_4$), centrifugation (1580 g, 5 min), ethyl acetate phase decanted, evaporated ($N_2$), residue dissolved (1 mL ACN-water, 10:90 v/v). | HPLC-UV | LOD (ng/g): 1.8, LOQ (ng/g): - | 68.0 |

Table 3. *Cont.*

| References | Analytes | Sample | Sample Preparation | Analytical Technique | LOD–LOQ, CC$_\alpha$–CC$_\beta$ | Recovery (%) |
|---|---|---|---|---|---|---|
| [61] | CAP, THI and FFC | shrimp | Matrix solid phase dispersion: SPE cartridge conditioned (2 mL MeOH and 2 mL water), frits and sorbent removed, sorbent placed in a beaker with 0.5 g homogenized shrimp (spiked with 400 μL of mixture of three antibiotics), blending, sonicated (10 min). SPE cartridges repacked (one frit at the bottom, then sorbent/spiked sample, second frit on top, compressed with glass stirring rod), cartridge washed (1 mL ultra pure water), sequential elution (1 mL ACN and then 1 mL MeOH).Evaporation (dryness, water bath, 40 °C, under stream of nitrogen), dry residue dissolved (400 μL aqueous solution of lamotrigine, 10 ng/mL), filtration (syringe filter, 0.2 μm). | HPLC | LOQ (μg/kg): 20 (all), CC$_\alpha$ (μg/kg): THI: 58.8, FFC: 1030.8, CAP: 59.2, CC$_\beta$ (μg/kg): THI: 64.6, FFC: 1046.8, CAP: 63.8 | THI: 81.3–114.5, FFC: 72.0–103.3, CAP: 89.1–120.6 |
| [71] | CAP | shrimp tissue | Samples spiked (isotopically labeled internal standard, *d*$^5$-Cap), ethyl acetate extraction, defatted (hexane), cleanup (SPE C$_{18}$). Elute evaporated, derivatized with Sylon BFT. | GC/MS-MS | LOQ (ng/g): 0.3 | 95.0–111.0 (ave.) |
| [72] | CAP, THI, FFC and FFC amine | shrimp | MISPE | LC | LOD (μg/kg): CAP: 0.016, THI: 0.093, FFC: 0.102, FFC amine: 0.029, LOQ (μg/kg): - | 92.4–98.8 (all) |
| [73] | CAP | fish and shrimp | Samples extracted (ethyl acetate), defatted (hexane), derivatized (Sylon BFT). | GC-MS | LOD (ng/g): 0.04 (all), LOQ (ng/g): 0.1 (all) | 69.9–86.3 (all) |
| [74] | CAP | aquatic products | Samples extracted with (ethyl acetate–NH$_4$OH, 98:2 *v/v*), cleanup (IAC). | HPLC-UV, HPLC-MS/MS | LOD (μg/kg): -, LOQ (μg/kg): 0.25 (all) | 92.0–97.3 (ave.) |
| [75] | CAP | chicken meat, fish meat, shrimp meat and honey | Analytes extracted (ethylacetate), defatted (n-hexane, LLE). | LC-MS/MS | - | 76.2 |
| [76] | CAP | honey and prawns | Sample shaken with buffer, centrifuged, applied to IAC | LC-MS/MS | LOD (μg/kg): 0.05, LOQ (μg/kg): - | 84.0%–108.0% |
| [77] | FFC and FFC amine | fish, shrimp, and swine muscle | Samples extracted, defatted (hexane), cleaned (SPE, Oasis MCX cartridges), eluate evaporated (dryness), derivatized. | GC-microwell electron capture detector | LOD (ng/g): FFC: 0.5, FFC amine: 1 (all), LOQ (ng/g): - | 94.1–103.4 (ave.) |
| [78] | CAP, THI and FFC | shrimp | Samples extracted (basic ethyl acetate), extracts defatted (L-L partition), cleaned (C$_{18}$ SPE cartridge). | LC-MS/MS | LOD (ng/g): CAP and THI: 0.01, FFC: 0.05, LOQ (ng/g): - | CAP: 73.9–96.0, THI: 78.6–99.5, FFC: 74.9–103.7 |
| [79] | CAP, THI, FFC and FFC amine | shrimp muscle and pork | Samples extracted with 2% basic ethyl acetate, L-L partition, SPE. | HPLC-MS/MS | LOD (μg/kg): CAP: 0.001, THI: 0.020, FFC: 0.002, FFC amine: 0.003 (all), LOQ (μg/kg): - | 78.17–99.86 |

**Table 4.** Overview of extraction methodologies for the determination of sulfonamides in shrimps.

| Reference | Analytes | Sample | Sample Preparation | Analytical Technique | LOD, LOQ, $CC_\alpha$, $CC_\beta$ | Recovery % |
|---|---|---|---|---|---|---|
| [62] | SDZ, SMR, SME, SMZ, SMX, SDM | Fish, Shrimp | Fish and shrimp samples (5.0 g), spiked at 3 levels, MISPE extraction, 1% acetic acid (10.0 mL), vortexed, sonicated (5 min), centrifuged ($5.0 \times 10^3$ g, 5 min), supernatant applied to MISPE/NISPE cartridges, washed (5% ACN 1.0 mL in water—1% acetic acid), elution (3 mL MeOH/acetic acid 9/1 $v/v$), evaporation with $N_2$ at 40 °C, dissolved with 0.5 mL 28% ACN in water, filtration (0.22 μm) | HPLC-UV | LOD (μg/kg): 8.4–10.9 LOQ (μg/kg): 22.4–27.7 | 85.5%–106.1% |
| [63] | SDZ, STZ, SMZ, SMX, SMP, SCPD, SDM, SMM, SPZ, SDX, SSZ, SCP, SMT, SQX | flatfish, jacopever, sea bream, common eel, blue crab, shrimp, abalone | Sample (1 g), ACN (5 mL), homogenized (1 min), extraction × 2 -sonication (10 min), centrifugation (4500× g, 10 min), supernatant collected, add 100 mg $C_{18}$, homogenization (30 s), powder dispersed, centrifugation (4500× g, 10 min), evaporation with $N_2$ to dryness, dissolved with $KH_2PO_4$ (5 mM, 1 mL), filtration (0.45 μm) | HPLC-PDA Confirmation with LC-MS/MS | LOD (μg/kg): 3–6 LOQ (μg/kg): 9–18 | 51.8%–89.7% |
| [64] | SN, SDZ, STZ, SPD, SMR, SMZ, SMTZ, SMP, SCP, SMM, SDX, SMX, SDM, SQX | catfish, shrimp, salmon | Samples (10 g), 0.2% $CH_3COOH$-MeOH-ACN (10 mL, 85:10:5), homogenized (30 s, 20000 rpm), ACN (90 mL), shaker (10 min), centrifugation (10 min), ACN (30 mL) for extraction, shaker, centrifugation, $CH_2Cl_2$ (60 mL), shake (3 min), leave 15 min, bottom layer to flask with boiling chips, extraction repeated, concentration to 2–3 mL, $CH_2Cl_2$:acetone (60:40, 5 mL). SPE on SCX cartridges preconditioned (2.5 mL acetone, 2.5 mL 0.2% acetic acid, 2.5 mL acetone), elution with acetone-0.4 M $CH_3COONH_4$ (50:50 $v/v$, 5 mL), evaporation with $N_2$ to 2 mL. | HPLC-FLD | LOQ (ng/g): 1 | 67.3%–90.5%. |
| [65] | SG, SDZ, SMT, SMM, SMX, SDM, SQX | Shrimp | Homogenizes shrimp (2 g), $Na_2$EDTA-McIlaine's buffer (10 mL), mixed, vortexed (5 min), sonicated, centrifuged (3500 rpm, 10 min), SPE with Oasis HLB (200 mg, conditioned with 5 mL Milli-Q water, 5 mL. $Na_2$EDTA-McIlvaine buffer solution), elution with MeOH (7 mL), evaporation with $N_2$, reconstituted with mobile phase (10 mL), filtered (0.45 μm). | HPLC-EC (BBD amperometric detection) | LOD (ng/mL): 1.2–3.4 LOQ (ng/mL): 4.1–11.3 | (Spiked samples 1.5, 5, 10 μg·g$^{-1}$) 81.7 to 97.5% |
| [67] | SDZ, SMR, SG, SSZ, SDM, SMM, SDX, SMX | Shrimp | Homogenized shrimp (2 g), $Na_2$EDTA—McIlvaine buffer, vortexed (5 min), sonicated, centrifuged (3500 rpm, 10 min), SPE Microcolumn Vertipak™ HCP (conditioned with MeOH, Milli-Q water, $Na_2$EDTA-McIlvaine buffer), supernatant (10 mL), supernatant (10 mL) loaded, elution with MeOH (7 mL), filtration (0.20 μm pore size). | UPLC-ECD | LOD (ng/mL): 1.162–2.900 LOQ (ng/mL): 3.336–20.425 ng·mL$^{-1}$ | - |

Four nitrofuran metabolites, 3-amino-2-oxazolidinone (AOZ), 3-amino-5-morpholino-methyl-1,3-oxazolidinone (AMOZ), semicarbazide (SEM) and 1-aminohydatoin (AHD), were determined in shrimp tissue. The samples were washed with methanol, HCl solution was then added, and derivatization was achieved with 2-NBA. After remaining overnight in an incubator, the samples were neutralized and extracted with ethyl acetate. The extract was evaporated to dryness under nitrogen flow, and the residue was reconstituted with reconstitution solvent (water, acetonitrile, glacial acetic acid). Extraction with hexane was performed, and the final aqueous solution was filtered, prior to HPLC analysis. The mobile phase consisted of water, glacial acetic acid (A) and acetonitrile, water, glacial acetic acid (B). The UV detection was achieved at 275 nm. The recoveries for spiked samples, for the four metabolites, were found between 107% and 115% [80].

Four nitrofuran metabolites, AOZ, AMOZ, SEM and AHD, were determined in shrimp samples with LC-IDMS/MS analysis. The samples were acidified with HCl, derivatized with 2-NBA and left in an incubator overnight. After being neutralized, the samples were extracted twice with ethyl acetate and the extract was evaporated to dryness under nitrogen flow. The residue was dissolved with HPLC grade water and filtered before LC analysis. A Symmetry $C_{18}$ (2.1 × 150 mm, 3.5 μm) analytical column and a Symmetry $C_{18}$ guard column (2.1 × 10 mm, 3.5 μm) were used. The mobile phase consisted of acetonitrile (A) and water with aqueous solution acetic acid (B). The CCα of the derivatized metabolites was between 0.08 μg·$kg^{-1}$ and 0.20 μg/kg. The CCβ was between 0.13 μg/kg and 0.85 μg/kg [81].

Four nitrofuran metabolites (furalizone—AOZ, furaltadone—AMOZ, nitrofurazone—SEM and nitrofurantoin—AHD) were determined in shrimp samples using LC-MS/MS analysis. The samples were firstly washed with methanol, HCl and 2 NBA were then added to them, and the samples were kept overnight in an incubator. After being neutralized with NaOH, ethyl acetate was added to the samples, and they were centrifuged. The supernatant was evaporated near to dryness, the residue was re-dissolved with methanol, and the final mixture was filtered before LC-MS/MS analysis. The CCα was between 0.12 μg/kg and 0.23 μg/kg. The CCβ was between 0.21 μg/kg and 0.38 μg/kg, and the recoveries were found to be 88%–110% [82].

An overview of the extraction methodologies for the determination of macrolides and nitrofurans in shrimps is presented in Table 5.

### 3.7. Multi-Residue Methods

Thirteen sulfonamides, 3 fluoroquinolones and 3 quinolones were extracted from the spiked samples with 10 mL of acetonitrile and the addition of 0.1 mL p-toluenesulfonic acid monohydrate (p-TSA), 0.1 mL N,N,N′,N′-tetramethyl-p-phenylenediamine dihydrochloride (TMPD) solution, 2 g NaCl and ceramic homogenizer pellet. The extract was evaporated to dryness under nitrogen stream at 50–55 °C, and the residue was re-dissolved in 2 mL of acetonitrile:formic acid:water (10:0.4:89.6 *v/v*). The extraction procedure yielded 98.0%–104.0% average recovery for all analytes. Analysis was carried out by LC-MS/MS system, separation was achieved by a YMC ODS-AQ 2 × 100 mm, 3 μm column, and the mobile phase consisted of 0.1% formic acid in water and acetonitrile delivered under gradient elution. This method does not require a SPE cleanup step because it is fast and inexpensive, and an analytical chemist can prepare and analyze 12–16 samples per working day [83].

Five different sample treatment methods were tested for the extraction of multiclass antibiotics and veterinary drugs: benzalkonium chloride, ethoxyquin, leucomalachite green, malachite green, mebendazole, sulfadiazine, sulfadimethoxine, sulfamethazine, sulfamethizole, sulfanilamide, sulfapyridine, sulfathiazole and trimethoprim. The sample preparation was as follows:

(1) Spiked samples were extracted with 10 mL of acetonitrile containing 1% acetic acid with the addition of 4 g of anhydrous magnesium sulfate and 1.75 g of sodium chloride. 250 mg of primary-secondary amine and 750 mg of anhydrous magnesium sulfate were added to 5 mL of the acetonitrile supernatant. The final extract was evaporated to near dryness, reconstituted with 20% (*v/v*) methanol in water to a final volume of 2 mL.

(2)   Spiked samples were extracted with 6 mL of trifluoroacetic (20%, $w/v$), and the extract was evaporated to near dryness and reconstituted with 20% ($v/v$) methanol in water to a final volume of 0.5 mL.

(3)   Spiked shrimp samples were mixed with 2 g of aminopropyl (Bondesil-NH$_2$), and the mixture was transferred to a SPE cartridge. The antibiotics were eluted twice with 5 mL of acetonitrile, and the eluates were evaporated to near dryness and reconstituted with 20% ($v/v$) methanol in water to a final volume of 1 mL.

(4)   Spiked shrimp samples were mixed with 3 mL of sulfuric acid 0.17 M, 0.158 g of sodium tungstate and 12 mL of acetonitrile in order to precipitate the proteins. An SPE cleanup step followed, with a C$_{18}$ cartridge preconditioned with 5 mL of methanol and 5 mL of water. The analytes were eluted with 1 mL of acetonitrile/water (30:70 $v/v$), 2 × 2 mL of ethyl acetate were added to the eluate, and the organic extracts were evaporated to near dryness and with 20% ($v/v$) methanol in water to a final volume of 1 mL.

(5)   In order to precipitate the proteins, spiked shrimp samples were mixed with 100 mL of 0.2% of metaphosphoric acid in acetonitrile, the mixture was filtered through a 0.45 μm filter, and the extract was evaporated to a final volume of 30 mL. A SPE cleanup step followed with an Oasis HLB SPE cartridge, and the analytes were eluted with 5 mL of acetonitrile. The eluate was evaporated to near dryness and reconstituted with 20% ($v/v$) methanol in water to a final volume of 1 mL.

The QuEChERS method was preferred. The extraction procedure yielded recoveries between 33.0%–118.0% for all analytes. Analysis was carried out by an LC–TOF/MS system, separation was achieved by a RR Zorbax Eclipse XDB-C$_{18}$ analytical column (50 mm × 4.6 mm and, 1.8 μm), and the mobile phase consisted of 0.1% formic acid and acetonitrile (gradient elution). The LOD ranged between 0.06–7.1 μg/kg for all analytes [84].

Forty-three multi-class veterinary drugs were extracted from the spiked samples with 5 mL of water and 15 mL of 1 vol. % formic acid in acetonitrile or acetonitrile. Four grams of magnesium sulfate, 1.5 g trisodium citrate dehydrate and 2 g sodium chloride were then added to the mixtures. The extract was used directly for LC-MS/MS analysis without further handling. Analysis was carried out by an LC-MS/MS system, separation was achieved by a Syncronis aQ (2.1 mm i.d. × 100 mm, 5 μm column, and the mobile phase consisted of 0.1% formic acid in 10 mmol/L ammonium acetate and acetonitrile under gradient conditions [85].

Lomefloxacin, enoxacin, sarafloxacin, enrofloxacin, sulfadiazine, sulfamethoxydiazine and sulfadimethoxypyrimidine were extracted from the spiked samples with accelerated solvent extraction and acetonitrile was the extraction solvent. The extract was evaporated to dryness under nitrogen stream at 45 °C, and the residue was re-dissolved in 1 mL of methanol. The extraction procedure yielded recoveries between 88.9%–94.8%, 88.0%–93.1%, 87.6%–95.7%, 88.0%–93.2%, 88.7%–91.0%, 86.7%–90.0%, 85.4%–88.8% for LOME, ENO, SAR, ENR, sulfadiazine, sulfamethoxydiazine and sulfadimethoxypyrimidine, respectively. Analysis was carried out by capillary zone electrophoresis system couples with a UV detector, separation was achieved by a uncoated fused silica capillary of i.d. 50 μm with total length of 48.5 cm (effective length 40 cm), capillary was filled with a borate buffer (25 mM, pH 8.8) containing methanol, and the analytes moved through the capillary by reversing the polarity (−25 V). The LOD was 0.025, 0.033, 0.025, 0.020, 0.013, 0.013, 0.013 μg/mL for LOME, ENO, SAR, ENR, sulfadiazine, sulfamethoxydiazine and sulfadimethoxypyrimidine, respectively. The accelerated solvent extraction provides rapid extraction procedures and lower solvent usage in comparison with the extraction procedures used in the literature [86].

Sulfadiazine, sulfamethazine, sulfamethazine, sulfachloropyridazine, sulfadimethoxine and sulfaquinoxaline were determined in shrimp samples after extraction with trichloroacetic acid and hydroxylamine hydrochloride. The samples were vortexed and centrifuged. Sodium succinate and NaOH were added to the supernatant, and it was then applied to Waters Oasis HLB cartridges. Elution was performed with MeOH and $CH_3CN$/MeOH. Into the eluate ammonium formate buffer, EDTA and ascorbic acid were added. After evaporation under vacuum until 0.8 mL and an addition of water/acetonitrile mix up to 1 mL, the final solution was vortexed and centrifuged. Half of the final solution was used for analysis, which was performed by LC-MS/MS. Phenyl column separated the analytes prior to analysis, and APCI was used as ionization source in negative mode. The estimated recovery ranged from over 40% to over 90%. The LOD of SQX was achieved at 20 ng/g and 10 ng/g for the other sulfonamides [87].

An overview of the extraction methodologies for the multi-class antibiotics analysis in shrimps is presented in Table 6.

**Table 5.** Overview of extraction methodologies for the determination of macrolides and nitrofurans in shrimps.

| Reference | Analytes | Sample | Sample Preparation | Analytical Technique | LOD, LOQ, CCα, CCβ | Recovery % |
|---|---|---|---|---|---|---|
| [68] | Macrolides—(erythromycin, tylosin, josamycin, spiromycin, neospiromycin, tilmicosin, gamithromycin, tildipirosin and oleandomycin) Lincosamides—Lincomycin, Pirlimycin, Clindamycin | Salmon, Shrimp, Tilapia | Homogenized sample (5 g), extraction with ACN (10 mL), water (1 mL), shaker (700 rpm), centrifugation (5 min, $400 \times g$ RCF), re-extraction with ACN, phosphate buffer (3 mL), shaker, centrifugation, the two supernatants centrifuged again (5 min, $6100 \times g$ RCF), SPE on Bond-Elut cartridge (pre-conditioned with water, 12% ACN), elution with methanolic $CH_3COONH_4$ (750 µL × 2), fat removal with water and hexane, vortexed, centrifuged (5 min, $1000 \times g$ RCF), evaporation with $N_2$ to volume < 0.75 mL, methanolic $CH_3COONH_4$ (50 µL), mixed, MeOH to volume 1 mL, centrifugation (15 min, $2130 \times g$, 5 °C). | LC-MS/MS | LOD (µg/kg): 0.5 | 47%–99% |
| [69] | Macrolides—ERY, ELAN, AZM, CLM, TIM, RXM, Quinolones—CIP, SPFX Amphenicols—CAP, TAP | Pork, Fish, Shrimp | Spiked samples, NaOH for hydrolysis of lipids (500 µL), extraction with ACN (20 mL), vortexed (15 min), centrifuged (5 min, 7000 rpm), supernatant with MMIPs (100 mg) mixed, magnetically removed, washed with ACN:water, elution with (10 mL) MeOH/50 mM $KH_2PO_4$ (pH 8), evaporation to dryness, residue reconstituted with mL ACN/25 mM $KH_2PO_4$ | UPLC-UV | LOD (µg/kg): 0.015–0.2 LOQ (µg/kg): 0.075–0.5 | - |
| [70] | Nitrofurans—Furazolidone, furaltadone, nitrofurazone, nitrofurantoin | Shrimp | Homogenized samples (2.5 g), added HCl aqueous and 2-NBA for derivatization, incubating overnight, neutralized with di-sodium hydrogen phosphate and NaOH, centrifugation (5 min, 4000 rpm), supernatants SPE Oasis HLB (conditioning with ethyl acetate, MeOH, Milli-Q water), cartridge washed with water, elution with ethyl acetate (6 mL), evaporation to dryness, redissolved with mobile phase (1 mL), filtered (0.20 µm) | UHPLC-QqQ-MS/MS | LOD (mg/kg)0.5–0.8, LOQ (mg/kg): 1 | - |
| [80] | Nitrofurans—nitrofuran metabolites, AOZ, AMOZ, SEM and AHD. | Shrimp | Homogenized sample (5 g), washed with MeOH (20 mL) mixing, centrifugation (10 min, 2500 rpm), washing, with MeOH and water, HCl (10 mL), derivatization with 2-NBA, incubated overnight, $Na_3PO_4$, $12H_2O$ solution added, neutralized with NaOH (2 M), extraction with ethyl acetate, evaporation to dryness, reconstituted with 500 µL reconstitution solvent, extracted free times with hexane, filtration (0.45 µm) | HPLC-UV | LOD (mg/kg): 2 | 107%–115% |
| [81] | Nitrofurans—AOZ, AMOZ, SEM and AHD | Shrimps | Sample (1 g) spiked at 2 µg/Kg, hydrolysis with HCl (5 mL), derivatization with 50 µL 2-NBA, overnight incubation, neutralizing with NaOH and phosphate buffer, extraction with ethyl acetate, evaporation to dryness, dissolution with HPLC grade water, filtration. | LC-IDMS/MS | CCα (µg/kg): 0.08–0.20 CCβ (µg/kg)0.13–0.85 | - |
| [82] | Nitrofurans—furalizone—AOZ, furaltadone—AMOZ, nitrofurazone—SEM and nitrofurantoin—AHD | Shrimps | Homogenized sample (1 g) washed with methanol, centrifuged (4 min 4000 rpm) repeated, HCl and 2-NBA to the sample, incubated overnight, neutralized with NaOH, ethyl acetate added (4 mL), centrifuged, extraction again, supernatant evaporation near dryness, residue dissolved with methanol, filtrated (0.45 µm) analysis, $AMOZ\text{-}d_5$ internal standard | LC-MS/MS | CCα (µg/kg): 0.12–0.23, CCβ (µg/kg): 0.21–0.38 | 88%–110% |

**Table 6.** Overview of extraction methodologies for the determination of multi-class antibiotics in shrimps.

| Reference | Analytes | Sample | Sample Preparation | Analytical Technique | LOD, LOQ, CCα, CCβ | Recovery (%) |
|---|---|---|---|---|---|---|
| [83] | 26 veterinary drugs: 13 SAs TRI, 3 FQ, 3 AQ, 3 TPM, 2 LC dyes metabolites, 1 hormone | fish and other aquaculture products (tilapia, catfish, eel, pangasius, sablefish, swai, salmon, trout, and shrimp | Tissue (4.0 g ± 0.03 g) weighed. SMZ-$^{13}$C solution (0.040 mL) added, EDTA-McIlvaine buffer (2.0 mL) added, mixed (50 s). ACN (10 mL), p-TSA (0.100 mL), TMPD solution (0.100 mL), NaCl (2.0 g) and ceramic homogenizer pellet added, shaken (5 min), centrifuged (6000 rpm, 5 °C, 5 min) Organic layer transferred, ACN (10 mL) added, shaken (5 min), centrifuged. ACN layers combined, evaporated (dryness, water bath, 50–55 °C, N$_2$). Residue reconstituted (2.0 mL the dissolution solution), mixed (30 s), sonicator (5 min), centrifuged (10,000 rpm, 5 °C, 5 min). 0.5 mL portion filtered (0.2 μm PTFE syringe filter). | LC-MS/MS | - | 98.0–104.0 (all, ave.) |
| [84] | BC, EQ, LMG, MG, MBZ, SDZ, SDM, SMZ, SMTZ, SN, SPD, STZ,TRO | shrimp | Five different sample treatment methodologies were tested: ACN extraction followed by cleanup by QuEChERS. Sample (10 g) homogenized, ACN containing 1% acetic acid (10 mL) added, shaken (1 min), anhydrous MgSO$_4$ (4 g) and NaCl (1.75 g) added, shaking repeated (1 min).Extract centrifuged (3700 rpm, 3 min), supernatant (5 mL) (ACN phase) transferred to centrifuge tube (with 250 mg PSA and 750 mg anhydrous MgSO$_4$), shaken (20 s), centrifuged (3700 rpm, 3 min), extract (2 mL) evaporated (near dryness, reconstituted (20% $v/v$ MeOH in water, to a final volume of 2 mL), filtered (0.45 μm PTFE filter). 1. Extraction with TCA: Shrimp (1 g), TCA solution added (6 mL, 20%, $w/v$), homogenized (ultrasonic bath, 30 s), centrifuged (5 min, 3700 rpm). Supernatant (3 mL) taken, evaporated (near dryness), dissolved in MeOH in water (20% $v/v$, to a final volume of 0.5 mL), filtered (0.45 μm PTFE filter). 2. Matrix solid phase dispersion (MSPD) procedure: Shrimp (1 g), (2 g, Bondesil-NH$_2$) added, mixture transferred to SPE cartridge containing 2 g Florisil, connected to vacuum system. Elution with ACN (2 × 5 mL), final extract evaporated (near dryness), reconstituted with MeOH in water (with 20% $v/v$, to a final volume of 1 mL), filtered (0.45 μm PTFE filter). | LC-TOFMS | LOD (μg/kg): BC-C12: 0.6, EQ: 7.1, LMG: 0.6, MG: 0.06, MBZ: 0.1, SDZ: 4.5, SDM: 0.3, SMZ: 0.1, SMTZ: 0.8, SN: 3.5, SPD: 0.5, STZ: 2.9, TRI: 0.7 | BC-C12: 53.0, EQ: 53.0, LMG: 90.0, MG: 118.0, MBZ: 118.0, SDZ: 82.0, SDM: 85.0, SMZ: 114.0, SMTZ: 33.0, SN: 115.0, SPD: 109.0, STZ: 81.0, TRI: 87.0 |

Table 6. *Cont.*

| Reference | Analytes | Sample | Sample Preparation | Analytical Technique | LOD, LOQ, CCα, CCβ | Recovery (%) |
|---|---|---|---|---|---|---|
| | | | 3. SPE-based method I: (a) Protein precipitation: Shrimp (1 g), sulfuric acid (3 mL, 0.17 M), sodium tungstate (0.158 g) and ACN (12 mL) added, mixture shaken, centrifuged. Supernatant (10 mL) filtered (0.45 µm PTFE filter). (b) SPE: Aliquot (3 mL) transferred to C₁₈ cartridge preconditioned with 5 mL MeOH and 5 mL water), washing (500 µL of water 500 µL ACN/water, elution (1 mL of ACN/water, 30:70). (c) LLE: SPE eluate, ethyl acetate (2 mL) added, shaking (30 s), organic phase separated, extraction repeated (2 mL ethyl acetate), extracts combined, evaporated (near dryness), reconstituted with MeOH in water (20% v/v, to a final volume of 1 mL), filtered (0.45 µm PTFE filter). 4. SPE-based method II: Shrimp (5 g), metaphosphoric acid in ACN (100 mL, 0.2%) added, mix filtered (0.45 µm filter), evaporated (N₂ stream, until 30 mL), Extract loaded onto Oasis HLB cartridge, washed (5 mL ACN:water, 20:80 v/v), eluted (5 mL ACN), evaporated (near dryness), reconstituted with MeOH in water (20% v/v, a final volume of 1 mL), filtered (0.45 µm PTFE). | | | |
| [85] | 43 multi-class veterinary drugs (sulfonamides, quinolones, coccidiostats and antiparasites) | milk, fish and shellfish (salmon, tiger shrimp, red sea bream and bastard halibut) | Sample (5 g), working standard solution (50 µL or 500 mL) added, waiting (>30 min), water (5 mL) and HCOOH in ACN (15 mL, 1 vol. %, Method A) or ACN alone (15 mL, 1 vol. %, Method B) added, homogenized, magnesium sulfate (4 g), trisodium citrate dehydrate(1.5 g) and NaCl (2 g) added, shaken (1 min), centrifuged (1800× *g*, 10 min), supernatant transferred, dilution extraction solvent, portion of the solution transferred to a microtube, centrifuged (16,000× *g*, 10 min). | LC-MS/MS | LOQ (µg/kg): 1–10 (all) | 48.5–113.6 (all) (Method A) 11.1–116.5 (all) (Method B) |
| [86] | LOME, ENO, SAR, ENR, SDZ, SMD, SDMP | shrimp and sardine | Sample (5 g) and diatomite (1.5 g) mixed, transferred into extraction cell (ACN extraction solvent). Extraction conditions: oven temperature 60 °C, 3 min heat-up time, pressure 10.3 MPa, two static cycles, static time 5 min, flush volume 40% of extraction cell volume. Extract purged (pressurized N₂, 90 s), evaporated (dryness, N₂, 45 °C), residue dissolved (MeOH, to 1 mL), filtered (0.45 µm). | CZE | LOD (µg/mL): LOME: 0.025, ENO: 0.033, SAR: 0.025, ENR: 0.020, SDZ: 0.013, SMD: 0.013, SDMP: 0.013 LOQ (µg/mL): LOME: 0.08, ENO: 0.10, SAR: 0.08, ENR: 0.07, SDZ: 0.04, SMD: 0.04, SDMP: 0.04 | LOME: 88.9–94.8, ENO: 88.0–93.1, SAR: 87.6–95.7, ENR: 88.0–93.2, SDZ: 88.7–91.0, SMD: 86.7–90.0, SDMP: 85.4–88.8 |

Table 6. *Cont.*

| Reference | Analytes | Sample | Sample Preparation | Analytical Technique | LOD, LOQ, CCα, CCβ | Recovery (%) |
|---|---|---|---|---|---|---|
| [87] | SDZ, SMR, SMZ, SCP, SDM, SQX, ENR, SAR, DIF, OXO, NAL, FLU, LMV, LVG, MG, GV, OTC, TOLSa | Shrimp | blended shrimp (2 g), 100 µL standard, TCA & NH₂OH-HCl added, homogenized, vortex (10 min), centrifugation (4000 rcf, 4 °C, 15 min), supernatant into solution sodium succinate (2.5 mL) and NaOH (10N, 280 µL)—pH 3.6 ± 0.1, Oasis HLB cartridge pre-conditioned (washed 3 mL ammonium formate buffer, 3 mL Milli-Q water, dried for 2 min), elution with MeOH (2 mL) CH₃CN/MeOH 1:1 *v*/*v* (1 mL), to the elute ammonium formate buffer (20 mM, pH 3.9), EDTA (50 µL, 0.1 M), ascorbic acid (1 mg/mL in MeOH), evaporation with N₂ till 0.8 mL, aliquot of 1:1 water/ACN added to fill 1 mL, centrifugation (14000 rpm, 10 min), analysis of the middle portion (~0.8 mL) | LC-ion trap-MS | 10 ng/g for SQX | 40%–90% |
| [88] | 21 veterinary drugs: SAs (SDZ, SMR, SMZ, SCP, SDM, SQX), TCs (OTC, TC, CTC), FQ (NOR, CIP, ENR, SAR, DIF, FLU, OXO,NAL) and cationic dyes (MG, GV, LMC, and LGV) | shrimp | Sample (2 g) extracted (×2, 2 different pH values), supernatant diluted (aqueous internal standard), online SPE automated sample cleanup. | HPLC-MS/MS | - | - |
| [89] | FQ, TCs, macrolides, lincosamides, SAs and others | livestock and fishery products | Extraction with two solutions of different polarity: highly polar compounds extracted with Na₂EDTA-McIlvaine's buffer (pH 7.0) and medium polar compounds were extracted with ACN containing 0.1% HCOOH. Cleanup with SPE polymer cartridge, first extracted solution applied to the cartridge (highly polar compounds retained), second extracted solution applied to the same cartridge, both highly and medium polar compounds eluted. | LC-MS/MS | LOQ (µg/kg): 0.1–5 | - |
| [90] | CAP and nitrofuran metabolites | shrimp | Extraction steps: neutralization of hydrolysates, addition of ACN for extraction, salting out of organic phase from the ACN-aqueous mixture | LC-MS/MS | - | 98.6–109.2 (all) |
| [91] | 33 FQ and SAs | eels and shrimps | Sample extracted with acidified ACN, cleaned-up (hexane), concentrated (evaporator). | HPLC-MS/MS | LOD (µg/kg): 1.0, LOQ (µg/kg): 2.0 | 66.0–123.0 |
| [92] | AMOZ, AOZ, AHD, SEM CAP | shrimp | single extraction procedure | LC-MS/MS | LOD (ng/g): nitrofuran metabolites: 0.5, CAP: 0.3 | - |

## 4. Conclusions

Increased aquaculture practice has resulted in increased levels of infections among species. Various classes of antibiotics including quinolones, tetracyclines, b-lactams, sulfonamides, *etc.* exhibit activity against both Gram-positive and Gram-negative bacteria; therefore, they are widely used in aquaculture to treat or prevent diseases.

However, the extended use of antibiotics in aquaculture has led to the demand for developing sensitive methods for their determination. The focus of this review has been to present the trends in microextraction techniques for the analysis of shrimps, as many different antibiotic classes are used in shrimp aquaculture worldwide, although some of them have been forbidden in other countries due to their dangerous side effects on humans.

Evidently, the analysis of antibiotics in shrimps still requires a significant amount of solvents and tedious extraction protocols due to the complex matrix; therefore, microextraction techniques are scarcely applied, indicating that there is still a lot of research to be done in this direction.

**Author Contributions:** The authors have equally contributed to the manuscript.

**Conflicts of Interest:** The authors declare no conflict of interest.

## Abbreviations

| | |
|---|---|
| ACN | Acetonitrile |
| AHD | 1-Aminohydatoin |
| AMOZ | 3-Amino-5-morpholino-methyl-1,3-Oxazolidinone |
| AOZ | 3-Amino-2-Oxazolidinone |
| APCI | Atmospheric Pressure Chemical Ionization |
| AQ | Acidic Quinolones |
| ASE | Accelerated Solvent Extraction |
| AVE | Average |
| AZM | Azithromycin |
| BC | Benzalkonium Chloride |
| BDD | Boron Doped Diamond |
| BG | Brilliant Green |
| BSA | $N,O$-Bis(trimethylsilyl)acetamide |
| CAP | Chloramphenicol |
| CCα | Decision Limit |
| CCβ | Detection Capability |
| CIN | Cinoxacin |
| CIP | Ciprofloxacin |
| CLM | Clarithromycin |
| CTC | Chlortetracycline |
| CV | Crystal Violet Cation |
| CWP | Coordinating Working Party |
| CZE | Capillary Zone Electrophoresis |
| $d^5$-Cap | $d^5$-Chloramphenicol |
| DC | Doxycycline |
| DES-CIP | Desethylene Ciprofloxacin |
| DI water | Deionized Water |
| DIF | Difloxacin |
| DMC | Demeclocycline |
| DNA | Deoxyribonucleic Acid |
| EDTA | Ethylenediaminetetraacetic Acid |

| | |
|---|---|
| ELAN | Elandomycin |
| ENR | Enrofloxacin |
| EQ | Ethoxyquin |
| ERY | Erythromycin |
| EU | European Union |
| FAO | Food and Agriculture Organization |
| FAO | Food and Agriculture Organization of The United Nations |
| FFC | Florfenicol |
| FLU | Flumequine |
| FQ | Fluoroquinolones |
| GC | Gas Chromatography |
| GC/MS-MS | Gas Chromatography-Mass Spectrometry |
| GC/NCI/MS | Gas Chromatography-Negative Chemical Ionization-Mass Spectrometry |
| GV | Gentian Violet |
| HLB | Hydrophilic-Lipophilic Balance |
| HPLC | High-Performance Liquid Chromatography |
| HPLC-CE | High-Performance Liquid Chromatography Cation-Exchange |
| HPLC-CL | High-Performance Liquid Chromatography-Chemiluminescenece Detection |
| HPLC-FLD | High-Performance Liquid Chromatography-Fluorescence Detection |
| HPLC-UV | High-Performance Liquid Chromatography-Ultraviolet Detection |
| IAC | Immunoaffinity Column |
| IS | Internal Standard |
| LC dye metabolites | Leuco Dye Metabolites |
| LC-ESI-MS/MS | Liquid Chromatography-Electrospray Ionization-Mass Spectrometry |
| LC-FLD | Liquid Chromatography-Fluorescence Detection |
| LC-FLD-MS | Liquid Chromatography-Fluorescence-Mass Spectrometry |
| LC-MS/MS | Liquid Chromatography-Tandem Mass Spectrometry |
| LC-TOFMS | Liquid Chromatography-Time-Of-Flight Mass Spectrometry |
| LC-UV | Liquid Chromatography-Ultraviolet Detection |
| LCV | Leucocrystal Violet |
| LDTD-MS/MS | Laser Diode Thermal Desorption-Mass Spectrometry |
| LGV | Leucogentian Violet |
| L-L partition | Liquid-Liquid Partition |
| LLE | Liquid-Liquid Extraction |
| LMG | Leucomalachite Green |
| LOD | Limit of Detection |
| LOME | Lomefloxacin |
| LOQ | Limit of Quantification |
| MARB | Marbofloxacin |
| MBZ | Mebendazole |
| MCX | Mixed Mode Cation Exchange |
| MDL | Method Detection Limit |
| MeCN | Acetonitrile |
| MeOH | Methanol |
| MG | Malachite Green Cation |
| MIP | Molecularly Imprinted Polymer |
| MISPE | Molecularly Imprinted Solid Phase Extraction |
| MNC | Minocycline |
| MQCA | 3-Methyl-quinoxaline-2-carboxylic Acid |
| MRLs | Maximum Residue Levels |
| MSPD | Matrix Solid Phase Dispersion |

| | |
|---|---|
| MT | Methyltestosterone |
| MTC | Methacycline |
| NAL | Nalidixic Acid |
| NBA | Nitrobenzaldehyde |
| Ni-DIA electrode | Nickel-Implanted Boron-Doped Diamond Thin Film Electrode |
| NIP | Non-Molecularly Imprinted Polymer |
| NOR | Norfloxacin |
| OFL | Ofloxacin |
| ORB | Orbifloxacin |
| OTC | Oxytetracycline |
| OXO | Oxolinic Acid |
| PABA | Para-Aminobenzoic Acid |
| PEF | Perfloxacin |
| PLE | Pressurized Liquid Extraction |
| PSA | Primary–Secondary Amine |
| PTFE | Polytetrafluoroethylene |
| *p*-TSA | *p*-Toluenesulfonic Acid Monohydrate |
| QCA | Quinoxaline-2-Carboxylic Acid |
| QuEChERS | Quick, Easy, Cheap, Effective, Rugged, and Safe |
| RNA | Ribonucleic Acid |
| RXM | Roxythromycin |
| SAR | Sarafloxacin |
| SAs | Sulfonamides |
| SCPD | Sulfachloropyridazine |
| SCPZ | Sulfachloropyrazine |
| SDB-RPS | Polystyrenedivinylbenzene-Reverse Phase Sorbent |
| SDM | Sulfadimethoxine |
| SDM | Sufladimethoxine |
| SDMP | Sulfadimethoxypyrimidine |
| SDX | Sulfadoxine |
| SDX | Sulfadoxine |
| SDZ | Sulfadiazine |
| SEM | Semicarbazide |
| SFE | Supercritical Fluid Extraction |
| SG | Sulfaguanidine |
| SLE | Solid Liquid Extraction |
| SMD | Sulfamethoxydiazine |
| SME | Sulfameter |
| SMM | Sulfamonomethoxine |
| SMP | Sulfamethoxypyridazine |
| SMR | Sulfamerazine |
| SMT | Sulfamethazine |
| SMTZ | Sulfamethizole |
| SMX | Sulfamethoxazole |
| SMZ | Sulfamethazine |
| SMZ-13C6 | Sulfamethazine-13C6 |
| SN | Sulfanilamide |
| SPD | Sulfapyridine |
| SPE | Solid Phase Extraction |
| SPZ | Sulfaphenazole |
| SQX | Sulfaquinoxaline |

| SSZ | Sulfisoxazole |
| SSZ | Sulfisoxazole |
| STZ | Sulfathiazole |
| Sylon BFT | {*N,O*-Bis(Trimethylsily) Trifluoroacetamide[BSTFA]-Trimethylchlorosilane [TMCS], 99 + 1} |
| TC | Tetracycline |
| TCA | Trichloroacetic Acid |
| TCs | Tetracyclines |
| THI | Thiamphenicol |
| TIM | Tilmicocin |
| TMPD | *N,N,N′,N′*-Tetramethyl-*P*-Phenylenediamine dihydrochloride |
| TOLSa | Toltrazurisulfone |
| TPM | Triphenylmethane Dyes |
| TRI | Trimethoprim |
| UPLC-MS/MS | Ultra-Performance Liquid Chromatography-Mass Spectrometry |
| UV | Ultra Violet |

## References

1. Food and Agriculture Organization of the United Nations. Available online: http://www.fao.org/fishery/cwp/handbook/j/en (accessed on 9 December 2015).
2. Cole, D.W.; Cole, R.; Gaydos, S.J.; Gray, J.; Hyland, G.; Jacques, M.L.; Powell-Dunford, N.; Sawhney, C.; Au, W.W. Aquaculture: Environmental, Toxicological, and Health Issues. *Int. J. Hyg. Environ. Health* **2009**, *212*, 369–377. [CrossRef] [PubMed]
3. Sapkota, A.; Sapkota, A.R.; Kucharski, M.; Burke, J.; McKenzie, S.; Walker, P.; Lawrence, R. Aquaculture Practices and Potential Human Health Risks: Current Knowledge and Future Priorities. *Environ. Int.* **2008**, *34*, 1215–1226. [CrossRef] [PubMed]
4. Gillett, R. *Global Study of Shrimp Fisheries*; FAO: Rome, Italy, 2008; Volume 475.
5. Rigos, G.; Troisi, G.M. Antibacterial Agents in Mediterranean Finfish Farming: A Synopsis of Drug Pharmacokinetics in Important Euryhaline Fish Species and Possible Environmental Implications. *Rev. Fish Biol. Fish.* **2005**, *15*, 53–73. [CrossRef]
6. Defoirdt, T.; Sorgeloos, P.; Bossier, P. Alternatives to Antibiotics for the Control of Bacterial Disease in Aquaculture. *Curr. Opin. Microbiol.* **2011**, *14*, 251–258. [CrossRef] [PubMed]
7. Bermúdez-Almada, M.C.; Espinosa-Plascencia, A. The Use of Antibiotics in Shrimp Farming. In *Health and Environment in Aquaculture*; Carvalho, E., Ed.; InTech: Rijeka, Croatia, 2010; pp. 199–214.
8. Mo, W.Y.; Chen, Z.; Leung, H.M.; Leung, A.O.W. Application of Veterinary Antibiotics in China's Aquaculture Industry and Their Potential Human Health Risks. *Environ. Sci. Pollut. Res.* **2015**. [CrossRef] [PubMed]
9. Samanidou, V.F.; Evaggelopoulou, E.N. Analytical Strategies to Determine Antibiotic Residues in Fish. *J. Sep. Sci.* **2007**, *30*, 2549–2569. [CrossRef] [PubMed]
10. Finch, R.G.; Greenwood, D.; Whitley, R.J.; Norrby, S.R. *Antibiotic and Chemotherapy*, 9th ed.; Elsevier: Amsterdam, The Netherlands, 2010.
11. Wagman, A.S.; Wentland, M.P. Quinolone Antibacterial Agents. In *Comprehensive Medicinal Chemistry II*; Elsevier: Amsterdam, The Netherlands, 2007; pp. 567–596.
12. Sousa, J.; Alves, G.; Abrantes, J.; Fortuna, A.; Falcão, A. Analytical Methods for Determination of New Fluoroquinolones in Biological Matrices and Pharmaceutical Formulations by Liquid Chromatography: A Review. *Anal. Bioanal. Chem.* **2012**, *403*, 93–129. [CrossRef] [PubMed]
13. Cheng, G.; Hao, H.; Dai, M.; Liu, Z.; Yuan, Z. Antibacterial Action of Quinolones: From Target to Network. *Eur. J. Med. Chem.* **2013**, *66*, 555–562. [CrossRef] [PubMed]
14. EUR-Lex. Available online: http://eur-lex.europa.eu/legal-content/EN (accessed on 9 December 2015).
15. Rang, H.P.; Ritter, J.M.; Flower, R.J.; Henderson, G. *RANG & DALE'S Pharmacology*, 8th ed.; Elsevier: Amsterdam, The Netherlands, 2016.
16. Nelson, M.L.; Ismail, M.Y. The Antibiotic and NonantibioticTetracyclines. In *Comprehensive Medicinal Chemistry II*; Elsevier: Amsterdam, The Netherlands, 2007; pp. 597–628.

17. Nicolaou, K.C.; Chen, J.S.; Edmonds, D.J.; Estrada, A.A. Recent Advances in the Chemistry and Biology of Naturally Occurring Antibiotics. *Angew. Chem. Int. Ed.* **2010**, *48*, 660–719. [CrossRef] [PubMed]

18. U.S. Food and Drug Administration. Available online: http://www.fda.gov/AnimalVeterinary/DevelopmentApprovalProcess/Aquaculture/ucm132954.htm (accessed on 9 December 2015).

19. Hauser, A.R. *Antibiotics Basics for Clinicians: The ABCs of Choosing the Right Antibacterial Agent*, 2nd ed.; Lippincott Williams & Wilkins: Philadelphia, PA, USA, 2007.

20. Manivasagan, P.; Venkatesan, J.; Sivakumar, K.; Kim, S.K. Marine Actinobacterial Metabolites: Current Status and Future Perspectives. *Microbiol. Res.* **2013**, *168*, 311–332. [CrossRef] [PubMed]

21. Kaneko, T.; Dougherty, T.J.; Magee, T.V. Macrolide Antibiotics. In *Comprehensive Medicinal Chemistry II*; Elsevier: Amsterdam, The Netherlands, 2007; pp. 520–539.

22. Aronson, J.K. *Meyler's Side Effects of Drugs: The International Encyclopedia of Adverse Drug Reactions and Interactions*; Elsevier: Amsterdam, The Netherlands, 2006; pp. 2248–2249.

23. Dost, K.; Jones, D.C.; Davidson, G. Determination of sulfomamides by packed column supercritical fluid chromatography with atmospheric pressure chemical ionization mass spectrometric detection. *Analyst* **2000**, *125*, 1243–1247. [CrossRef] [PubMed]

24. FAO/WHO. *Evaluation of Certain Veterinary Drug Residues in Food*; 40th Report of the Joint FAO/WHO Expert Committee on Food; World Health Organization: Geneva, Switzerland, 1993.

25. Garrido-French, A.; Plaza-Bolanos, P.; Aguilera-Luiz, M.M.; Martinez-Vidal, J.L. Recent advances in the analysis of veterinary drugs and growth promoting agents by chromatographic techniques. In *Chromatography Types, Techniques and Methods*; Quintin, T.J., Ed.; Nova Science Publishers, Inc.: New York, NY, USA, 2009; pp. 1–102.

26. Delatour, T.; Gremaud, E.; Mottier, P.; Richoz, J.; Vera, F.A.; Stadler, R.H. Preparation of stable isotope-labeled 2-nitrobenzaldehyde derivatives of four metabolites of nitrofuran antibiotics and their comprehensive characterization by UV, MS, and NMR techniques. *J. Agric. Food Chem.* **2003**, *51*, 6371–6379. [CrossRef] [PubMed]

27. Tarbin, J.A.; Potter, R.A.; Stolker, A.A.M.; Berendsen, B. Single-residue quantitative and confirmatory methods. In *Chemical Analysis of Antibiotic Residues in Food*; Wiley: Hoboken, NJ, USA, 2011; pp. 227–262.

28. Sriket, P.; Benjakul, S.; Visessanguan, W.; Kijroongrojana, K. Comparative Studies on Chemical Composition and Thermal Properties of Black Tiger Shrimp (*Penaeus monodon*) and White Shrimp (*Penaeus vannamei*) Meats. *Food Chem.* **2007**, *103*, 1199–1207. [CrossRef]

29. Danyi, S.; Widart, J.; Douny, C.; Dang, P.K.; Baiwir, D.; Wang, N.; Tu, H.T.; Tung, V.T.; Phuong, N.T.; Kestemont, P.; *et al.* Determination and Kinetics of Enrofloxacin and Ciprofloxacin in Tra Catfish (*Pangasianodon Hypophthalmus*) and Giant Freshwater Prawn (*Macrobrachium Rosenbergii*) Using a Liquid Chromatography/mass Spectrometry Method. *J. Vet. Pharmacol. Ther.* **2011**, *34*, 142–152. [CrossRef] [PubMed]

30. Pearce, J.N.; Burns, B.G.; van de Riet, J.M.; Casey, M.D.; Potter, R.A. Determination of Fluoroquinolones in Aquaculture Products by Ultra-Performance Liquid Chromatography-Tandem Mass Spectrometry (UPLC-MS/MS). *Food Addit. Contam. A Chem. Anal. Control Expo. Risk Assess.* **2009**, *26*, 39–46. [CrossRef] [PubMed]

31. Wan, G.H.; Cui, H.; Pan, Y.L.; Zheng, P.; Liu, L.J. Determination of Quinolones Residues in Prawn Using High-Performance Liquid Chromatography with Ce(IV)-Ru(bpy)$_3{}^{2+}$-HNO$_3$ Chemiluminescence Detection. *J. Chromatogr. B Anal. Technol. Biomed. Life Sci.* **2006**, *843*, 1–9. [CrossRef] [PubMed]

32. Poapolathep, A.; Jermnak, U.; Chareonsan, A.; Sakulthaew, C.; Klangkaew, N.; Sukasem, T.; Kumagai, S. Dispositions and Residue Depletion of Enrofloxacin and Its Metabolite Ciprofloxacin in Muscle Tissue of Giant Freshwater Prawns (*Macrobrachium Rosenbergii*). *J. Vet. Pharmacol. Ther.* **2009**, *32*, 229–234. [CrossRef] [PubMed]

33. Takeda, N.; Gotoh, M.; Matsuoka, T. Rapid Screening Method for Quinolone Residues in Livestock and Fishery Products Using Immobilised Metal Chelate Affinity Chromatographic Clean-up and Liquid Chromatography-Fluorescence Detection. *Food Addit.Contam. A Chem. Anal. Control Expo. Risk Assess.* **2011**, *28*, 1168–1174. [CrossRef] [PubMed]

34. Schröder, U.; MacHetzki, A. Determination of Flumequine, Nalidixic Acid and Oxolinic Acid in Shrimps by High-Performance Liquid Chromatography with Fluorescence Detection. *Eur. Food Res. Technol.* **2007**, *225*, 627–634. [CrossRef]

35. Karbiwnyk, C.M.; Carr, L.E.; Turnipseed, S.B.; Andersen, W.C.; Miller, K.E. Determination of Quinolone Residues in Shrimp Using Liquid Chromatography with Fluorescence Detection and Residue Confirmation by Mass Spectrometry. *Anal. Chim. Acta* **2007**, *596*, 257–263. [CrossRef] [PubMed]

36. Lohne, J.J.; Andersen, W.C.; Clark, S.B.; Turnipseed, S.B.; Madson, M.R. Laser Diode Thermal Desorption Mass Spectrometry for the Analysis of Quinolone Antibiotic Residues in Aquacultured Seafood. *Rapid Commun. Mass Spectrom.* **2012**, *26*, 2854–2864. [CrossRef] [PubMed]

37. Costi, E.M.; Sicilia, M.D.; Rubio, S. Supramolecular Solvents in Solid Sample Microextractions: Application to the Determination of Residues of Oxolinic Acid and Flumequine in Fish and Shellfish. *J. Chromatogr. A* **2010**, *1217*, 1447–1454. [CrossRef] [PubMed]

38. Wang, J.; Dai, J.; Meng, M.; Song, Z.; Pan, J.; Yan, Y.; Li, C. Surface Molecularly Imprinted Polymers Based on Yeast Prepared by Atom Transfer Radical Emulsion Polymerization for Selective Recognition of Ciprofloxacin from Aqueous Medium. *J. Appl. Polym. Sci.* **2014**. [CrossRef]

39. Andersen, W.C.; Roybal, J.E.; Gonzales, S.A.; Turnipseed, S.B.; Pfenning, A.P.; Kuck, L.R. Determination of Tetracycline Residues in Shrimp and Whole Milk Using Liquid Chromatography with Ultraviolet Detection and Residue Confirmation by Mass Spectrometry. *Anal. Chim. Acta* **2005**, *529*, 145–150. [CrossRef]

40. Treetepvijit, S.; Preechaworapun, A.; Praphairaksit, N.; Chuanuwatanakul, S.; Einaga, Y.; Chailapakul, O. Use of Nickel Implanted Boron-Doped Diamond Thin Film Electrode Coupled to HPLC System for the Determination of Tetracyclines. *Talanta* **2006**, *68*, 1329–1335. [CrossRef] [PubMed]

41. Venkatesh, P.; Kumar, N.A.; Prasad, R.H.; Krishnamoorthy, K.B.; Prasath, K.H.; Soumya, V. LC-MS/MS Analysis of Tetracycline Antibiotics in Prawns (*Penaeus monodon*) from South India Coastal Region. *JOPR J. Pharm. Res.* **2012**, *6*, 48–52. [CrossRef]

42. Liu, Y.; Yang, H.; Yang, S.; Hu, Q.; Cheng, H.; Liu, H.; Qiu, Y. High-Performance Liquid Chromatography Using Pressurized Liquid Extraction for the Determination of Seven Tetracyclines in Egg, Fish and Shrimp. *J. Chromatogr. B Anal. Technol. Biomed. Life Sci.* **2013**, *917–918*, 11–17. [CrossRef] [PubMed]

43. Zheng, L.; Wu, Y.; Li, Y.; Li, L. Determination of the Residues of 3-Methyl-Quinoxaline-2-Carboxylic Acid and Quinoxaline-2-Carboxylic Acid in Animal Origin Foods by High Performance Liquid Chromatography-Tandem Mass Spectrometry. *Chin. J. Chromatogr.* **2013**, *30*, 660–664. [CrossRef]

44. Schneider, M.J.; Vazquez-Moreno, L.; Bermudez-Almada, M.C.; Guardado, R.B.; Ortega-Nieblas, M. Multiresidue Determination of Fluoroquinolones in Shrimp by Liquid Chromatography-Fluorescence-Mass Spectrometry. *J. AOAC Int.* **2005**, *88*, 1160–1166. [PubMed]

45. Dufresne, G.; Fouquet, A.; Forsyth, D.; Tittlemier, S.A. Multiresidue Determination of Quinolone and Fluoroquinolone Antibiotics in Fish and Shrimp by Liquid Chromatography/tandem Mass Spectrometry. *J. AOAC Int.* **2007**, *90*, 604–612. [PubMed]

46. Chonan, T.; Fujimoto, T.; Inoue, M.; Tazawa, T.; Ogawa, H. Multiresidue Determination of Quinolones in Animal and Fishery Products by HPLC. *J. Food Hyg. Soc. Jpn.* **2008**, *49*, 244–248. [CrossRef]

47. Sun, Y.; Xu, X.; Zhu, X.; Mou, Z.; Ge, Y.; Wu, S.; Du, N. Optimization of Sample Preparation for the Determination of Fluoroquinolone Antibiotics Residues in Aquatic Products by UPLC-MS/MS. *J. Harbin Inst. Technol.* **2013**, *45*, 52–57.

48. Chang, C.S.; Wang, W.H.; Tsai, C.E. Simultaneous Determination of 18 Quinolone Residues in Marine and Livestock Products by Liquid Chromatography/tandem Mass Spectrometry. *J. Food Drug Anal.* **2010**, *18*, 87–97.

49. Chang, C.S.; Wang, W.H.; Tsai, C.E. Simultaneous Determination of Eleven Quinolones Antibacterial Residues in Marine Products and Animal Tissues by Liquid Chromatography with Fluorescence Detection. *J. Food Drug Anal.* **2008**, *16*, 87–96.

50. Zhao, S.; Jiang, H.; Li, X.; Mi, T.; Li, C.; Shen, J. Simultaneous Determination of Trace Levels of 10 Quinolones in Swine, Chicken, and Shrimp Muscle Tissues Using HPLC with Programmable Fluorescence Detection. *J. Agric. Food Chem.* **2007**, *55*, 3829–3834. [CrossRef] [PubMed]

51. Siqueira, S.R.R.; Donato, J.L.; de Nucci, G.; Reyes, F.G.R. A High-Throughput Method for Determining Chloramphenicol Residues in Poultry, Egg, Shrimp, Fish, Swine and Bovine Using LC-ESI-MS/MS. *J. Sep. Sci.* **2009**, *32*, 4012–4019. [CrossRef] [PubMed]

52. Douny, C.; Widart, J.; de Pauw, E.; Maghuin-Rogister, G.; Scippo, M.L. Determination of Chloramphenicol in Honey, Shrimp, and Poultry Meat with Liquid Chromatography-Mass Spectrometry: Validation of the Method According to Commission Decision 2002/657/EC. *Food Anal. Methods* **2013**, *6*, 1458–1465. [CrossRef]

53. Tyagi, A.; Vernekar, P.; Karunasagar, I.; Karunasagar, I. Determination of Chloramphenicol in Shrimp by Liquid Chromatography-Electrospray Ionization Tandem Mass Spectrometry (LC-ESI-MS-MS). *Food Addit. Contam. A Chem. Anal. Control Expo. Risk Assess.* **2008**, *25*, 432–437. [CrossRef] [PubMed]

54. Barreto, F.; Ribeiro, C.; Hoff, R.B.; Costa, T.D. Determination and Confirmation of Chloramphenicol in Honey, Fish and Prawns by Liquid Chromatography-tandem Mass Spectrometry with Minimum Sample Preparation: Validation according to 2002/657/EC Directive. *Food Addit.Contam. A* **2012**, *29*, 550–558. [CrossRef] [PubMed]

55. Polzer, J.; Hackenberg, R.; Stachel, C.; Gowik, P. Determination of Chloramphenicol Residues in Crustaceans: Preparation and Evaluation of a Proficiency Test in Germany. *Food Addit. Contam.* **2006**, *23*, 1132–1140. [CrossRef] [PubMed]

56. Ashwin, H.M.; Stead, S.L.; Taylor, J.C.; Startin, J.R.; Richmond, S.F.; Homer, V.; Bigwood, T.; Sharman, M. Development and Validation of Screening and Confirmatory Methods for the Detection of Chloramphenicol and Chloramphenicol Glucuronide Using SPR Biosensor and Liquid Chromatography-tandem Mass Spectrometry. *Anal. Chim. Acta* **2005**, *529*, 103–108. [CrossRef]

57. Shi, X.; Wu, A.; Zheng, S.; Li, R.; Zhang, D. Molecularly Imprinted Polymer Microspheres for Solid-Phase Extraction of Chloramphenicol Residues in Foods. *J. Chromatogr. B Anal. Technol. Biomed. Life Sci.* **2007**, *850*, 24–30. [CrossRef] [PubMed]

58. Tao, Y.; Zhu, F.; Chen, D.; Wei, H.; Pan, Y.; Wang, X.; Liu, Z.; Huang, L.; Wang, Y.; Yuan, Z. Evaluation of Matrix Solid-Phase Dispersion (MSPD) Extraction for Multi-Fenicols Determination in Shrimp and Fish by Liquid Chromatography-Electrospray Ionisation Tandem Mass Spectrometry. *Food Chem.* **2014**, *150*, 500–506. [CrossRef] [PubMed]

59. Liu, W.L.; Lee, R.J.; Lee, M.R. Supercritical Fluid Extraction *in Situ* Derivatization for Simultaneous Determination of Chloramphenicol, Florfenicol and Thiamphenicol in Shrimp. *Food Chem.* **2010**, *121*, 797–802. [CrossRef]

60. Stidl, R.; Cichna-Markl, M. Sample Clean-up by Sol-Gel Immunoaffinity Chromatography for Determination of Chloramphenicol in Shrimp. *J. Sol Gel Sci. Technol.* **2007**, *41*, 175–183. [CrossRef]

61. Samanidou, V.F.; Makrygianni, E.A. Ultrasound-Assisted Matrix Solid Phase Dispersion for the HPLC-DAD Analysis of Amphenicols in Shrimps. *Sample Prep.* **2015**, *2*, 66–73. [CrossRef]

62. Shi, X.; Meng, Y.; Liu, J.; Sun, A.; Li, D.; Yao, C.; Lu, Y.; Chen, J. Group-selective molecularly imprinted polymer solid phase extraction for the simultaneous determination of six sulfonamides in aquaculture products. *J.Chromatogr. B* **2011**, *879*, 1071–1076. [CrossRef] [PubMed]

63. Won, S.Y.; Lee, C.H.; Chang, H.S.; Kim, S.O.; Lee, S.H.; Kim, D.S. Monitoring of 14 sulfonamide antibiotic residues in marine products using HPLC-PDA and LC-MS/MS. *Food Control* **2011**, *22*, 1101–1107. [CrossRef]

64. Gehring, T.A.; Griffin, B.; Williams, R.; Geiseker, C.; Rushing, L.G.; Siitonen, P.H. Multiresidue determination of sulfonamides in edible catfish, shrimp and salmon tissues by high-performance liquid chromatography with postcolumn derivatization and fluorescence detection. *J. Chromatogr. B* **2006**, *840*, 132–138. [CrossRef] [PubMed]

65. Sangjarusvichai, H.; Dungchai, W.; Siangproh, W.; Chailapakul, O. Rapid separation and highly sensitive detection methodology for sulfonamides in shrimp using a monolithic column coupled with BDD amperometric detection. *Talanta* **2009**, *79*, 1036–1041. [CrossRef] [PubMed]

66. Done, H.Y.; Halden, R.U. Reconnaissance of 47 antibiotics and associated microbial risks in seafood sold in the United States. *J. Hazard. Mater.* **2015**, *282*, 10–17. [CrossRef] [PubMed]

67. Thammasoontaree, N.; Rattanarat, P.; Ruecha, N.; Siangproh, W.; Rodthongkum, N.; Chailapakul, O. Ultra-performance liquid chromatography coupled with graphene/polyaniline nanocomposite modified electrode for the determination of sulfonamide residues. *Talanta* **2014**, *123*, 115–121. [CrossRef] [PubMed]

68. Dickson, L.C. Performance characterization of a quantitative liquid chromatography-tandem mass spectrometric method for 12 macrolide and lincosamide antibiotics in salmon, shrimp and tilapia. *J. Chromatogr. B* **2014**, *967*, 203–210. [CrossRef] [PubMed]

69. Zhou, Y.; Zhou, T.; Jin, H.; Jing, T.; Song, B.; Zhou, Y.; Mei, S.; Lee, Y. Rapid and selective extraction of multiple macrolide antibiotics in foodstuff samples based on magnetic molecularly imprinted polymers. *Talanta* **2015**, *137*, 1–10. [CrossRef] [PubMed]

70. Alarcón-Flores, M.I.; Romero-González, R.; Vidal, J.L.M.; Frenich, A.G. Multiclass determination of phytochemicals in vegetables and fruits by ultra high performance liquid chromatography coupled to tandem mass spectrometry. *Food Chem.* **2013**, *141*, 1120–1129. [CrossRef] [PubMed]

71. Yang, X.; Zhou, R.; Zhao, Y.F.; Wu, Y.N. Analysis of Chloramphenicol in Shrimps by Gas Chromatography Tandem Mass Spectrometry. *Wei Sheng Yan Jiu* **2005**, *34*, 584–587. [PubMed]

72. Shi, X.; Song, S.; Sun, A.; Liu, J.; Li, D.; Chen, J. Characterisation and Application of Molecularly Imprinted Polymers for Group-Selective Recognition of Antibiotics in Food Samples. *Analyst* **2012**, *137*, 3381–3389. [CrossRef] [PubMed]

73. Ding, S.; Shen, J.; Zhang, S.; Jiang, H.; Sun, Z. Determination of Chloramphenicol Residue in Fish and Shrimp Tissues by Gas Chromatography with a Microcell Electron Capture Detector. *J. AOAC Int.* **2005**, *88*, 57–60. [PubMed]

74. Zhang, Q.J.; Peng, T.; Chen, D.D.; Xie, J.; Wang, X.; Wang, G.M.; Nie, C.M. Determination of Chloramphenicol Residues in Aquatic Products Using Immunoaffinity Column Cleanup and High Performance Liquid Chromatography with Ultraviolet Detection. *J. AOAC Int.* **2013**, *96*, 897–901. [CrossRef] [PubMed]

75. Tang, Y.Y.; Chung, Y.J.; Shih, Y.C.; Hwang, D.F. Determination of Chloramphenicol Residues in Foods by Liquid Chromatography-Electrospray Tandem Mass Spectrometry. *Taiwan. J. Agric. Chem. Food Sci.* **2010**, *48*, 239–247.

76. Mackie, J.; Marley, E.; Donnelly, C. Immunoaffinity Column Cleanup with LC/MS/MS for the Determination of Chloramphenicol in Honey and Prawns: Single-Laboratory Validation. *J. AOAC Int.* **2013**, *96*, 910–916. [CrossRef] [PubMed]

77. Zhang, S.; Sun, F.; Li, J.; Cheng, L.; Shen, J. Simultaneous Determination of Florfenicol and Florfenicol Amine in Fish, Shrimp, and Swine Muscle by Gas Chromatography with a Microcell Electron Capture Detector. *J. AOAC Int.* **2006**, *89*, 1437–1441. [PubMed]

78. Peng, T.; Li, S.; Chu, X.; Cai, Y.; Li, C. Simultaneous Determination of Residues of Chloramphenicol, Thiamphenicol and Florfenicol in Shrimp by High Performance Liquid Chromatography-Tandem Mass Spectrometry. *Fenxi Huaxue* **2005**, *33*, 463–466.

79. Tao, X.; Huang, H.; Liao, J.; Gao, P.; Huang, G.; Zeng, D. Simultaneous Determination of Residues of Chloramphenicol, Thiamphnicol, Florfenicol and Florfenicol Amine in Shrimp Muscle and Pork by High Performance Liquid Chromatography-Tandem Mass Spectrometry. *J. Chin. Inst. Food Sci. Technol.* **2014**, *14*, 232–238.

80. Fernando, R.; Munasinghe, D.M.S.; Gunasena, A.R.C.; Abeynayake, P. Determination of nitrofuran metabolites in shrimp muscle tissue by liquid chromatography-photo diode array detection. *Food Control* **2015**, in press. [CrossRef]

81. Douny, C.; Widart, J.; de Pauw, E.; Silvestre, F.; Kestemont, P.; Tu, H.T.; Phuong, N.T.; Maghuin-Rogister, G.; Scippo, M.L. Development of an analytical method to detect metabolites of nitrofurans: Application to the study of furazolidone elimination in Vietnamese black tiger shrimp (*Penaeus monodon*). *Aquaculture* **2013**, *376–379*, 54–58. [CrossRef]

82. Hossain, M.B.; Ahmed, S.; Rahman, M.F.; Kamaruzzam, B.Y.; Jalal, K.C.A.; Amin, S.M.N. Method Development and Validation of Nitrofuran Metabolites in Shrimp by Liquid Chromatographic Mass Spectrometric System. *J. Biol Sci.* **2013**, *13*, 33–37.

83. Storey, J.M.; Clark, S.B.; Johnson, A.S.; Andersen, W.C.; Turnipseed, S.B.; Lohne, J.J.; Burger, R.J.; Ayres, P.R.; Carr, J.R.; Madson, M.R. Analysis of Sulfonamides, Trimethoprim, Fluoroquinolones, Quinolones, Triphenylmethane Dyes and Methyltestosterone in Fish and Shrimp Using Liquid Chromatography-Mass Spectrometry. *J. Chromatogr. B* **2014**, *972*, 38–47. [CrossRef] [PubMed]

84. Villar-Pulido, M.; Gilbert-López, B.; García-Reyes, J.F.; Martos, N.R.; Molina-Díaz, A. Multiclass Detection and Quantitation of Antibiotics and Veterinary Drugs in Shrimps by Fast Liquid Chromatography Time-of-Flight Mass Spectrometry. *Talanta* **2011**, *85*, 1419–1427. [CrossRef] [PubMed]

85. Nakajima, T.; Nagano, C.; Kanda, M.; Hayashi, H.; Hashimoto, T.; Kanai, S.; Matsushima, Y.; Tateishi, Y.; Sasamoto, T.; Takano, I. Single-Laboratory Validation Study of Rapid Analysis Method for Multi-Class Veterinary Drugs in Milk, Fish and Shellfish by LC-MS/MS. *ShokuhinEiseigakuZasshi* **2013**, *54*, 335–344. [CrossRef]

86. Wang, N.; Su, M.; Liang, S.; Sun, H. Sensitive Residue Analysis of Quinolones and Sulfonamides in Aquatic Product by Capillary Zone Electrophoresis Using Large-Volume Sample Stacking with Polarity Switching Combined with Accelerated Solvent Extraction. *Food Anal. Methods* **2015**, *9*, 1020–1028. [CrossRef]

87. Li, H.; Kijak, P.J.; Turnipseed, S.B.; Cui, W. Analysis of veterinary drug residues in shrimp: A multi-class method by liquid chromatography-quadrupole ion trap mass spectrometry. *J. Chromatogr. B* **2006**, *836*, 22–38. [CrossRef] [PubMed]

88. Li, H.; Kijak, P.J. Development of a Quantitative Multiclass/multiresidue Method for 21 Veterinary Drugs in Shrimp. *J. AOAC Int.* **2011**, *94*, 394–406. [PubMed]

89. Kanda, M.; Nakajima, T.; Hayashi, H.; Hashimoto, T.; Kanai, S.; Nagano, C.; Matsushima, Y.; Tateishi, Y.; Yoshikawa, S.; Tsuruoka, Y.; *et al*. Multi-Residue Determination of Polar Veterinary Drugs in Livestock and Fishery Products by Liquid Chromatography/Tandem Mass Spectrometry. *J. AOAC Int.* **2015**, *98*, 230–247. [CrossRef] [PubMed]

90. An, H.; Parrales, L.; Wang, K.; Cain, T.; Hollins, R.; Forrest, D.; Liao, B.; Paek, H.C.; Sram, J. Quantitative Analysis of Nitrofuran Metabolites and Chloramphenicol in Shrimp Using Acetonitrile Extraction and Liquid Chromatograph-Tandem Mass Spectrometric Detection: A Single Laboratory Validation. *J. AOAC Int.* **2015**, *98*, 602–608. [PubMed]

91. Wang, Z.; Leng, K.; Sun, W.; Ning, J.; Zhai, Y. Simultaneous Determination of 33 Quinolone and Sulfonamide Residues in Eels and Shrimps by High Performance Liquid Chromatography-Tandem Mass Spectrometry. *Chin. J. Chromatogr.* **2009**, *27*, 138–143.

92. El-Demerdash, A.; Song, F.; Reel, R.K.; Hillegas, J.; Smith, R.E. Simultaneous Determination of Nitrofuran Metabolites and Chloramphenicol in Shrimp with a Single Extraction and LC-MS/MS Analysis. *J. AOAC Int.* **2015**, *98*, 595–601. [PubMed]

![separations logo] *separations*

MDPI

*Review*

# Trends in Microextraction-Based Methods for the Determination of Sulfonamides in Milk

**Maria Kechagia and Victoria Samanidou \***

Laboratory of Analytical Chemistry, University of Thessaloniki, Thessaloniki GR 54124, Greece;
marikech92@yahoo.gr
\* Correspondence: samanidu@chem.auth.gr; Tel.: +30-231-099-7698

Academic Editor: Frank L. Dorman
Received: 1 May 2017; Accepted: 12 June 2017; Published: 23 June 2017

**Abstract:** Sulfonamides (SAs) represent a significant category of pharmaceutical compounds due to their effective antimicrobial characteristics. SAs were the first antibiotics to be used in clinical medicine to treat a majority of diseases, since the 1900s. In the dairy farming industry, sulfa drugs are administered to prevent infection, in several countries. This increases the possibility that residual drugs could pass through milk consumption even at low levels. These traces of SAs will be detected and quantified in milk. Therefore, microextraction techniques must be developed to quantify antibiotic residues, taking into consideration the terms of Green Analytical Chemistry as well.

**Keywords:** sulfonamides; determination; extraction; microextraction; milk; green analytical chemistry

---

## 1. Introduction

The term "sulfonamide" derives from para-amino-benzene-sulfonamide (sulfanilamide) and it is also known as streptocid. The structure is similar to para-aminobenzoic acid (PABA), which is demanded by microorganisms (such as bacteria) for dihydrofolic and folic acid synthesis. SAs (SAs) or alternative sulfa drugs have been known since the middle of the twentieth century and were proved to have antibacterial properties in 1935, thus they are some of the oldest antimicrobial drugs. Nowadays, SAs are widely used as antibiotics in veterinary medicine either in single formulation or synergistically with other antibiotics (such as tetracyclines (TCs), SAs, quinolones (Qs), fluoroquinolones (FQs) and trimethoprim (TMP). Their use aims to protect the animals from infectious diseases. In addition, their administration as additives to animal feed can promote growth that results in the rise of the productivity of livestock, despite the fact that the use of antibiotics is prohibited in various places around the world. For example, the concentration of SAs in meat produced in Denmark was on average 4.82 mg/kg (pork), 17.2 mg/kg (cattle), 0.033 mg/kg (broilers) and 58.5 mg/kg (fish). Human health can be influenced by the consumption of meat and dairy products (such as eggs, milk and cheese) which contain amounts of SAs. The use of SAs as additives is forbidden, aiming to ensure safety in human health. Furthermore, SAs are widely used as antibacterial drugs because they are of low cost and active against a broad spectrum of microorganisms. SAs can act, despite their bacteriostatic and chemotherapeutic activity, against infections caused by gram-positive and gram-negative bacteria and protozoa. As it is observed, every human is a passive consumer of sulfa-drugs, which are obtained from the treatment of diseases in animals. The systematic and long-term intake of SAs through the food could be characterized as dangerous and in some cases toxic, appearing in the form of allergic reactions, suppression of enzyme activity and alteration of the intestinal microflora [1,2].

It should be noted that only 40 of the 10,000 sulfanilamide derivatives that have been synthesized are used in medical and veterinary practice. SAs have similarities in their structure, as all of them carry the same molecule with the addition of diverse radicals at the R position [2].

*Separations* **2017**, *4*, 23

170

The structures of some of the most commonly used SAs that are mentioned in this review are presented in Figure 1.

Instructions for the withdrawal period have to be followed in order not be found as residues in milk, eggs, meat, tissues and other livestock products after drug administration [1,3].

On the other hand, milk is one of the most widely consumed foods and it is a rich source of protein and calcium, especially in children's growth. With regards to its analysis, milk is considered as a complex matrix, which contains water, proteins, lactose, fats, minerals and vitamins. According to the European Union (EU), the combined total residues of all substances within the sulfonamide group should not exceed 100 μg/kg [4]. This is the maximum residue level (MRL) which should be in force for all target tissues (muscle, fat, liver and kidney) and for milk coming from bovine, ovine and caprine. The same MRLs were established in the USA and Canada [3]. Also, the regulation of *Codex Alimentarius* for the MRLs of the above drugs is referred to sulfadimidine and is set as 100 μg/kg for all target tissues, except for milk that is set to 25 μg/L [5].

Several methods have been developed and validated for the determination of SAs in different matrices such as environmental, biological or food by applying various techniques including photometry, spectrophotometry, gas chromatography (GC), gas chromatography tandem mass spectrometry (GC-MS/MS), capillary electrophoresis-ultraviolet detection (CE-UV), and high performance liquid chromatography (HPLC) coupled with ultraviolet (UV) detection, fluorescence detection (FD) and mass spectrometry (MS) [6]. All of the developed methods require various sample preparation procedures. Sample preparation is the most demanding step of the analysis in comparison with the other two steps, such as sampling or measurement. At this critical step, the analytes need to be successfully isolated and the sample should reach a capable form for analysis. Sample preparation is crucial to produce accurate results. Thus, this step requires special attention, and it is a time-consuming procedure as well. There are many well-established sample preparation techniques and this field is very interesting for researchers. Therefore, novel techniques are introduced and used in different matrices. In recent years, the trend for every novel technique is to be environmentally friendly according to the principles of Green Analytical Chemistry (GAC) [7].

During recent decades, several extraction techniques have been used for sample pre-treatment. For the analysis of solid samples, the most applied techniques are Soxhlet (SOX) and pressurized solvent extraction techniques (e.g., supercritical fluid extraction (SFE), accelerated solvent extraction (ASE) and subcritical water extraction (SWE)), and the well-known liquid–liquid extraction (LLE) for the analysis of liquid samples. However most of the conventional techniques (SOX, SPE and LLE) seem to have significant drawbacks. They are time-consuming and complicated, consume large amounts of sample and organic solvents and are difficult to be automated.

In 1990, a novel technique, known as solid phase microextraction (SPME), was introduced by Pawliszyn and co-workers. SPME uses a fused silica fiber, which is coated with a sorbent (the fiber is incorporated in a chromatographic syringe) to extract the target analytes which subsequently are directly transferred into GC or HPLC. The technique has significant advantages. It is fast, simple, solvent free and it is compatible with analyte separation and detection by a chromatographic system (directly in gas chromatography, or via an interface in high-performance liquid chromatography) [8]. Due to the plethora of advantages, the SPME was extensively used for sample preparation of environmental and food samples. The environmental samples include water, air, soil and sediments, whereas food applications are based on fruits, vegetables, fats, oils, wine, meat and dairy products [9].

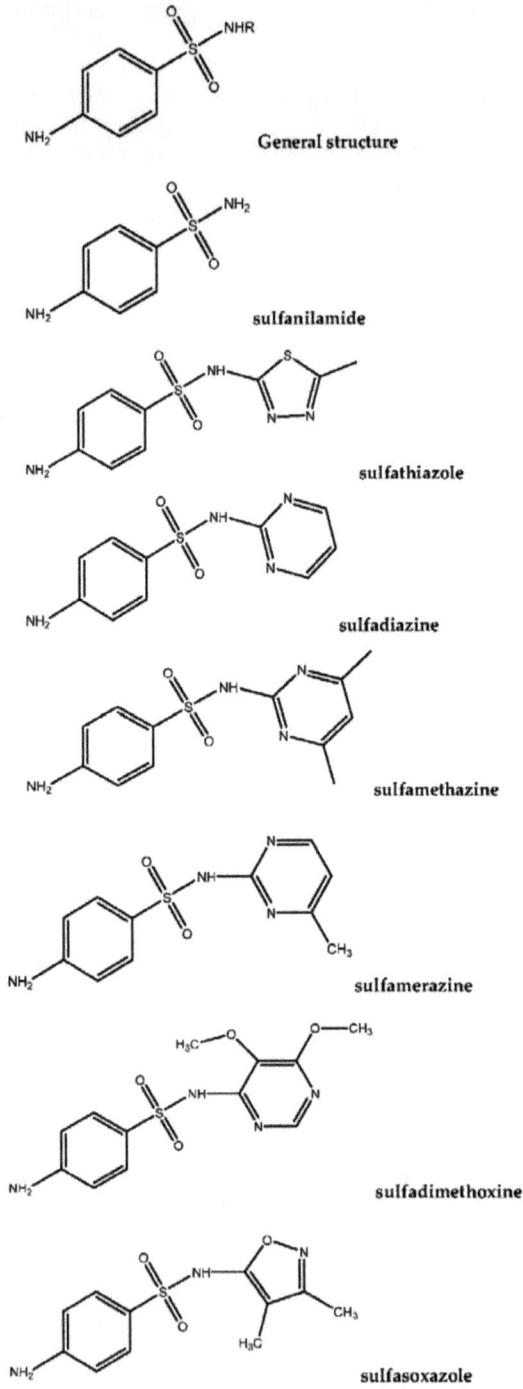

**Figure 1.** Chemical structures of sulfonamides (SAs).

However, the use of SPME fibers involves some drawbacks such as:

- their maximum operating temperatures are in the range between 240–280 °C
- they are not stable with the organic solvents due to swelling
- they break easily
- the possibility of stripping of coatings due to analyst's handling errors.

A modification of SPME was introduced in 1999 by Baltussen. This novel technique is called stir bar sorptive extraction (SBSE) and uses a stir bar consisting of a magnet covered with glass which in turn is coated by a layer (typically 0.5–1 mm) of sorptive material (usually polydimethylsiloxane-PDMS) for the extraction. Furthermore, microextraction by packed sorbent (MEPS) was developed and introduced as a further miniaturization version of SPE. In MEPS, SPE's conventional polymeric cartridge was replaced by a stainless steel, miniaturized version termed the barrel insert and needle (BIN), which could hold any of a great number of sorbents, such as those used in SPE [7,10].

In the meantime, liquid phase microextraction (LPME) was introduced in order to overcome significant drawbacks of liquid phase extraction modes. In LPME, the amount of solvents is smaller in comparison with LLE; only some µL are required, whereas LLE consumes hundreds of mL. It is simple and cheap, as well as adaptable with capillary electrophoresis (CE), HPLC and GC.

Three modes of LPME can be applied. These include single drop microextraction (SDME), hollow fiber liquid-phase microextraction (HF-LPME) and dispersive liquid–liquid microextraction (DLLME). Extraction occurs into a small amount (usually 1–100 µL) of organic solvent (acceptor phase) from an aqueous matrix containing the analytes (donor phase) [11].

The essential feature of the above extraction techniques is the elimination of large amounts of organic solvents due to the fact that organic solvents are toxic and hazardous for the environment and human health. This complies with the principles of GAC, following the trend of using solvent-less or better described as solvent-free extraction methods.

The introduction of new sorbent materials in sample preparation is also of significant importance and has been widely investigated in order to prepare materials with higher adsorption capacity and selectivity, as well as to expand the availability of cheaper, more easily synthesized sorbents. The combination of microextraction techniques with the new sorbent materials is also base to the GAC demands [12,13].

The purpose of this review was to focus on the most recently developed microextraction techniques for the determination of SAs in milk.

## 2. The Demand of Microextraction Techniques

As already mentioned, sample preparation is the most demanding step in any analytical workflow. The main purpose of sample preparation is to transfer the analytes of interest from a complex matrix to a compatible medium for further determination by an analytical technique. In addition, sample preparation often includes procedures such as clean up, analyte enrichment and derivatization. So, it is clear that this step is time consuming and usually requires the use of large organic solvent volumes and the waste of reagents and consumables. For these reasons, the trend is the introduction of more "green" and micro-techniques in sample preparation.

The idea of sustainable ecological development was introduced in 1987 in a report of the World Commission on Environment and Development. The term green chemistry was mentioned by P. Anastas in 1991 at the US Environmental Protection Agency (EPA). As a result, in 1993, the comprehensive US Green Chemistry Program was established, which included cooperation among many governmental agencies and research institutions. While Anastas and co-workers were elaborating the ideas of green chemistry, the first paradigms of green analytical chemistry were introduced. GAC, introduced in 1999, became a whole part of chemical nomenclature, and numerous reviews and original studies have been published in this topic. The principles of green chemistry and by extension of GAC are presented in Table 1 [14].

Table 1. The twelve principles of Green Analytical Chemistry (GAC).

| No. | Principle of GAC |
|---|---|
| 1 | Direct analytical approaches should be preferred in order to avoid sample preparation |
| 2 | Minimum sample size and minimum number of samples are the main goals of this principle |
| 3 | In situ measurements are considered necessary |
| 4 | Energy saving and reagent reduction is achieved by process integration |
| 5 | Automated and miniaturized methods should be developed and applied |
| 6 | Derivatization is not preferable |
| 7 | Production of a large volume of analytical waste should be avoided and proper management of analytical waste should be provided |
| 8 | Multi-analyte or multi-parameter methods are preferred versus methods using single-analyte |
| 9 | The use of energy should be reduced |
| 10 | Reagents ensured from renewable sources should be preferred |
| 11 | Toxic reagents should be minimized or replaced |
| 12 | The operator should work under safe conditions |

Concerning green analytical methods, the goals to be achieved include:

- elimination or reduction of the use of chemical substances

  (such as solvents, reagents, preservatives, additives for pH adjustment)
- elimination of energy consumption
- proper management of analytical waste
- increased safety for the operator

The demands of GAC are automatization, no derivatization, and no sample treatment in the step of sample preparation. The latter is not possible in most cases, so sample preparation by a microextraction technique is the next best choice. Microextraction techniques arose as a development of conventional extraction techniques. The term microextraction means that all modes of these techniques require small volumes of extraction, which becomes under described conditions [13,14].

With a quick review of the literature, it is obvious that various methods have been developed for the determination of the SAs in several food matrices. The most often applied techniques for the determination of SAs in milk are HPLC coupled with different detectors, such as ultraviolet [15], diode-array [16], or mass spectroscopy [17].

A significant number of contributions can be found in literature with regards to sample preparation of milk for the determination of SAs. These include either traditional techniques or modern ones.

Solid phase extraction is widely used either in the classical approach or in an alternative way, based on the use of modern adsorbent materials. The use of the commercial SPE presents some disadvantages:

1. Although there is a wide range of chemistries, many choices for manipulating solvent and pH conditions, optimization is time consuming. In many cases, several steps are required.
2. The cost per sample is higher than that of simple liquid–liquid extraction (LLE).

Novel microexraction techniques were introduced to overcome these drawbacks. The new techniques require less time and labor than the multi-step procedures of SPE. These include SPME, SBSE, magnetic solid phase microextraction (MSPE), and other greener approaches. For example, a new in-tube solid-phase microextraction technique was introduced by Wen Y. et al. in 2005. The aim for this sample preparation technique is the determination of five SA residues in milk with HPLC-UV. The on-line in-tube SPME used poly-(methacrylic acid-ethylene glycol dimethacrylate) monolithic capillary as extraction media. The method is easily applied and environmentally friendly, following the demands of GAC [18].

## 3. Microextraction Techniques for the Determination of SAs in Milk

Some green microextraction techniques that have been developed and applied to determine SAs in milk will be presented in the next few paragraphs. The reported techniques are summarized in Table 2.

**Table 2.** Microextraction techniques for the determination of SAs in milk.

| Analyte | Extraction Type | Determination | LOD | Recovery | Reference |
|---|---|---|---|---|---|
| sulfachloropyridazine, sulfadiazine, sulfadimethoxine, sulfamethazine, sulfamethoxazole, sulfamethoxypyridazine, sulfaquinoxaline, sulfathiazole, sulfisoxazole | MSPE | HPLC-DAD | 7–14 µg/L | 81.88%–114.9% | [16] |
| sulfadiazine, sulfamerazine, sulfamethazine sulfamethizole, sulfamethoxazole, sulfadimethoxine | SBSE | HPLC-MS/MS | 0.9–10.5 µg/L | 68%–115% | [17] |
| sulfamerazine, sulfamethizole, sulfadoxine, sulfamethoxazole, sulfisoxazole | MSPE | HPLC-UV | 1.16–1.59 µg/L | 62.0%–104.3% | [19] |
| sulfadiazine, sulfamethazine, sulfamonomethoxine, sulfamethoxazole, sulfaquinoxaline | SBSE | HPLC-DAD | 4.29–26.3 µg/L | 54.8%–126% | [20] |
| sulfapyridine, sulfadiazine, sulfachloropyridazine, sulfadoxin, sulfamethoxazole, sulfadimethoxin, sulfamethizol, sulfameter, sulfamethazine | DLLME | HPLC-FL | 0.60–1.21 µg/L | 90.8%–104.7% | [21] |
| (same analytes with the above) | QuEChERS | HPLC-FL | 1.15–2.73 µg/L | 83.6%–104.8% | [21] |
| sulfamethazine | MIP-silica column | HPLC-UV | 7.9 µg/L | 79.3%–87.4% | [22] |
| sulfamethazine, sulfamethoxazole, sulfadizaine, sulfaquinoxaline, sulfametoxydiazine, sulfadimethoxine, sulfamethizole | RA-MIP | HPLC-UV | 0.8–2.7 µg/L | 93%–107% | [23] |
| sulfamethazine, sulfisoxazole, sulfadimethoxine | FPSE | HPLC-UV | - | 22.98%–49.5% | [24] |
| sulfadimidine, sulfachloropyridazine, sulfamonomethoxine, sulfachloropyrazine | M-G-PTE | LC-UV | 0.004–0.012 µg/L | 90.1%–113.5% | [25] |
| sulfamethazine , sulfamethoxypyridazine, sulfamethoxydiazine, sulfamethoxazole, sulfadimethoxine, sulfaphenazole | IL-based MADLLME | HPLC-FL | 0.018–0.031 µg/L | 97.3%–107.9% | [26] |

### 3.1. Magnetic Solid Phase Extraction

MSPE uses magnetic particles (MPs) as absorbents and in recent years has attracted the interest of analytical scientists. Magnetic separation was introduced by Robinson and co-workers in 1993, although the term MSPE was introduced for the first time in analytical chemistry by Safarikova and Safarik in 1999. The MSPE includes immersion of the MPs to the sample solution. The magnetic core of these MPs is coated with silica or alumina oxides according to the sol–gel technique. After adsorption of analytes on the surface of the MPs, separation of the latter from aqueous solution is achieved by applying an external magnetic field. Target analytes are subsequently desorbed by a suitable eluent and determined by a suitable analytical technique. The procedure of the MSPE technique is illustrated in Figure 2.

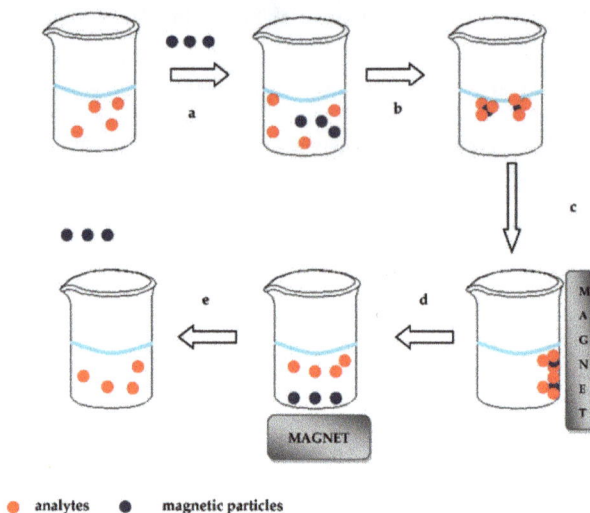

**Figure 2.** Magnetic solid phase extraction (MSPE) procedure. (**a**) addition of magnetic; (**b**) adsorption; (**c**) magnetic separation; (**d**) solvent exchange-elution; and (**e**) separation.

In comparison to the commercial SPE, the application of MSPE simplifies the sample preparation procedure because no pre-packed columns are used. Moreover, separation can be faster due to fewer necessary steps [10,16].

Micromagnetic particles, nanomagnetic particles and magnetic molecular imprinted polymers are used as absorbents in MSPE. The application of magnetic solid phase extraction reduces the total time of analysis and provides simultaneous isolation and enrichment of the analytes. In addition, MSPE reduces the use of organic solvents and thus the accumulation of toxic and dangerous wastes, which is acceptable by the principles of GAC [10].

In 2010, Gao Q. et al. [27] reported a MSPE technique which achieved the extraction of eleven SAs from milk. A magnetite/silica/poly (methacrylic acid-coethylene glycol dimethacrylate) ($Fe_3O_4$/$SiO_2$/P(MAA-*co*-EGDMA)) sorbent material was synthesized and characterized by elemental analysis, electron microscopy, X-Ray diffraction and Fourier-transformed infrared spectroscopy. A quantity of 50 mg of the sorbent material was inserted into a vial, and the magnetic particles were preconditioning with methanol and water. Then, the sample was added to the vial, and vortexed for 30 s. Subsequently, the sorbent with adsorbed SAs was separated rapidly from the solution under a strong external magnetic field. After the removal of supernatant solution, SAs were eluted from the magnetic composite by 1 mL of acetone containing 5% ammonium hydroxide solution (*v*/*v*) with the assistance of vortex mixing for 30 s. An external magnet was used for the separation of magnetic composite from the solution. The sample solution was injected into a LC-MS/MS system for further analysis [27].

A method applying MSPE of SAs from milk samples was developed by Li Y. and his team in 2015, using a graphene-based magnetic nanocomposite ($CoFe_2O_4$-graphene, $CoFe_2O_4$-G). The target analytes included sulfamerazine, sulfamethizole, sulfadoxine, sulfamethoxazole and sulfisoxazole. After the extraction of the absorbent, the analytes were determined by HPLC. In order to achieve the optimal extraction efficiency, several parameters were investigated, such as sample pH, extraction time, the amount of the magnetic material ($CoFe_2O_4$-G) and elution solvent. As it is mentioned in this work, pH is an important parameter in the MSPE technique because it could affect the speciation of SAs. Therefore, after an optimization study in the range of pH values from 2.0 to 12.0, it was found that higher extraction efficiency was achieved at pH 4.0. The optimum time of the extraction was selected at 20 min. A quantity of 15.0 mg of the $CoFe_2O_4$-G was selected as optimal. The key parameter of a successful extraction with high recovery was the elution solvent. Several solvents were investigated and finally a volume of 0.5 mL of MeOH solution containing 5% ($v/v$) acetic acid was selected. Among the advantages of this research is the fact that the $CoFe_2O_4$-G nanocomposite could be re-used after washing with acetone and ultrapure water successively. Moreover, the MSPE procedure presents easy operation, sensitivity and efficiency [19].

Ibarra I. and co-workers in 2014 published a method for the simultaneous determination of nine SAs in milk samples with HPLC, according to the EU MRLs, using a magnetic solid phase extraction consisting of a silica-based magnetic absorbent ($Fe_3O_4$-$SiO_2$-phenyl modified). The advantages were that the developed method was faster than traditional preparation techniques such as SPE and also demanded minimum sample pretreatment and reduced volumes of organic solvents, thus being cost effective. Moreover, the analytical results provide sensitivity and accuracy [16].

## 3.2. Stir Bar Sorptive Extraction

SBSE is another novel technique which fulfills the requirements of GAC, and was introduced by Baltussen and co-workers in 1999. This technique uses a polymer coated bar in which the sorptive material is usually the polydimethylsiloxane (PDMS), as shown in Figure 3. The bar is inserted in the vial containing the sample, and equilibrium of analyte concentration between sorbent and sample matrix is reached by stirring. When the extraction is completed, the bar is removed and transferred to a clean vial where the target compounds are desorbed thermally or in liquid mode. Analysis can be subsequently performed by liquid or gas chromatography using various detection techniques such as ultraviolet, mass spectrometry or fluorescence, etc. [10,28,29].

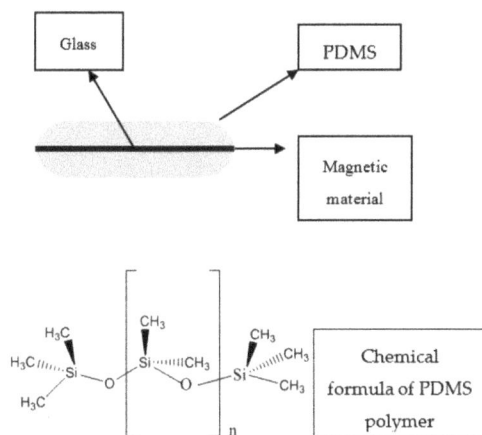

**Figure 3.** Analytical device used for stir bar sorptive extraction: classic case of polydimethylsiloxane (PDMS) as sorptive material.

SBSE takes place in two distinctive procedures: sorption/extraction and desorption. During sorption, the polymer-coated stir bar comes in contact with analytes, either by immersion or in the headspace (HS). HS mode is more preferable for volatile compounds. After the extraction step, the stir bar is removed, rinsed with distilled water, wiped with a paper tissue and submitted to desorption procedure. This can be accomplished either thermally (thermal desorption (TD)) or by means of a suitable solvent system (liquid desorption (LD)). TD is typically followed by Gas Chromatographic analysis, while LD is followed by HPLC, CE or GC. TD is usually used for thermally stable volatile or semi-volatile compounds, whereas LD is preferred for semi-volatile, non-volatile and thermo-labile compounds.

Besides PDMS, which is the most commonly-used SBSE sorbent, other materials that are either commercially available or developed in the lab can be used. Some of them include:

- polyurethane foams
- silicone materials
- poly(ethylene glycol)-modified silicone
- poly(dimethylsiloxane)/polypyrrole
- poly(phthalazine ether sulfone ketone)
- polyvinyl alcohol
- polyacrylate
- carbon nanotube-poly(dimethylsiloxane) (CNT-PDMS)
- alkyl-diol-silica (ADS) restricted access materials
- molecularly imprinted polymers (MIPs)
- sorbents obtained with sol–gel techniques
- monolithic materials, or
- cyclodextrin.

SBSE is a promising sample preparation technique, showing reliability in terms of enrichment capacity, outstanding performance, great sensitivity and selectivity for ultra-trace extraction of non-polar to medium-polar organic compounds from complex matrices with a great variety of applications [28,29].

The above-mentioned technique has been already used for the determination of SAs in milk samples.

A method of $C_{18}$-SBSE HPLC-MS/MS was proposed for the determination of six SAs (sulfadiazine, sulfamerazine, sulfamethazine, sulfamethizole, sulfamethoxazole and sulfadimethoxine) in milk and milk powder samples by Yu C. and Hu B. Adhesion method was used for the preparation of $C_{18}$ silica coated stir bar, using two types of adhesive glue, PDMS sol and epoxy glue. The results showed that the $C_{18}$-coated stir bar prepared by PDMS sol as adhesive glue is more robust than the one prepared by epoxy glue when liquid desorption was used, with regards to lifetime and organic solvent compatibility. Several parameters, such as extraction and desorption time, ionic strength, sample pH and stirring speed were investigated in order to achieve the optimized performance. Optimum parameters lead to a sensitive, accurate, rapid, and inexpensive method, which can be used as an alternative for trace SAs analysis in milk and milk powder samples [17].

Huang X. and co-workers suggested a simple, rapid, and sensitive method for the determination of five SAs in milk. The proposed method was applied by SBSE coupling to HPLC with diode array detection. Stir bar used poly (vinylimidazole-divinylbenzene) monolithic (VIDB) as a polymer-coated material. To achieve optimum performance with the application of SBSE-VIDV, several parameters, including extraction and desorption time, desorption solvent, ionic strength and pH value of sample matrix, were investigated. The preparation of milk samples was simple and the manufactured SBSE-VIDB showed higher selectivity to SAs than commercial SBSE-PDMS [20].

### 3.3. QuEChERS Approach

A popular procedure, recently developed, called QuEChERS (Quick, Easy, Cheap, Effective, Rugged and Safe), has been widely applied. The advantages of QuEChERS include simplicity, with short steps, and effective cleaning for complex samples, by avoiding a great number of steps. The methodology includes two basic steps: (1) extraction based on partitioning via salting-out extraction involving the equilibrium between an aqueous and an organic layer; and (2) dispersive SPE that involves further clean-up using combinations of $MgSO_4$ and different sorbents, such as $C_{18}$ or primary and secondary amine (PSA), to remove interference [30].

Qin Y. et al. have developed a liquid chromatographic method tandem mass spectrometry for the determination of SAs, Tilmicosin and Avermectins residues in animal matrices including milk and using QuEChERS as the sample preparation procedure. In comparison with traditional techniques such as LLE, this approach is less time-consuming and environmentally friendly, as the use of solvents is limited [31].

In addition, a quick method was developed for the simultaneous screening and quantification of 90 veterinary drugs in milk (including SAs) by Zhang Y. and co-workers in 2015. The determination was achieved by ultra-performance liquid chromatography coupled to quadrupole time-of-flight mass spectrometry (UPLC-QTOF-MS). A modified QuEChERS method was applied and proved to be fast and easy with good repeatability. The proposed method is accurate and gives reliable results for drug concentrations below or above the MRL [32].

### 3.4. Dispersive Liquid–Liquid Microextraction

Assadi et al. was the first to mention DLLME as an alternative LLE technique that uses 1 µL volume of extraction solvent, along with a few mL of dispersive solvents. In this technique, a cloudy solution is formed when an appropriate mixture of extraction and dispersive solvents is injected into an aqueous sample. The extraction solvent is dispersed into the bulk aqueous solution. Centrifugation is applied prior to the determination of the analytes in the lower phase (in the bottom of a conical tube) by conventional analytical techniques. The main advantages of DLLME are simplicity of operation, quickness, low cost, great recovery, high enrichment factor and very short extraction time (usually a few seconds) [11].

An ultrasound-assisted ionic liquid/ionic liquid-dispersive liquid–liquid microextraction (UA-IL/IL-DLLME) HPLC method was developed by Gao S. et al. [33] The method was proposed for the extraction, separation and determination of six SAs in infant formula milk powder samples. Hydrophobic ionic liquids were used for extraction, while hydrophilic ionic liquids were used as dispersive solvents. The formation of cloudy solutions, which was achieved by fine drops of $[C_6MIM][PF_6]$, facilitated the extraction procedure. The above method was applied to the analysis of infant formula milk powder samples and the recoveries of the analytes ranged from 90.4% to 114.8% [33].

DLLME and QuEChERS were applied for the determination of nine SAs in milk. The quantification was achieved by HPLC with fluorescence detection. This study was proposed by Arroyo-Manzanares N. et al. in 2014. [21] The DLLME and QuEChERS have been proposed as extraction modes due to the fact that they are environmentally friendly, use a reduced amount of organic solvents and are in agreement with the new trends of GAC. For the optimization, several parameters such as extraction and disperser solvent (for DLLME) and value of pH were investigated. The developed methods were accurate and were successfully applied to the extraction and determination of SAs in milk samples [21].

### 3.5. The Molecularly Imprinted Polymers Approach

MIPs play an important role in the field of extraction. There are numerous applications using MIPs as absorbents for a wide variety of substances in complex matrices applying different extraction or microextraction techniques. Among these, there are some applications with regards to the extraction and determination of SAs in milk.

MIPs are artificial materials synthesized by polymerization of functional and cross-linking monomers in the presence of the target molecule (template). According to this procedure, MIPs transfer the molecular imprint of the template and could recognize it, with high selectivity among other molecules which are similar to the template. Due to the fact that they present high selectivity and other advantages such as easy preparation, chemical and thermal stability and low manufacturing cost, MIPs have been suggested and used in many applications [10].

Su S. and Zhang M. developed a novel method for the determination of sulfamethazine (SMZ) in milk. A new polymer material was synthesized, and the MIP layer was grafted on the silica surface. For its preparation, the authors used sulfamethazine (SMZ) as the template, methacrylic acid (MAA) as the functional monomer and ethylene dimethacrylate (EDMA) as cross-linker. As it is mentioned, during synthesis, the initiator-transfer in combination with the agent-terminator (called iniferter) were immobilized on a silica surface using chemical reagents, which shows good availability. The imprinting polymerization was initialized by the silica-supported iniferter under the UV radiation. Two chromatographic columns were investigated, which included a MIP-silica (3h-MIP-Sil) and its non-imprinted polymer-grafted silica (NIP-silica) as the stationary phases. The results show that the MIP-silica column has better selectivity to the template molecule in comparison with the NIP-silica phase. Furthermore, with the use of this stationary phase, better column efficiency and low backpressure were observed. Finally, the constructed SMZ-MIP-silica was applied for the determination of SMZ in milk samples [22].

A novel restricted access-molecularly imprinted material (RA-MIP) with selectivity for SAs was synthesized using the initiator-transfer agent-terminator method, a "living"/controlled radical polymerization technique by Xu W. et al. Two layers with different functions on the silica support, grafted, were used for the preparation of the material. To perform a "grafting from" polymerization, iniferter was immobilized on the surface of silica, followed by grafting of the internal sulfamethazine imprinted polymer and the external poly(glycidyl methacrylate) [poly(GMA)]. The hydrophilic structures were formed on the external layer of the material by the hydrolysis of the linear poly(GMA) for protein removal. The RA-MIP-SG was used as a pre-column for the determination of SAs in milk. The manufactured material has the properties of a MIP and a RAM material and can be used in selective extraction and sample clean-up in SAs analysis in milk. Thus, a simple direct-injection HPLC method was established using the RAM-MIP grafted silica for sample online pretreatment [23].

The above methods promise minimization of the analytical steps with the use of MIP either as a stationary phase or as pre-column material. Thus, the methods are approaching the principles of GAC.

*3.6. Fabric Solid Phase Extracton*

Fabric solid phase extraction (FPSE) is a novel microextraction technique introduced by Kabir and Furton in 2014. It can be used as a solvent-free or solvent-minimized technique in various applications. FPSE includes the advantages of sol–gel technology as well as the rich surface chemistry of cellulose fabric substrates. The result is a manufactured robust microextraction device with high sample capacity, quick extraction equilibrium, and very high solvent and chemical stability. The ability of FPSE to extract (without any sample preparation) target analytes directly from a raw sample matrix containing particulates, biomasses, and debris is outstanding. In comparison with other applications for the determination of SAs in milk, it is obvious that FPSE is a simple, quick and green sample preparation procedure. In the application of FPSE, many unnecessary steps such as filtration, protein precipitation, solvent evaporation and sample reconstitution are omitted, making total time of sample preparation as short as possible. Also, the consumption of organic solvents is reduced comparatively with the other developed techniques. Summing it up, the above technique covers the requirements of GAC. The main steps of the FPSE approach are presented in Figure 4 [24].

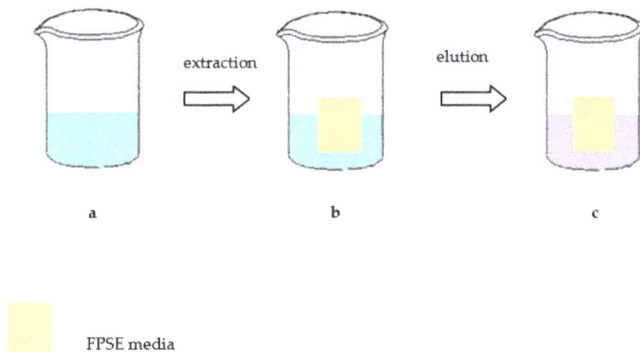

**Figure 4.** Schematic representation of fabric solid phase extraction (FPSE) process. (a) sample; (b) addition of the FPSE media in the sample; (c) elution with a suitable elution system.

FPSE was successfully applied for the extraction of three SAs (sulfamethazine, sulfisoxazole and sulfadimethoxine) in milk by Samanidou V. and co-workers, using a highly polar sol–gel poly(ethylene glycol) (sol–gel PEG) coated FPSE medium, followed by analysis using HPLC with UV detection. The optimization of fabric solid phase extraction conditions, which include the diluents of conditioning, sample loading conditions, elution solvent, elution time and sonication, showed that the sol–gel PEG material yielded the higher recovery of the three examined SAs. The FPSE procedure is applied at four steps. The FPSE medium was immersed in a mixture of 1 mL $CH_3OH$ and 1 mL ACN for 5 min and then rinsed with 2 mL of deionized water. Subsequently, a quantity of 1 g milk spiked with 0.5 mL of standard solution was transferred in a new, clean glass vial in which the FPSE media was added with the magnetic stir bar. Magnetic stirring was performed for 30 min. Finally, the FPSE media was inserted in a clean vial with 250 µL MeOH for 8 min and then in another clean vial with 250 µL ACN for 5 min. The sample was injected to HPLC after filtration. The FPSE media could be reused up to 30 times. In conclusion, the above method is convenient, reliable, and fast and it can be easily applied in any food testing laboratory [24].

### 3.7. Miniaturized Graphene-Based Pipette Tip Extraction

A miniaturized graphene-based pipette tip extraction (M-G-PTE) method coupled with liquid chromatography-ultraviolet detection was developed for the determination of four SAs (sulfadimidine, sulfachloropyridazine, sulfamonomethoxine, and sulfachloropyrazine) in bovine milk by Yan H. et al. [25] In comparison with other adsorbents such as $C_{18}$, HLB, SCX, PCX, and multi-walled carbon nanotubes, the manufactured M-G-PTE device showed better recovery and selectivity for SAs. Several parameters were investigated for the optimization of the pretreatment, such as the amount of grapheme in M-G-PTE cartridge, the type and volume of washing, and elution sorbents. The procedure of the extraction is very interesting. Two dried and clean pipette tips (100 µL and 1.0 mL) were used for assembling the pipette tip cartridge. Then, 3 mg of the graphene adsorbent was packed in the smaller tip using degreased cotton at both ends to stabilize the adsorbent. The configuration of the pipette that was used is presented in Figure 5. The tip of the larger pipette was cut and was connected with the packed tip. The sorbent was conditioned successively with 1.0 mL methanol and 1.0 mL water. Then, 2.0 mL of the sample was loaded, washed with 1.0 mL water and eluted with 0.5 mL 5% ammonia-methanol. The eluent was evaporated to dryness at 40 °C under vacuum and redissolved in 100 µL of mobile phase prior to HPLC analysis. The proposed method, beyond being simple and economic, also uses reduced amounts of organic solvents [25].

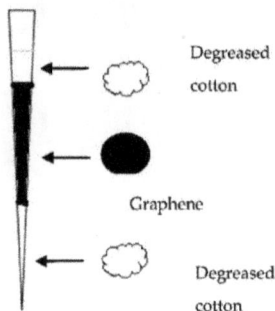

**Figure 5.** Pipette used in miniaturized graphene-based pipette tip extraction (M-G-PTE).

### 3.8. Ionic Liquid Based Microextraction

Ionic liquids (ILs) can be used instead of organic solvents as a more green approach. ILs have unique physicochemical properties, such as negligible vapor pressure, miscibility with water and organic solvents, good solubility for organic and inorganic compounds, and high thermal stability. Therefore, they can promise a greener extraction than the commonly used organic solvents such as chlorobenzene, chloroform, and carbon tetrachloride, which were typically highly toxic and not environmentally friendly. The ILs have been already used in DLLME as extraction solvents with good results.

Xu X. et al. [26] developed an ionic liquid-based microwave-assisted dispersive liquid–liquid microextraction (IL-based MADLLME) followed by HPLC-FD for the determination of six SAs in various liquid samples including milk. The aim of this work was to simplify the analytical step, to reduce the consumption of toxic solvents and improve the sensitivity. Extraction, derivatization and preconcentration were carried out by adding methanol (disperser), fluorescamine solution (derivatization reagent) and ionic liquid (extraction solvent) into the sample. Several experimental parameters, such as the type and volume of extraction solvent, the type and volume of disperser, amount of derivatization reagent, microwave power, microwave irradiation time, pH of sample solution, and ionic strength were investigated and optimized [26].

### 3.9. Dispersive Micro-Solid Phase Extraction

Dispersive micro-solid phase extraction is a miniaturized technique which is based on the dispersion of sorbents (micro- or nano-) in the sample solutions. The isolation/extraction step includes centrifugation, filtration or using an external magnetic media (for the removal of magnetic sorbents). The operation of dispersive micro-SPE is similar to the classical SPE approach. The main difference is that a small amount of sorbent (in a range of μg to mg) is added to the sample solution without conditioning. The phenomenon of dispersion minimizes the extraction time, as the analytes provide better and rapid interaction with the sorbent. Several commercial or in-house made types of sorbents have been applied in dispersive extraction, such as functionalized silica, multi-walled carbon nanotubes, graphene, graphene oxide, modified magnetic NPs, polymers, MIPs, silica, graphene, surfactants, carbon nanotubes, ionic liquids and metal/metal oxides [34].

A metal-organic framework/graphite oxide (MIL-101(Cr)@GO) was synthesized (using the hydrothermal method) and introduced by Jia X. and co-workers in 2017. This novel material was applied as sorbent in dispersive micro-solid phase extraction, for the determination of SAs in milk samples. Several parameters were investigated for the extraction, such as type of sorbents, the effect of pH, the amount of MIL-101(Cr)@GO, ionic strength, adsorption time, desorption solvent and desorption time. A UPLC-MS/MS method was developed and validated for the determination of analytes. The proposed method is characterized by advantages such as easy and quick modification, minimized use of organic solvents and stability of the sorbent [35].

Finally, another dispersive micro solid-phase extraction approach coupled with liquid chromatography-high resolution mass spectrometry (LC-HRMS) was introduced for the analysis of 24 SAs in milk by Hu S. et al. [36]. In this study, a commercially available polymer cation exchange PCX powder sorbent was used. The extraction efficiency was evaluated by several parameters, including the pH of sample solution, the amount of PCX, the desorption solvent, and volume. This rapid, selective and environmentally green method was validated under the optimized conditions [36].

## 4. Conclusions

Several methods are developed for the determination of SAs in different matrices and especially in milk, which is widely consumed. In recent years, the trend is to develop microextraction methods for the determination of residual antibiotics in food of animal origin due to potential hazards for human health. Microextraction techniques are preferable towards conventional techniques due to the fact that they are environmentally friendly, easy in operation, use new novel sorbent materials and the majority of them are less time and reagent consuming. The purpose of this article was to review the most recently developed microextraction techniques for the determination of SAs in milk.

**Conflicts of Interest:** The authors declare no conflict of interest.

## References

1. Martins, T.A.; Melo, J.; Barreto, F.; Hoff, R.B.; Jank, L.; Bittencourt, M.S.; Arsand, J.B.; Schapoval, E.E.S. A simple, fast and cheap non-SPE screening method for antibacterial residue analysis in milk and liver using liquid chromatography–tandem mass spectrometry. *Talanta* **2014**, *129*, 374–383. [CrossRef] [PubMed]
2. Dmitrienko, G.S.; Kochuk, V.E.; Apyari, V.V.; Tolmacheva, V.V.; Zolotov, Y.A. Recent advances in sample preparation techniques and methods of sulfonamides detection—A review. *Anal. Chim. Acta* **2014**, *850*, 6–25. [CrossRef] [PubMed]
3. Dmitrienko, G.S.; Kochuk, V.E.; Tolmacheva, V.V.; Apyari, V.V.; Zolotov, Y.A. Determination of the total content of some sulfonamides in milk using solid-phase extraction coupled with off-line derivatization and spectrophotometric detection. *Food Chem.* **2015**, *188*, 51–56. [CrossRef] [PubMed]
4. European Commission Decision 2002/657/EC. Available online: http://www.fao.org/fao-who-codexalimentarius/en/ (accessed on 20 November 2016).
5. Shao, M.; Zhang, X.; Li, N.; Shi, J.; Zhang, H.; Wang, Z.; Zhang, H.; Yu, A.; Yu, Y. Ionic liquid-based aqueous two-phase system extraction of sulfonamides in milk. *J. Chromatogr. B* **2014**, *961*, 5–12. [CrossRef] [PubMed]
6. Spietelum, A.; Marcinkowski, L.; Guardi, M.; Namieśnik, J. Recent developments and future trends in solid phase microextraction techniques towards green analytical chemistry. *J. Chromatogr. A* **2013**, *1321*, 1–13. [CrossRef] [PubMed]
7. Płotka-Wasylka, J.; Szczepanska, N.; Guardia, M.; Namiesnik, J. Miniaturized solid-phase extraction techniques. *Trends Anal. Chem.* **2015**, *73*, 19–38. [CrossRef]
8. Ouyang, G.; Pawlizyn, J. Recent developments in SPME for on-site analysis and monitoring. *Trends Anal. Chem.* **2006**, *25*, 692–703. [CrossRef]
9. Merkle, S.; Kleeberg, K.K.; Fritsche, J. Recent Developments and Applications of Solid Phase Microextraction (SPME) in Food and Environmental Analysis—A Review. *Chromatography* **2015**, *2*, 293–381. [CrossRef]
10. Baltussen, E.; Sandrs, P. Stir Bar Sorptive Extraction SBSE (SBSE), a Novel Extraction Technique for Aqueous Samples: Theory and Principles. *J. Microcolumn Sep.* **1999**, *11*, 693–749. [CrossRef]
11. Sarafraz-Yazdi, A.; Amiri, A. Liquid-phase microextraction. *Trends Anal. Chem.* **2010**, *20*, 1–14. [CrossRef]
12. Fumes, H.B.; Silva, M.R.; Andrade, F.N.; Nazario, C.E.D.; Lanças, M.F. Recent advances and future trends in new materials for sample preparation. *Trends Anal. Chem.* **2015**, *71*, 9–25. [CrossRef]
13. Tobiszewski, M.; Mechlinska, A.; Zygmunt, B.; Namiesnik, J. Green analytical chemistry in sample preparation for determination of trace organic pollutants. *Trends Anal. Chem.* **2009**, *28*, 943–951. [CrossRef]

14. Galuszka, A.; Zdzislaw, M.; Namiesnik, J. The 12 principles of green analytical chemistry and the SIGNIFICANCE mnemonic of green analytical practices. *Trends Anal. Chem.* **2013**, *50*, 78–84. [CrossRef]
15. Chung, H.H.; Lee, J.B.; Chung, Y.H.; Lee, K.G. Analysis of sulfonamide and quinolone antibiotic residues in Korean milk using microbial assays and high performance liquid chromatography. *Food Chem.* **2009**, *113*, 297–301. [CrossRef]
16. Ibarra, I.; Miranda, J.M.; Rodriguez, J.A.; Nebot, C.; Cepeda, A. Magnetic solid phase extraction followed by high-performance liquid chromatography for the determination of sulphonamides in milksamples. *Food Chem.* **2014**, *157*, 511–517. [CrossRef] [PubMed]
17. Yu, C.; Bin, H. C$_{18}$-coated stir bar sorptive extraction combined with high performance liquid chromatography-electrospray tandem mass spectrometry for the analysis of sulfonamides in milk and milk powder. *Talanta* **2012**, *90*, 77–84. [CrossRef] [PubMed]
18. Wen, Y.; Zhang, M.; Zhao, Q.; Feng, Y.Q. Monitoring of Five Sulfonamide Antibacterial Residues in Milk by In-Tube Solid-Phase Microextraction Coupled to High-Performance Liquid Chromatography. *J. Agric. Food Chem.* **2005**, *53*, 8468–8473. [CrossRef] [PubMed]
19. Li, Y.; Xu, W.; Li, Z.; Zhong, S.; Wang, W.; Wang, A.; Chen, J. Fabrication of CoFe$_2$O$_4$-grapheme nanocomposite and its application in the magnetic solid phase extraction of sulfonamides from milk samples. *Talanta* **2015**, *144*, 1279–1286. [CrossRef] [PubMed]
20. Huang, X.; Qiu, N.; Yuan, D. Simple and sensitive monitoring of sulfonamide veterinary residues in milk by stir bar sorptive extraction based on monolithic material and high performance liquid chromatography analysis. *J. Chromatogr. A* **2009**, *1216*, 8240–8245. [CrossRef] [PubMed]
21. Arroyo-Manzanares, N.; Gracia, G.L.; Garcia-Campana, M.A. Alternative sample treatments for the determination of sulfonamides in milk by HPLC with fluorescence detection. *Food Chem.* **2014**, *143*, 459–464. [CrossRef] [PubMed]
22. Su, S.; Zhang, M.; Li, B.; Zhang, H.; Dong, X. HPLC determination of sulfamethazine in milk using surface-imprinted silica synthesized with iniferter technique. *Talanta* **2008**, *76*, 1141–1146. [CrossRef] [PubMed]
23. Su, S.; Xu, W.; Jiang, P.; Wang, H.; Dong, X.; Zhang, M. Determination of sulfonamides in bovine milk with column-switching high performance liquid chromatography using surface imprinted silica with hydrophilic external layer as restricted access and selective extraction material. *J. Chromatogr. A* **2010**, *1217*, 7198–7207. [CrossRef]
24. Samanidou, V.; Kabir, A.; Furton, G.K.; Karageorgou, E.; Manousi, N. Fabric phase sorptive extraction for the fast isolation of sulfonamides residues from raw milk followed by high performance liquid chromatography with ultraviolet detection. *Food Chem.* **2016**, *196*, 428–436. [CrossRef]
25. Yan, H.; Sun, N.; Liu, S.; Row, H.K.; Song, Y. Miniaturized graphene-based pipette tip extraction coupled with liquid chromatography for the determination of sulfonamide residues in bovine milk. *Food Chem.* **2014**, *158*, 239–244. [CrossRef] [PubMed]
26. Wang, Z.; Xu, X.; Su, R.; Zhao, X.; Liu, Z.; Zhang, Y.; Li, D.; Li, X.; Zhang, H. Ionic liquid-based microwave-assisted dispersive liquid-liquid microextraction and derivatization of sulfonamides in river water, honey, milk, and animal plasma. *Anal. Chim. Acta* **2011**, *707*, 92–99. [CrossRef]
27. Gao, Q.; Luo, D.; Ding, J.; Feng, Y.Q. Rapid magnetic solid-phase extraction based on magnetite/silica/poly(methacrylic acid-co-ethylene glycol dimethacrylate) composite microspheres for the determination of sulfonamide in milk samples. *J. Chromatogr. A* **2010**, *1217*, 5602–5609. [CrossRef] [PubMed]
28. Nogueira, J.M.F. Stir-bar sorptive extraction: 15 years making sample preparation more environment-friendly. *Trends Anal. Chem.* **2015**, *71*, 214–223. [CrossRef]
29. Nogueira, J.M.F. Novel sorption-based methodologies for static microextraction analysis: A review on SBSE and related techniques. *Anal. Chim. Acta* **2012**, *757*, 1–10. [CrossRef] [PubMed]
30. González-Curbelo, M.A.; Socas-Rodríguez, B.; Herrera-Herrera, A.V.; González-Sálamo, J.; Hernández-Borges, J.; Rodríguez-Delgado, M.A. Evolution and applications of the QuEChERS method. *Trends Anal. Chem.* **2015**, *71*, 169–185. [CrossRef]

31. Qin, Y.; Jatamunua, F.; Zhang, J.; Han, Y.; Zon, N.; Shan, J.; Jiang, Y.; Pac, C. Analysis of Sulfonamides, Tilmicosin and Avermectins Residues in Typical Animal Matrices with multi-Plug Filtration Cleanup by Liquid Chromatography–Tandem Mass Spectrometry Detection. *J. Chromatogr. B* **2017**, *1053*, 27–33. [CrossRef] [PubMed]

32. Zhang, Y.; Li, X.; Liu, X.; Zhang, J.; Cao, Y.; Shi, Z.; Sun, H. Multi-class, multi-residue analysis of trace veterinary drugs in milk by rapid screening and quantification using ultra-performance liquid chromatography–quadrupole time-of-flight mass spectrometry. *J. Dairy Sci.* **2015**, *98*, 8433–8444. [CrossRef] [PubMed]

33. Gao, S.; Yang, X.; Yu, W.; Liu, Z.; Zhang, H. Ultrasound-assisted ionic liquid/ionic liquid-dispersive liquid-liquid microextraction for the determination of sulfonamides in infant formula milk powder using high-performance liquid chromatography. *Talanta* **2012**, *99*, 875–882. [CrossRef] [PubMed]

34. Khezeli, T.; Daneshfar, A. Development of dispersive micro-solid phase extraction based on micro and nano sorbents. *Trends Anal. Chem.* **2017**, *89*, 99–118. [CrossRef]

35. Jia, X.; Zhao, P.; Ye, X.; Zhang, L.; Wang, T.; Chen, Q.; Hou, X. A novel metal-organic framework composite MIL-101(Cr)@GO as an efficient sorbent in dispersive micro-solid phase extraction coupling with UHPLC-MS/MS for the determination of sulfonamides in milk samples. *Talanta* **2017**, *169*, 227–238. [CrossRef] [PubMed]

36. Hu, S.; Zhao, M.; Xi, Y.; Mao, Q.; Zhan, X.; Chen, D.; Yan, P. Nontargeted Screening and Determination of Sulfonamides: A Dispersive Micro Solid-Phase Extraction Approach to the Analysis of Milk and Honey Samples Using Liquid Chromatography-High-Resolution Mass Spectrometry. *J. Agric. Food Chem.* **2017**, *65*, 1984–1991. [CrossRef] [PubMed]

*separations*

MDPI

Article

# Application of Carbon Nanotubes Modified Coatings for the Determination of Amphetamines by In-Tube Solid-Phase Microextraction and Capillary Liquid Chromatography

Ana Isabel Argente-García, Yolanda Moliner-Martínez, Esther López-García, Pilar Campíns-Falcó and Rosa Herráez-Hernández *

Department of Analytical Chemistry, University of Valencia, Dr. Moliner 50, Burjassot, Valencia 46100, Spain; arai@uv.es (A.I.A.-G.); yolanda.moliner@uv.es (Y.M.-M.); estloga3@uv.es (E.L.-G.); pilar.campins@uv.es (P.C.-F.)
* Correspondence: rosa.herraez@uv.es; Tel.: +34-96-3544978; Fax: +34-96-3544436

Academic Editor: Victoria F. Samanidou
Received: 17 December 2015; Accepted: 6 February 2016; Published: 1 March 2016

**Abstract:** In this study, polydimethylsiloxane (PDMS)-coated capillary columns (TRB-5 and TRB-35), both unmodified and functionalized with single-wall carbon nanotubes (SWCNTs) or multiwall carbon nanotubes (MWCNTs), have been tested and compared for the extraction of amphetamine (AMP), methamphetamine (MET) and ephedrine (EPE) by in-tube solid-phase microextraction (IT-SPME). Prior to their extraction, the analytes were derivatized with the fluorogenic reagent 9-fluorenylmethyl chloroformate (FMOC). For separation and detection capillary chromatography with fluorimetric detection has been used. The presence of carbon nanotubes in the extractive coatings enhanced the extraction efficiencies and also significantly improved the chromatographic profiles, thus resulting in a reliable option for the analysis of these drugs. As an example of application, a new method is proposed for the analysis of the tested amphetamines in oral fluid using a TRB-35 capillary column functionalized with MWCNTs. The proposed conditions provided suitable selectivity and reproducibility (CV $\leqslant$ 6%, $n$ = 3) at low $\mu$g/mL levels, and limits of detection of 0.5–0.8 $\mu$g/mL.

**Keywords:** carbon nanotubes (CNTs); in-tube solid-phase microextraction (IT-SPME); derivatization; amphetamines; capillary liquid chromatography

## 1. Introduction

Among the novel microextraction techniques developed in the last decades, in-tube solid-phase microextraction (IT-SPME) has emerged as one of the most attractive options, as demonstrated by the increasing number of publications that use it for the analysis of a variety of analytes and matrices [1]. IT-SPME typically uses a polymeric-coated capillary column coupled to a liquid chromatograph. The target compounds can be extracted by repeated draw/injection cycles of the samples using a programmable sample injector until the analytes reach partition equilibrium between the coating of the capillary and the sample. Sample volumes are typically in the 100–500 µL range, although they are usually mixed with an aliquot of an organic solvent or a buffer before the extraction. Alternatively, the extractive capillary can be used as the loop of the injector valve (in-valve IT-SPME), and a volume of sample as large as necessary (up to several mL) is passed through the capillary until the amount of analytes extracted is sufficient to reach the required analytical responses. The former option has been extensively used for the analysis of drugs and biomarkers in biological fluids, whereas the in-valve IT-SPME microextraction approach is particularly well-suited for the extraction of traces of organic pollutants from water samples [1–3].

In IT-SPME the extraction efficiency depends on a variety of factors such as the capillary dimensions, and the type and thickness of the extractive coating. Although segments of commercially available open tubular capillary columns such as those used in gas chromatography (GC) can be used for many applications, improved selectivity and extraction efficiency can be expected through the development of new extractive phases [4,5]. Examples of alternative phases developed for IT-SPME are polypyrrole-based coatings, restricted access materials, immunosorbents or molecularly imprinted polymers (MIPs), as well as monolithic packings [6].

Nanostructured coatings are becoming popular in the context of extraction and microextraction mainly because they offer high surface-to-volume ratios, thus increasing the extraction efficiencies. In this sense, the employment of sorbents with carbon nanotubes (CNTs) appears as one of the most attractive options. CNTs have been extensively tested as sorbents for solid phase extraction (SPE) and fiber SPME because they can interact with a variety of compounds through different mechanisms such as hydrogen bonding, π–π stacking interactions, electrostatic and van der Waals forces, and hydrophobic interactions [7,8]. However, only a few studies have been reported on the utilization CNTs modified phases for IT-SPME. Liu *et al.* immobilized –COOH functionalized MWCNTs onto the external surface of a fused silica fiber using epoxy resin glue. The fibre was then inserted into a PEEK tube which was used as the injection loop in an in-valve IT-SPME configuration. The proposed device was applied to the extraction of aniline compounds from water [9]. In recent studies, we have reported the use of different polydimethylsiloxane (PDMS)-based capillary columns modified covalently with CNTs. The new coatings provided improved efficiencies for the extraction of some pollutants from water samples [10,11].

Enhancing the extraction capability of the capillary coating may be the key factor to extend the applicability of in-valve IT-SPME to those analytical problems in which the volume of sample is limited. A typical example is the identification and determination of drugs in oral fluid. Although there are several methods available for drugs in plasma and urine samples, there is an increasing interest in the analysis of alternative biological matrices such oral fluid or sweat [12,13]. In particular, oral fluid is the matrix preferred for the investigation of illicit drug consumption from drivers suspected for driving under the influence of drugs. This is because sample collection is non-invasive and it can be done under supervision. Therefore, the risk of sample adulteration is minimized. Moreover, drug concentrations in oral fluid are generally higher than in other biological fluids, especially after oral intake. The main difficulty encountered in the analysis of these fluids is the low amount of sample available. Thus, most of the analytical schemes proposed for the analysis of plasma and urine can hardly be applied to these alternative fluids.

In the present study, the extraction capabilities of CNTs modified PDMS-based capillaries for amphetamines have been explored. Different PDMS-based extractive capillaries both unmodified and modified with CNTs have been investigated. As indicated above, these coatings have been previously applied to the extraction of organic pollutants from water samples [10,11]. In the present work, we have evaluated for the first time the potential utility of such extractive phases in order to extend the IT-SPME methodology to drug analysis in oral fluid. To our knowledge CNTs modified coating have never been used for the extraction of drugs by in-valve IT-SPME. Amphetamine (AMP), methamphetamine (MET) and ephedrine (EPE) have been selected as model compounds because, according to the literature, amphetamines (including amphetamine-derived designer drugs) are among the most frequently detected drugs in oral fluid from drivers [13]. Routine analysis of amphetamines in biological samples are performed on a daily basis in clinical, forensic and toxicological laboratories [14].

Immunoassays are generally used for screening tests, whereas chromatographic methods are required for quantitative purposes. Chromatographic procedures involve extensive sample treatment for analyte isolation and preconcentration [12,13]. The determination of amphetamines by liquid chromatography (LC) usually entails a previous chemical derivatization to make the analytes more amenable for chromatography and/or to enhance the sensitivity. According to the results obtained in previous studies, in the present work the fluorogenic reagent FMOC has been selected for derivatization (see Figure 1) [15,16]. The amphetamine derivatives formed have been processed by in-valve IT-SPME coupled online to capillary liquid chromatography with fluorimetric detection. The analytical performance and possible applications of the proposed approach are discussed.

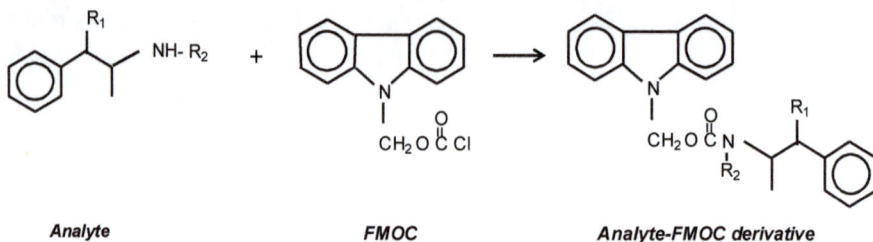

| | R₁ | R₂ |
|---|---|---|
| ANF | H | H |
| MET | CH₃ | H |
| EPE | CH₃ | OH |

**Figure 1.** Scheme of the reaction between FMOC and the amphetamines tested.

## 2. Experimental Section

### 2.1. Reagents and Solutions

All the reagents were of analytical grade. SWCNTs, MWCNTs, 3-aminopropyl triorthoxysilane (APTS), 1,3-dicyclohexylcarbodiimide, glutaraldehyde, AMP sulphate, MET hydrochloride and EPE hydrochloride were obtained from Sigma (St. Louis, MO, USA), and FMOC was purchased from Aldrich (Stenheim, Germany). Sodium hydrogen carbonate was purchased Probus (Badalona, Spain). Acetonitrile was of HPLC grade (Romil, Cambridge, UK); methanol and acetone were purchased from Scharlau (Barcelona, Spain). Sodium hydroxide, nitric acid (60%), sulphuric acid (98%) and dimethylformamide were purchased from Panreac (Barcelona, Spain).

Stock standard solutions of AMP, MET and EPE (1000.0 µg/mL) were prepared in water. Working solutions of these compounds were prepared by dilution of the stock solutions with water. Stocks solutions of FMOC (10 mM) were prepared weekly by dissolving the pure compound in acetonitrile. Solutions used for derivatization (0.1 mM) were prepared daily by dilution of the 10 mM FMOC stock solutions with acetonitrile. The hydrogencarbonate buffer (4%, *w/v*) was prepared by dissolving the appropriate amount of sodium hydrogen carbonate in water, and then by adjusting the pH to 10.6 with 5 M NaOH.

Ultrapure water was obtained from a Nanopure II (Sybron, Barnstead, UK) system. All solutions were stored in the dark at 4 °C.

### 2.2. Apparatus and Chromatographic Conditions

The chromatographic system consisted of an isocratic capillary pump, a high-pressure six-port valve (Reodyne, Rohnert Park, CA, USA), a LC-Net II/ADC interface and a fluorescence detector (Micro 21PU-01, Jasco Corporation, Tokyo, Japan). The detector was coupled to a data system (Jasco ChromNAV Chromatography Data System) (Jasco Corparation, Tokyo, Japan) for data acquisition and calculation. The excitation and emission wavelengths were 285 nm and 320 nm, respectively. A Zorbax SB-C18 (35 mm × 0.5 mm i.d., 3.5 µm) column (Agilent, Waldbronn, Germany) was used for the separation of the analytes. The mobile-phase was a mixture of acetonitrile-water, and the flow rate was 20 μL/min.

### 2.3. IT-SPME

The set-up used in the present study corresponds to that developed for in-valve IT-SPME [1,2]. The stainless steel loop of the injection valve was replaced by the extractive capillary. In this study segments of PDMS-based open tubular columns (0.32 mm i.d., 3 μm coating thickness) were used as extractive capillaries. The columns tested were TRB-35 (35% diphenyl-65% polydimethylsiloxane) and TRB-5 (5% diphenyl-95% polydimethylsiloxane), both purchased from Teknokroma (Barcelona, Spain). Segments of 30–50 cm these columns were directly tested for IT-SPME or functionalized with SWCNTs or MWCNTs.

The procedures used for the functionalization of the columns were extensively described in [10]. Briefly, the 0.025 g of SWCNTs were previously carboxylated with 5 mL of a mixture of sulfuric and nitric acids (3:1, $v/v$); MWCNTs were carboxylated by adding 80 mL of a mixture of sulfuric and nitric acids (3:1, $v/v$) to 0.020 g of MWCNTs; The PDMS columns were activated by passing successively 2 M NaOH (24 h), water (5 min), (2%, $v/v$) APTS in anhydrous acetone (15 min), water (5min), methanol (5 min) and 10% glutaraldehyde prepared in 50 mM borate buffer of pH 9.0 (10 min). Then, suspensions of the carboxylated CNTs (c-CNTs) 5 mg/mL prepared in dimethylformamide containing 1,3-dicyclohexylcarbodiimide were passed through the activated capillaries for 30 min in order to immobilize the c-CNTs into the PDMS-based coating. Finally, unreacted c-CNTs were eliminated by flushing water through the capillaries. The total time required for preparation of the capillaries after activation of the PDMS phase was about 2 h. Capillary connections to the valve were facilitated by the use of 2.5 cm sleeve of 1/16 in. polyether ether ketone (PEEK) tubing (Teknokroma); 1/16 in PEEK nuts and ferrules (Teknokroma) were used to complete the connections.

Aliquots of the solutions to be processed (20–50 μL) were manually loaded into the extractive capillary by means of a 100 μL-precision syringe. Next, 20 μL of water were flushed through the capillary in order to eliminate the solution remaining into it. Finally, the valve was manually rotated so the analytes were desorbed from the coating of the extractive capillary with the mobile-phase, and transferred to the analytical column for separation and detection.

### 2.4. Derivatization of the Amphetamines

Conditions for the derivatization of the tested amphetamines were selected according to previous works [14,15]. Aliquots of the standard solutions of the amphetamines (125 μL) were placed into 2 mL glass vials, and mixed with 125 μL of carbonate buffer and with 250 μL of a solution of the derivatization reagent (0.1 mM of FMOC). After a reaction time of 5.0 min, 50 μL of the resulting mixture were loaded into the extractive capillary of the IT-SPME device.

Each sample was derivatized in duplicate and all assays were carried out at ambient temperature.

### 2.5. Analysis of Real Samples

Samples (standard solutions of the analytes or oral fluid) were placed in 2-mL glass vials; then, one of the tips of a cotton swab was immersed into the sample, so that the cotton was totally wetted by the sample. The swab was then removed and introduced into a 2-mL glass vial containing 250 μL of the carbonate buffer and 250 μL of the 0.1 mM FMOC solutions. The swab was left inside the derivatization solution for 5 min, so the analytes passed from the cotton to the solution and the reaction occurred. After the reaction period, the swab was discarded, and an aliquot of the solution was removed by means of a 100 μL syringe and loaded into the IT-SPME capillary. Drug-free oral fluid samples were obtained from volunteers after informed consent. Cotton swaps were purchased from a local market; the amount of cotton on each tip was of about 0.03 g.

Unless otherwise stated, each sample was derivatized in duplicate and all assays were carried out at ambient temperature.

## 3. Results and Discussion

### 3.1. Optimization of the IT-SPME and Chromatographic Conditions

Initially, experiments were carried out in order to optimize conditions for the transfer of the FMOC derivatives from the extractive capillary of the IT-SPME device to the analytical column, as well as for the subsequent chromatographic separation. In these studies, unmodified capillaries were used for IT-SPME. The mobile-phase was a mixture acetonitrile-water and flow-rate was 20 µL/min.

When IT-SPME is combined on-line with capillary liquid chromatography, peak widths of the analytes are significantly higher than those typically achieved by conventional capillary chromatography, which is due to the inclusion of the extractive capillary in the chromatography system. A reduction of the extractive capillary dimensions would improve the resolution, but at the expense of a lower sensitivity (lower amount of analyte extracted). On the other hand, under a conventional capillary chromatographic scheme, the resolution could be improved by using lower mobile-phase flow rates. However, when an IT-SPME device is added to the chromatographic system, a reduction in the mobile-phase flow rate would increase the time of residence of the analytes in the extractive capillary causing extra peak broadening [17]. In other words, when in-valve IT-SPME is coupled on-line to capillary liquid chromatography, a compromise has to be reached between optimal conditions for the IT-SPME and for the chromatographic separation. In the present study, conditions were selected to achieve suitable separation of the three analytes of interest in the minimum time of analysis. Indeed, extending the proposed methodology to other amphetamine-like compounds would require different separation conditions (for example, a longer column).

Two mobile-phase compositions were tested, 50:50 and 70:30 acetonitrile:water ($v/v$). The results obtained demonstrated that both compositions were adequate to desorb the FMOC derivatives from the extractive capillary. However, best chromatographic separation was achieved by using a mixture acetonitrile:water 70:30 ($v/v$) (see Figure S1), which was the eluent composition selected for further work. The dead times observed in the chromatograms were higher than those expected for a conventional capillary chromatographic separation due to the inclusion of the extractive device. It has to be noted that the EPE-FMOC derivative eluted on the tail of the FMOC peak; for this reason, peak areas for this analyte were calculated throughout the study by the tangent skimming integration method.

On the other hand, the effect of the capillary length on the analytical responses (peak areas) was evaluated by processing solutions of the same concentration of amphetamines with TRB-35 capillaries of 30, 40 and 50 cm. In principle, an increment in the capillary length increases the surface area available for extraction, and thus, the amount of analyte extracted also increased. A slight increment on the analyte peak areas was observed when extending the capillary length from 30 cm to 40 cm. Unfortunately, the amount of unwanted products extracted also increases. In the present case, the chromatographic profile observed for an extractive capillary of 50-cm length was unsuitable due to the large amount of unreacted FMOC extracted and transferred to the analytical column (see Figure S2). Consequently, capillaries longer than 40 cm were not considered for further assays.

### 3.2. Evaluation of the Extraction Efficiencies

The extraction efficiency was evaluated for unmodified TRB-5 and TRB-35 capillaries, as well as for TRB-5 capillaries after their functionalization with c-SWCNTs, and TRB-35 capillaries functionalized with c-SWCNTs or with c-MWCNTs. The length of TRB-5 capillaries was 36 cm, whereas all TRB-35 capillaries were 40-cm length. The mean peak areas obtained for the tested capillaries are depicted in Figure 2; it has to be mentioned that AMP and MET peak areas were calculated by a valey-to-valey integration approach.

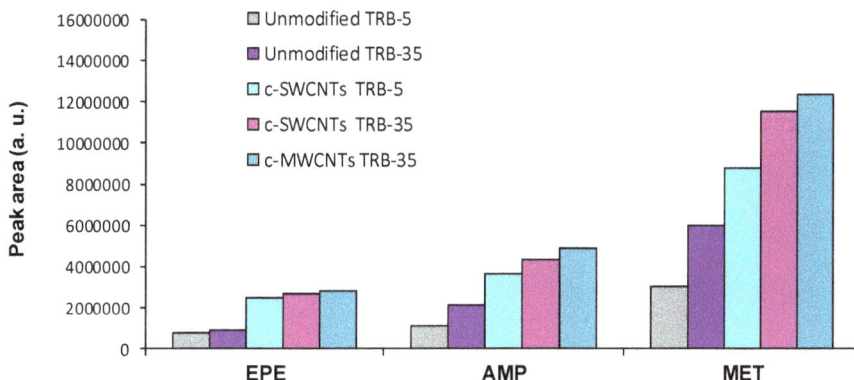

**Figure 2.** Peak areas obtained for a standard solution of the amphetamines tested (8.0 µg/mL, each) with different extractive capillaries: unmodified TRB-5 (36 cm), unmodified TRB-35 (40 cm), c-SWCNTs functionalized TRB-5 (36 cm), c-SWCNTs functionalized TRB-35 (40 cm) and c-MWCNTs functionalized TRB-35 (40 cm). For other experimental details, see text.

In order to compare the extraction efficiencies, the preconcentration rates (established as the peak area ratios between two given capillaries) were also calculated [10]. The results are summarized in Table 1.

**Table 1.** Preconcentration rates (PRs) values found for the extractive capillaries tested.

| Capillaries Compared | PR | | |
|---|---|---|---|
| | EPE | AMP | MET |
| Unmodified TRB-35/Unmodified TRB-5 | 1.2 | 2.0 | 2.0 |
| c-SWCNT functionalized TRB-5/Unmodified TRB-5 | 3.3 | 3.3 | 2.9 |
| c-SWCNT functionalized TRB-35/Unmodified TRB-35 | 2.9 | 2.2 | 1.9 |
| c-SWCNT functionalized TRB-35/c-SWCNTs functionalized TRB-5 | 1.1 | 1.2 | 1.3 |
| c-MWCNT- functionalized TRB-35/c-SWCNTs functionalized TRB-5 | 1.0 | 1.1 | 1.1 |

First, the solutions containing the three amphetamines were assayed onto the two unmodified capillaries tested. Higher analytical responses for the three analytes were obtained with the TRB-35 phase, as it can be seen in Table 1. It can be deduced that a higher percentage of diphenyl groups in the extractive phase lead to higher extraction rates. This suggests that the extraction involves $\pi$–$\pi$ interactions with the FMOC derivatives, which is consistent with the fact that the extracted compounds possess three aromatic rings (see Figure 1).

The effect of the presence of CNTs in the extractive coating was investigated for the TRB-5 and TRB-35 capillaries. In Figure 3 are depicted the chromatograms obtained for an unmodified TRB-5 capillary and the same capillary functionalized with c-SWCNTs. As observed, the presence of c-SWCNTs in the extractive phase affected not only the peak areas of the FMOC derivatives, but also their retention times. Higher retention times were observed with the c-SWCNTs coated phase, which indicates that the interaction between the extractive phase and the FMOC derivatives is stronger in the presence of CNTs, most probably by a $\pi$–$\pi$ mechanism [10]. This can also explain the higher peak areas observed with the c-SWCNTs functionalized coating (see Figure 2); the analyte responses were about three times higher with the modified coating (see Table 1).

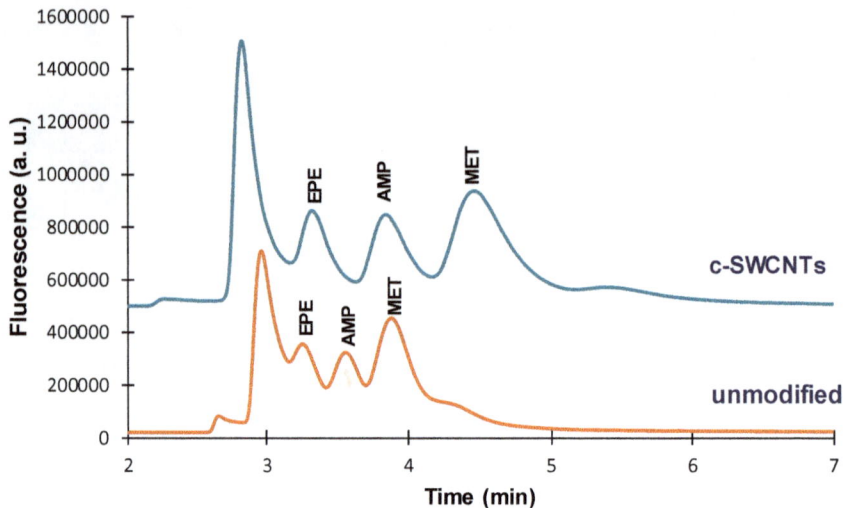

**Figure 3.** Chromatograms obtained for a standard solution of the amphetamines tested (8.0 µg/mL, each) with the unmodified TRB-5 and the c-MWCNTs functionalized TRB-5 capillaries. For other experimental details, see text.

Similar results were observed for the PDMS-TRB-35 capillaries. This is illustrated in Figure 4, which shows the chromatograms obtained for unmodified TRB-35 capillary and the same capillary after the introduction of c-MWCNTs. However, the increment in the analytical signals was more moderate than with the TRB-5 capillaries, especially for AMP and MET (Table 1). It is interesting to note that, although the retention times slightly increased in the presence of c-MWCNTs, the chromatographic peaks are narrower. As a result, not only the sensitivity but also the resolution improved with the introduction of the CNTs.

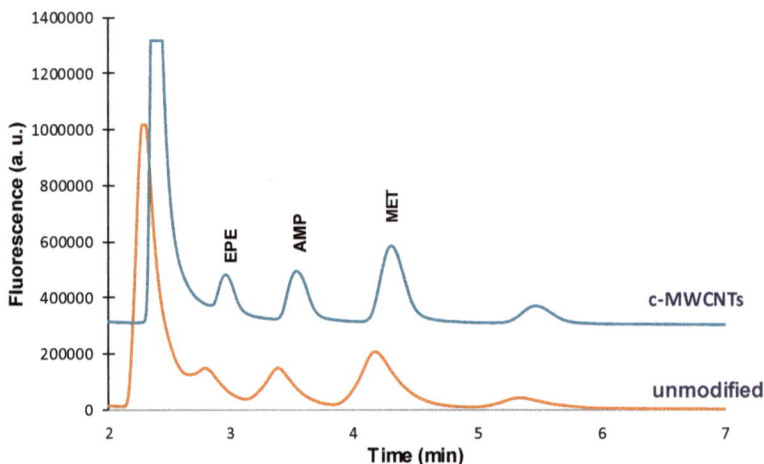

**Figure 4.** Chromatograms obtained for a standard solution of the amphetamines tested (8.0 µg/mL, each) with the unmodified TRB-35 and the c-MWCNTs functionalized TRB-35 capillaries. For other experimental details, see text.

On the other hand, the effect of the type of CNTs on the extractive phase was evaluated for a TRB-35 capillary. Figure 5 shows that both c-SWCNTs and c-MWCNTs modified coatings provided similar chromatographic profiles, although the peak areas were slightly higher for the c-MWCNTs functionalized capillary. The chromatographic profiles obtained when using the TRB-35 capillaries were better than those observed with the TRB-5 capillaries (Figures 3–5).

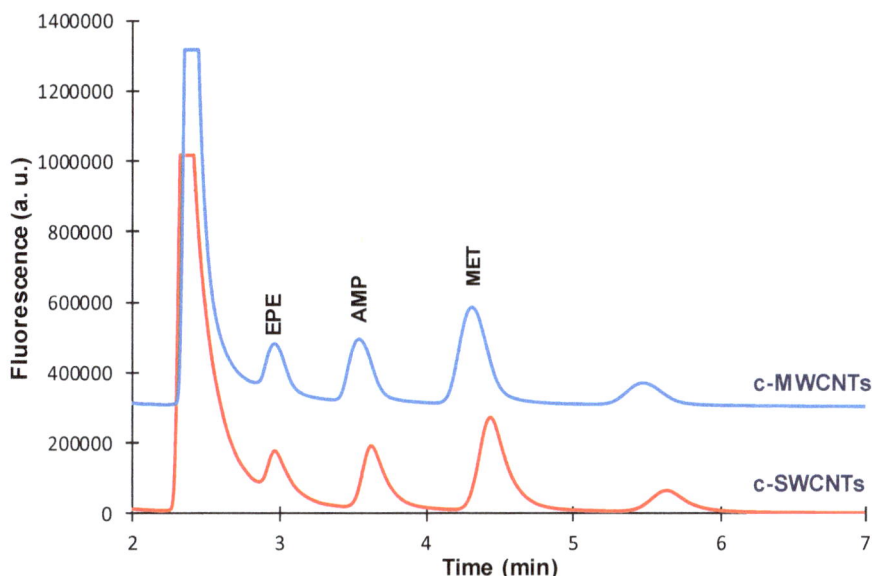

**Figure 5.** Chromatograms obtained for a standard solution of the amphetamines tested (8.0 μg/mL, each) with the c-SWCNTs and the c-MWCNTs functionalized TRB-35 capillaries. For other experimental details, see text.

Finally, the loading capacity was examined by processing solutions of the tested analytes at concentrations ranging from 2.0 to 10.0 μg/mL with different capillaries. The results are depicted in Figure 6. As it can be deduced from this figure, a linear relation between concentration and peaks areas was observed within the tested concentration interval with the TRB-35 columns. However, poor linearity was observed for the TRB-5 columns regardless the presence of CNTs in the extractive coating. For the TRB-5 capillaries, better correlation was observed between peak areas and concentration within the 2.0–8.0 μg/mL range (see Table S1), which indicates that their loading capacity was lower than that of the TRB-35 capillaries.

Taking into account the absolute peak areas and chromatographic profiles, as well as the linear working interval, a c-MWCNTs coated TRB-35 capillary of 40 cm length was selected as the best option for further experiments.

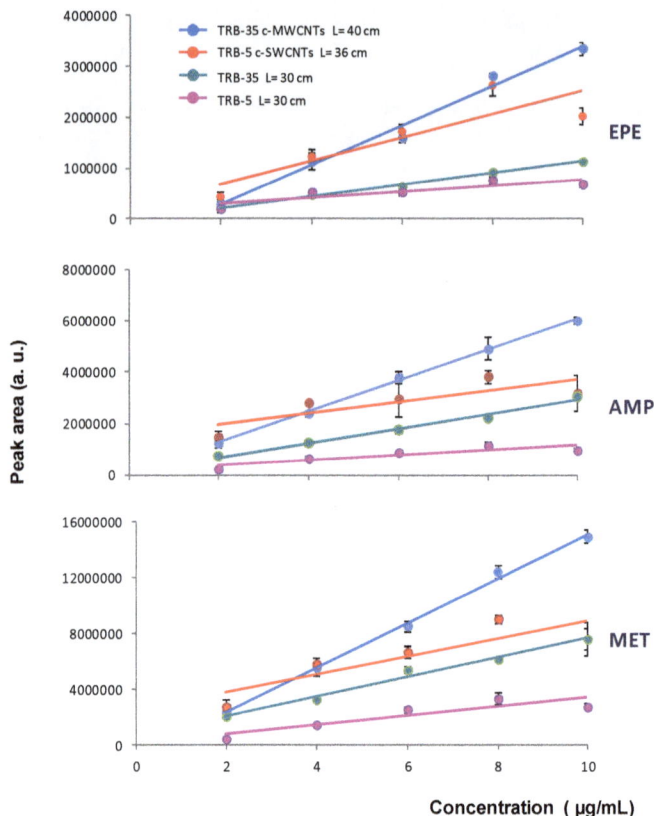

**Figure 6.** Peak areas obtained for different concentrations of the amphetamines tested within the 2.0–10.0 μg/mL concentration range with different extractive capillaries. For other experimental details, see text.

### 3.3. Application to the Quantification of Amphetamines in Oral Fluid Samples

In the analysis of drugs in oral fluid, samples are usually collected by spitting or by using sweat wipes, sponges or swabs [12,13]. Owing to its high viscosity the direct loading of oral fluid into the IT-SPME capillary was unsuitable, even after dilution with the buffer and FMOC solutions. For this reason, in the present study cotton swabs were used to collect oral fluid. In order to simplify the analytical procedure and to avoid excessive dilution of the analytes, the swabs were directly immersed into vials containing the derivatization solution. In such a way, the analytes were simultaneously extracted from the swabs and derivatized. Finally, aliquots of the resulting solutions were removed from the vials and processed by IT-SPME with a c-MWCNTs coated TRB-35 capillary. The reliability of the proposed approach was first tested with standard solutions of the analytes.

Preliminary experiments were carried out to estimate the volume of the sample that could be absorbed with the swabs. Volumes of the working solutions ranging from 50 to 200 μL were introduced into 2-mL glass vials, and then swabs were put into contact with these solutions. It was found that the swabs used in the present study absorbed about 125 μL of the samples. According to this observation, quantitative studies were carried out by placing aliquots 125 μL of the standard solutions containing the amphetamines in glass vials; then, the tips of the swabs were put into contact with the working solutions until the liquid was totally absorbed. Next, the swabs were immersed into another vial which contained the derivatization solution (250 μL of the carbonate buffer and 250 μL of the 0.1 mM FMOC solutions). Finally, aliquots of the resulting solutions were processed by IT-SPME and chromatographed.

The effect of volume of the extracts loaded into the IT-SPME capillary was tested within the 10–50 μL interval. A volume of 20 μL was found to be the best option, as higher volumes did not significantly increase the signal, probably due to the high percentage of acetonitrile in the extracts, whereas the peak corresponding to the excess of FMOC interfered with the measurement of the FMOC-EPE peak. Under such conditions, good linearity and reproducibility was found within the tested concentration intervals (1.0–10.0 μg/mL for AMP and MET and 2.0–10.0 μg/mL for EPE) (see Table 2).

**Table 2.** Linearity and reproducibility achieved with the proposed extraction/derivatization-IT-SPME-capillary liquid chromatographic method for AMP, MET and EPE within the 1.0–10.0 μg/mL (2.0–10.0 μg/mL for EPE) concentration interval (values obtained with the c-MWCNTs coated TRB-35 capillary of 40 cm length).

| Compound | Linearity, $y = ax + b$ ($n = 8$) * | | | Reproducibility **, Relative Standard Deviation (RSD) (%) ($n = 3$) |
|---|---|---|---|---|
| | $a$ | $b$ | $R^2$ | |
| EPE | $(229 \pm 8) \times 10^3$ | $(-18 \pm 4) \times 10^4$ | 0.991 | 5 |
| AMP | $(402 \pm 15) \times 10^3$ | $(-39 \pm 8) \times 10^4$ | 0.991 | 3 |
| MET | $(606 \pm 22) \times 10^3$ | $(-46 \pm 11) \times 10^4$ | 0.991 | 3 |

* Established from four concentrations assayed in duplicate; ** established at a concentration of 4.0 μg/mL.

According to the above results, the conditions finally selected for the analysis of amphetamines in oral fluid were those indicated in Figure 7. A small peak was observed at a retention time slightly lower than that of the EPE-FMOC derivative. This peak was not detected when effecting the derivatization in solution; thus, this peak was due to a compound present in the cotton. Nevertheless, it could be differentiated from the peak corresponding to the EPE derivative (see Figure S3).

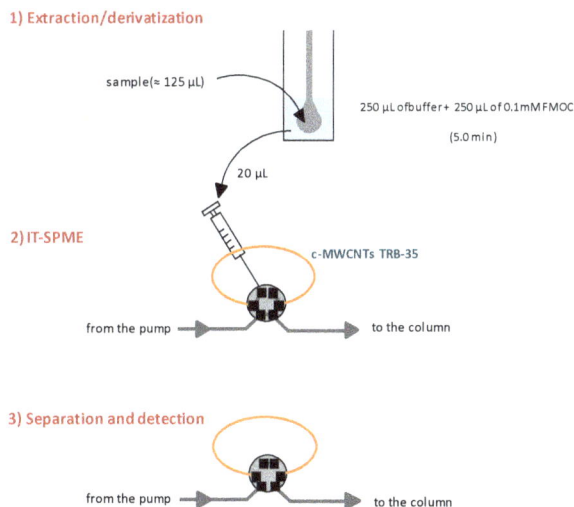

**Figure 7.** Scheme of the extraction/derivatization-IT-SPME-capillary liquid chromatographic method used for the analysis of amphetamines in oral fluid. The extractive capillary was a c-MWCNT functionalized TRB-35 capillary (40 cm length). For other experimental details, see text.

In Figure 8 are shown the chromatograms obtained for a sample of oral fluid and the same sample spiked with a mixture of the analytes. The chromatograms obtained for oral fluid were similar to those obtained for standard solutions subjected to the same extraction/derivatization procedure (Figure S3). It was then concluded that the selectivity provided by the c-MWCNTs coated TRB-35 capillary was suitable. In addition, no significant differences were observed in the chromatograms obtained for samples obtained from different donors. The limits of detection (LODs), established as the concentration of amphetamines that yielded a signal-to-noise ratio of 3, were 0.8 µg/mL for EPE, and 0.5 µg/mL for AMP and MET. The reproducibility was also satisfactory, being the interday coefficients of variation 6% for EPE and AMP, and 3% for MET ($n$ = 3). Finally, no additional cleaning of the extractive capillary with an organic solvent after each run was necessary as during the transfer and chromatographic separation steps the capillary was flushed with an eluent containing a high percentage of acetonitrile (70%).

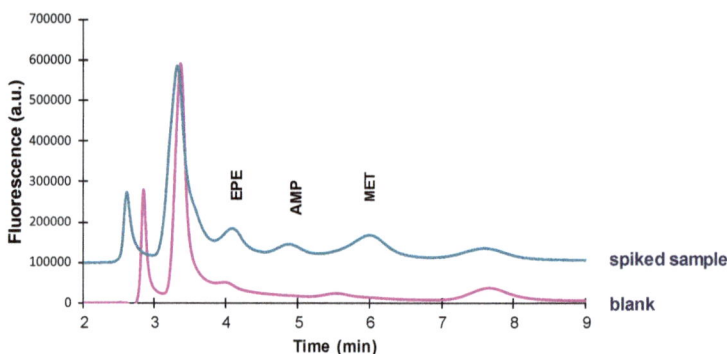

**Figure 8.** Chromatograms obtained with the proposed extraction/derivatization-IT-SPME-capillary liquid chromatographic method for an oral fluid sample, and the same sample spiked with the tested amphetamines (8.0 µg/mL, each). For other experimental details, see text.

*3.4. Utility*

Different strategies have been proposed for sample treatment prior to the chromatographic analysis of drugs in oral fluid. Extraction of the analytes into an organic solvent (toluene, butyl chloride, chloroform) has been the option traditionally used for the extraction of amphetamines [14,18,19], often in combination with evaporation of the extracts to dryness in order to concentrate the extracts to be chromatographed. Solid-phase extraction followed by solvent evaporation has also been proposed [20]. These methodologies allow the detection of the analytes at low ng/mL levels by GC, although the resulting procedures are labor-intensive. More recently, different methods have been reported that involve SPME with fibres [21,22] again prior to GC; LODs of 14–27 ng/mL were reported for MET. Very recently, microextraction with a $C_{18}$ sorbent packed into the needle of a syringe has been proposed for the analysis of illicit drugs in oral fluid by LC coupled to mass spectrometry (MS) (LC-MS/MS) [23]. In this case, samples were treated to precipitate the proteins, and the supernatant was passed through the sorbent by repetitive aspirating/dispensing cycles, being the LOD reported for AMP 1 ng/mL.

Compared with previously described procedures, the main advantage of the present methodology is simplicity. After sample collection, the only operations required are the immersion of the swabs into the derivatization solution and, after 5 min, the loading of an aliquot of the resulting solution into the IT-SPME device. The proposed method is also advantageous in terms of time of analysis and solvent and reagents consumption. Although the LODs are about one order of magnitude higher than those reported by other authors, the proposed method can be considered adequate for the measurement of amphetamines in oral fluid of abusers. For example, concentrations amphetamine found in oral fluid samples during random tests programs were in the 0.3–8.8 µg/mL range [24,25]. Nevertheless, for other types of studies, more sensitive methodologies may be required.

## 4. Conclusions

The present study shows that IT-SPME is a valid option for the on-line extraction of amphetamines derivatized with FMOC, and for their subsequent separation by capillary liquid chromatography. The efficiencies of the IT-SPME have been examined for different PDMS-based extractive capillaries, as well as for the same capillaries after their functionalization with CNTs. The introduction of CNTs in the extractive coatings has a positive effect not only on the extraction efficiencies but also on the chromatographic profiles. In all instances, highest responses were obtained for TRB-35 capillaries, which can be explained by a π–π interaction mechanism. In this sense, the use of FMOC (which introduces two aromatic rings in the compounds to be extracted) may be advantageous over other derivatization reagents typically used for amphetamines [15]. On the other hand, the chromatograms observed for TRB-35 capillaries functionalized with c-SWCNTs and c-MWCNTs were quite similar, although the analytical responses obtained with the latter capillary were slightly higher. According to these results, a new approach has been described for the analysis of amphetamines derivatives using a TRB-35 capillary functionalized with c-MWCNTs.

The described approach combines the simplicity of in-valve IT-SPME with enhanced sensitivity, and can be considered an alternative for the analysis of amphetamines. As an illustrative example of application, in the present study oral fluid samples (≈125 µL) collected with cotton swabs have directly been derivatized and processed by IT-SPME-capillary liquid chromatography. The proposed method is very simple and cost-effective, and provides suitable selectivity, sensitivity and reproducibility at low µg/mL concentration levels, which are concentrations typically found in oral fluid of abusers.

**Supplementary Materials:** The following are available online at www.mdpi.com/2227-9075/3/1/7/s1.

**Acknowledgments:** The authors are grateful to the Spanish Ministerio de Economia y Competitividad/FEDER for the financial support received (project CTQ2014-53916-P) and to the Generalitat Valenciana (PROMETEO 2012/045). A. Argente-García expresses her gratitude to the Spanish Ministerio de Educación, Cultura y Deporte for her FPU grant.

**Author Contributions:** The manuscript was written through contributions of all authors and all authors have given approval to the final version.

**Conflicts of Interest:** The authors declare no conflict of interest.

## References

1. Moliner-Martínez, Y.; Herráez-Hernández, R.; Verdú-Andres, J.; Molins-Legua, C.; Campíns-Falcó, P. Recent advances in in-tube solid-phase microextraction. *Trends Anal. Chem.* **2015**, *71*, 205–213. [CrossRef]

2. Kataoka, H.; Saito, K. Recent advances in SPME techniques in biomedical analysis. *J. Pharm. Biomed. Anal.* **2011**, *54*, 926–950. [CrossRef] [PubMed]

3. Queiroz, M.E.C.; Melo, M.P. Selective capillary coating materials for in-tube solid-phase microextraction coupled to liquid chromatography to determine drugs and biomarkers in biological samples: A review. *Anal. Chim. Acta* **2014**, *826*, 1–11. [CrossRef] [PubMed]

4. Mehdinia, A.; Aziz-Zanjani, M.O. Recent advances in nanomaterials utilized in fiber coatings for solid-phase microextraction. *Trends Anal. Chem.* **2013**, *42*, 205–215. [CrossRef]

5. Weng, Y.; Chen, L.; Li, J.; Liu, D.; Chen, L. Recent advances in solid-phase sorbents for samples preparation prior to chromatographic analysis. *Trends Anal. Chem.* **2014**, *59*, 26–41.

6. González-Fuenzalida, R.A.; Moliner-Martínez, Y.; Verdú-Andrés, J.; Molins-Legua, C.; Herráez-Hernández, R.; Jornet-Martínez, N.; Campíns-Falcó, P. Microextraction with phases containing nanoparticles. *Bioanalysis* **2015**, *17*, 2163–2170. [CrossRef] [PubMed]

7. Zhang, B.-T.; Zheng, X.; Li, H.-F.; Lin, J.-M. Application of carbon-based nanomaterials in sample preparation: a review. *Anal. Chim. Acta* **2013**, *784*, 1–17. [CrossRef] [PubMed]

8. Liang, X.; Liu, S.; Wang, S.; Guo, Y.; Jiang, S. Carbon-based sorbents: carbon nanotubes. *J. Chromatogr. A* **2014**, *1357*, 53–67. [CrossRef] [PubMed]

9.  Liu, X.-Y.; Ji, Y.-S.; Zhang, H.-X.; Liu, M.-C. Highly sensitive analysis of substituted aniline compounds in water samples by using oxidized multiwalled carbon nanotubes as an in-tube solid-phase microextraction medium. *J. Chromatogr. A* **2008**, *1212*, 10–15. [CrossRef] [PubMed]

10. Jornet-Martínez, N.; Serra-Mora, P.; Moliner-Martínez, Y.; Herráez-Hernández, R.; Campíns-Falcó, P. Evaluation of Carbon Nanotubes Functionalized Polydimethylsiloxane Based Coatings for In-Tube Solid Phase Microextraction Coupled to Capillary Liquid Chromatography. *Chromatography* **2015**, *2*, 515–528. [CrossRef]

11. Moliner-Martínez, Y.; Serra-Mora, P.; Verdú-Andrés, J.; Herráez-Hernández, R.; Campíns-Falcó, P. Analysis of polar triazines and degradation products in waters by in-tube solid-phase microextraction and capillary chromatography: and environmentally friendly method. *Anal. Bioanal. Chem.* **2015**, *407*, 1485–1497. [CrossRef] [PubMed]

12. Samyn, L.; Laloup, M.; de Boek, G. Bioanalytical procedures for determination of drugs of abuse in oral fluid. *Anal. Bioanal. Chem.* **2007**, *388*, 1437–1453. [CrossRef] [PubMed]

13. Lledo-Fernandez, C.; Banks, C.E. An overview of quantifying and screening drugs of abuse in biological samples: Past and present. *Anal. Methods* **2011**, *3*, 1227–1245. [CrossRef]

14. Kankaanpää, A.; Gunnar, T.; Ariniemi, K.; Lillsunde, P.; Mykkänen, S.; Seppälä, T. Single-step procedure for gas chromatography-mass spectrometry screening and quantitative determination of amphetamine-type stimulants and related drugs in blood, serum, oral fluid and urine samples. *J. Chromatogr. B* **2004**, *810*, 57–68. [CrossRef]

15. Herráez-Hernández, R.; Campíns-Falcó, P.; Sevillano-Cabeza, A. On-line derivatization into precolumns for the determination of drugs by liquid chromatography and column switching: determination of amphetamines in urine. *Anal. Chem.* **1996**, *68*, 734–739. [CrossRef] [PubMed]

16. Cháfer-Pericás, C.; Campíns-Falcó, P.; Herráez-Hernández, R. Application of solid-phase microextraction combined with derivatization to the determination of amphetamines by liquid chromatography. *Anal. Biochem.* **2004**, *333*, 328–335. [CrossRef] [PubMed]

17. Moliner-Martínez, Y.; Molins-Legua, C.; Verdú-Andrés, J.; Herráez-Hernández, R.; Campíns-Falcó, P. Advantages of monolithic over particulate columns for multiresidue analysis of organic pollutants by in-tube solid-phase microextraction coupled to capillary liquid chromatography. *J. Chromatogr. A* **2011**, *1218*, 6256–6262. [CrossRef] [PubMed]

18. Drummer, O.H.; Gerostamoulos, D.; Chu, M.; Swann, P.; Boorman, M.; Cairns, I. Drugs in oral fluid in randomly selected drivers. *Forensic Sci. Int.* **2007**, *170*, 105–110. [CrossRef] [PubMed]

19. Meng, P.; Wang, Y. Small volume liquid extraction of amphetamines in saliva. *Forensic Sci. Int.* **2010**, *197*, 80–84. [CrossRef] [PubMed]

20. Concheiro, M.; de Castro, A.; Quintela, O.; Cruz, A.; López-Rivadulla, M. Determination of illicit and medicinal drugs and their metabolites in oral fluid and preserved oral fluid by liquid chromatography-tandem mass spectrometry. *Anal. Bioanal. Chem.* **2008**, *391*, 2329–2338. [CrossRef] [PubMed]

21. Djozan, D.; Baheri, T. Investigation of pencil leads fiber efficiency for SPME of trace amount of methamphetamine from human saliva prior to GC-MS analysis. *J. Chromatogr. Sci.* **2010**, *48*, 224–228. [CrossRef] [PubMed]

22. Djozan, D.; Farajzadeh, M.A.; Sorouraddin, S.M.; Baheri, T. Synthesis and application of high selective monolithic fibers based on molecularly imprinted polymer for SPME of trace methamphetamine. *Chromatographia* **2011**, *73*, 975–983. [CrossRef]

23. Montesano, C.; Simeoni, M.C.; Curini, R.; Sergi, M.; lo Sterzo, C.; Compagnone, D. Determination of illicit drugs and metabolites in oral fluid by microextraction of packed sorben coupled with LC-MS/MS. *Anal. Bioanal. Chem.* **2015**, *407*, 3647–3658. [CrossRef] [PubMed]

24. Samyn, N.; van Haeren, C. On-site testing of saliva and sweat with drugwipe and determination of concentrations of drugs of abuse in saliva, plasma and urine of suspected users. *Int. J. Legal Med.* **2000**, *113*, 150–154. [CrossRef] [PubMed]

25. Samyn, N.; de Boeck, G.; Wood, M.; Lamers, C.; de Waard, D.; Brookhuis, K.; Verstraete, A.; Riedel, W. Plasma, oral fluid and sweat wipe ecstasy concentrations in controlled and real life conditions. *Forensic Sci. Int.* **2002**, *128*, 90–97. [CrossRef]

*separations*

MDPI

*Communication*

# Design of a Molecularly Imprinted Stir-Bar for Isolation of Patulin in Apple and LC-MS/MS Detection

**Patricia Regal \*, Mónica Díaz-Bao, Rocío Barreiro, Cristina Fente and Alberto Cepeda**

Department of Analytical Chemistry, Nutrition and Bromatology, Faculty of Veterinary Science, Universidade de Santiago de Compostela, 27002 Lugo, Spain; monica.diaz@usc.es (M.D.-B.); rocio.barreiro@usc.es (R.B.); cristina.fente@usc.es (C.F.); alberto.cepeda@usc.es (A.C.)
\* Correspondence: patricia.regal@usc.es; Tel.: +34-982-285-900

Academic Editor: Victoria F. Samanidou
Received: 31 December 2016; Accepted: 20 March 2017; Published: 24 March 2017

**Abstract:** Mycotoxins are a very diverse group of natural products produced as secondary metabolites by fungi. Patulin is produced by mold species normally related to vegetable-based products and fruit, mainly apple. Its ingestion may result in agitation, convulsions, edema, intestinal ulceration, inflammation, vomiting, and even immune, neurological or gastrointestinal disorders. For this reason, the European Commission Regulation (EC) 1881/2006 established a maximum content for patulin of 10 ppb in infant fruit juice, 50 ppb for fruit juice for adults and 25 ppb in fruit-derived products. In this work, a rapid and selective method based on magnetic molecularly imprinted stir-bar (MMISB) extraction has been developed for the isolation of patulin, using 2-oxindole as a dummy template. The final extraction protocol consisted of simply pouring in, stirring and pouring out samples and solvents from a beaker with the MMISB acting inside. The magnetic device provided satisfactory recoveries of patulin (60%–70%) in apple samples. The successful MMISB approach has been combined with high performance liquid chromatography coupled to tandem mass spectrometry (HPLC-MS/MS) to determine patulin.

**Keywords:** patulin; molecularly imprinted polymers (MIP); stir-bar; apple; HPLC-MS/MS

## 1. Introduction

Mycotoxins are low-molecular-weight natural products, very diverse in terms of structure and abilities, produced as secondary metabolites by fungi. Patulin (PAT) is an important mycotoxin produced by over 30 genera of mold such as Penicillium, Aspergillus, and Byssochlamys. In particular, *Penicillium expansum* is recognized as the main source of PAT and it has been commonly associated with apple rot [1]. These molds grow easily in damaged fruit or in derived products such as juices, if storage conditions are deficient. Some of the most serious health effects of PAT ingestion in humans are agitation, convulsions, edema, intestinal ulceration, inflammation and vomiting [2]. The toxicity of these molecules has led to the set-up of strict regulations in many countries for their control in food and feed, and the consequent establishment of official legislation. The establishment of maximum limits in some food products resulted in an increasing demand of sensitive, selective and effective analytical methods. The European Commission Regulation (EC) 1881/2006 established a maximum content for PAT of 10 ppb in infant fruit juices, 50 ppb for fruit juices for adults and 25 ppb in fruit-derived products [3].

For the determination of PAT, thin layer chromatography was firstly used. Nowadays, the official analytical method adopted by the Association of Official Analytical Chemists (AOAC) for the analysis of PAT in food is high performance liquid chromatography (HPLC) with ultraviolet (UV) detection, using an extraction with ethyl acetate and clean-up with sodium carbonate [4]. The main drawback of UV detection is the poor resolution between PAT and other co-extracted compounds such as hydroxymethylfurfural. To overcome this interference, liquid chromatography may be combined with mass spectrometric determination [1]. As an additional problem, PAT is vulnerable to the alkaline conditions of the previously mentioned extraction method. Purification by solid-phase extraction (SPE) has been frequently applied as an alternative procedure [5]. In the last years, molecularly imprinted polymers (MIPs) started to be used, and are becoming promising materials for extracting different analytes present in food and beverages [6]. However, mycotoxins are usually too toxic or too expensive to be used as template molecules in MIP preparation. Template bleeding may be an additional problem of these polymers, especially when dealing with very low detection levels. To overcome these limitations, dummy templates can be applied during MIP synthesis [7]. Also, magnetic materials can provide fast and simple methods of extraction and have already demonstrated their effectiveness to extract patulin [8].

In the present work, a rapid and selective method based on an in-house designed magnetic molecularly imprinted stir-bar (MMISB) has been developed for the isolation of PAT. For MIP synthesis, a structural analogue of PAT, 2-oxindole, was used as a dummy template (Figure 1) [9]. The molecularly imprinted polymer was grafted on the silanized surface of a glass-covered stir-bar using an adaptation of typical protocols used in grafting techniques. The applicability of this novel stirring bar for the extraction of PAT has been tested in spiked apple samples using HPLC-MS/MS for detection.

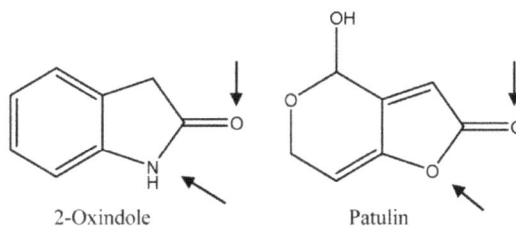

2-Oxindole                    Patulin

**Figure 1.** Chemical structures of patulin and dummy template used for molecularly imprinted polymers (MIP) synthesis. The potential sites for intermolecular interactions between template or patulin and the functional monomer methacrylic acid (MAA) are indicated by arrows.

## 2. Materials and Methods

### 2.1. Materials

The standard for PAT was purchased from Sigma-Aldrich Chemical Company (Madrid, Spain). The dummy template 2-oxindole, methacrylic acid (MAA), divinylbenzene 80% (DVB-80), ethylene glycol dimethacrylate (EGDMA), and the initiator 2,2′-azobis-(2-methyl-butyronitril) (AIMN) were from Sigma-Aldrich. The 3-(methacryloxy) propyltrimethoxysilane was purchased from Sigma-Aldrich Chemical Company. HPLC grade solvents were supplied by Merck (Madrid, Spain).

MAA and EGDMA were freed from stabilizers by distillation under reduced pressure and AIMN was recrystallized from methanol prior to use. DVB-80 was freed from stabilizers by passing through a small column packed with neutral alumina (Aldrich).

### 2.2. Apparatus

The polymerization was carried out into a temperature controllable incubator (Stuart Scientic, Redhill, Surrey, UK). Separation was performed in an 1100 series HPLC system from Agilent Technologies (Santa Clara, CA, USA). A Luna 3 μm C18 (150 × 2 mm) column from Phenomenex (Torrance, CA, USA) was used. The mobile phase was water and methanol with 0.1% formic acid, mixed in isocratic mode at 70% and 30%, respectively; the analytical run lasted for 10 min at 250 μL·min$^{-1}$.

A Q-Trap 2000 mass spectrometer with ESI Source from AB Sciex (Toronto, ON, Canada) was used, working in negative mode for PAT and in positive mode for 2-oxindole. For quantification of PAT, the most intense MRM transition was monitored along with a second transition for identity confirmation: 153 > 109 and 153 > 81, respectively.

### 2.3. Design of Molecularly Imprinted Stir-Bars (MMISB) for Patulin Extraction

To achieve a stir-bar grafted with molecularly imprinted polymer on its surface, a chemical coating protocol adapted from the work of Turiel and Martin-Esteban was used [10]. First, a commercial glass-covered magnetic stir-bar remained in a combination of methanol and hydrochloric acid (1:1, $v/v$), stirring for 30 min for clean-up of the glass surface. Next, the surface was silanized with 2% 3-(methacryloxy) propyltrimethoxysilane in toluene for 1 h. Finally, the stir-bar was rinsed methanol and let dry under $N_2$ stream.

Once the silanized glass surface was ready, the chemical coating process took place as follows: in a glass tube of 4 mL, the bar was placed and covered with the pre-polymerization mixture. The polymerization mixture was prepared with 2-oxindole, MAA and EGDMA, ratio 1:4:20, dissolved in the porogen solvent toluene/methanol (90:10) at 40 wt % and initiator 2 wt %. The stirring bar immersed in the monomeric recipe was allowed to polymerize at 60 °C for 12 h, or until the appearance of the polymer was completely white.

After (bulk) polymerization, the glass tube was broken to release the polymer-coated stir bar (Figure 2). Then, the bars were placed in a beaker and covered with water and methanol (50/50, $v/v$) to test the adhesion to the surface and the resistance of the polymer, using a magnetic stirrer. Next, the template molecule was removed by Soxhlet extraction for 12 h with methanol/acetic acid (1/1, $v/v$) solution. To optimize the conditions of use of this MMISB for the extraction of PAT, different loading, washing and elution solution were tested. A magnetic non-imprinted stir bar (MNISB) was prepared in parallel, without the addition of template.

**Figure 2.** Magnetic molecularly imprinted stir bar (MMISB) developed for the extraction of patulin; a screw cap was placed below the stir-bar to illustrate its size (Reproduced with permission from [11]. MDPI under CC BY 4.0, 2015.

### 2.4. Application to Real Samples

The extraction protocol started adding water to cover 3 g of diced apple in a 50 mL plastic tube. Apple samples were spiked with water containing PAT at 50 ng·g$^{-1}$ (MRL to for fruit juices in adults) and introduced into an ultrasound bath for 30 min. Samples were shaken for 10 min, centrifuged at 5000 g for 10 min and the supernatant was transferred to a baker with the molecularly imprinted stir-bar (MMISB). A conventional SPE protocol was applied, consisting of loading, washing and elution steps with different solvents.

The analyte was loaded within the pores (active sites) of the MMISB under stirring for 30 min. After this sorptive extraction, water was removed with the aid of external magnet and the stir-bar was washed with water for 5 min. Then, to elute the retained patulin from the MMISB, 5 mL methanol/acetic acid (75/25) were added and the solution was kept stirring in a beaker for 30 min.

Finally, the eluate was evaporated under nitrogen stream at 30 °C and re-dissolved in 100 µL of mobile phase. Twenty microliters were immediately injected into the chromatographic system for analysis. Recoveries were calculated using HPLC-MS/MS.

## 3. Results and Discussion

Nowadays, the official analytical method adopted by AOAC International is HPLC with UV detection, using clean-up with ethyl acetate and sodium carbonate. However, the diverse drawbacks of this method (poor stability of patulin under alkaline extraction, poor resolution between patulin and co-extracted hydroxymethylfurfural) have originated interest in alternative options, such as LC methods coupled to mass spectrometry [1]. In the last years, purification with molecularly imprinted polymers and magnetic materials started to be used and they are becoming promising materials in analytical chemistry and, more specifically, in mycotoxins determination [7,8,12,13]. Imprinted polymers have also showed potential for detoxification purposes in large-scale environmental applications [14].

### 3.1. Molecularly Imprinted Stir-Bars (MMISB) for Patulin Extraction

In this study, 2-oxindole, a compound structurally related to PAT, was selected as a dummy template for the design of a MIP selective towards PAT. This analogue of patulin has already been used successfully to synthesize a MIP capable of selectively binding PAT molecules [9]. The removal of the template leaves binding sites within the polymeric matrix that are complementary in shape and functionality to the template and to the target molecule of PAT (Figure 1). Dummy templates are usually selected to overcome bleeding problems. Additional advantages of these templates may include lower cost and toxicity. In this context, a dummy-template approach was preferred to avoid the manipulation of patulin during MIP synthesis, as it could be hazardous to the personnel working in the laboratory. Patulin has been classified as having acute toxicity (oral, dermal, inhalation) for humans, while 2-oxindole has not been classified as hazardous (Sigma-Aldrich website). As for the costs, relatively high amounts of template are normally required to synthesize imprinted polymers. In that respect, 2-oxindole is much cheaper than patulin. To provide just one example, 5 mL of PAT correspond to the price of 25 g of 2-oxindole in the Spanish market (Sigma-Aldrich website).

The polymer coated on the surface of the magnetic stir-bars was prepared by a non-covalent approach, based on the formation of the non-covalent interactions of MAA and the dummy template. MAA was selected because of its high capability to act both as a hydrogen bond and a proton donor and as a hydrogen bond acceptor [15]. The selected polymerization technique was bulk polymerization, because it does not require sophisticated instrumentation and the reaction conditions can be easily controlled. Furthermore, it is the most widely used method for the preparation of imprinted polymers [16]. In the design of any pre-polymerization mixture, the selection of the cross-linker and porogen solvent are two key factors, as they would determine the aspect, strength and even the color and porosity of the final polymer. DVB-80 and EGDMA were therefore tested separately and in combination with various solvent mixtures, in order to achieve the best option for MIP synthesis. The tested combinations were methanol and/or acetonitrile with different percentages of toluene, as follows: 0%, 20%, 50%, 70% and 90% toluene in methanol and 0%, 20%, 50%, 70% and 90% toluene in acetonitrile. The different combinations of monomer-solvent were introduced in an injection vial and allowed to polymerize under bulk conditions for 24 h at 60 °C. After polymerization, the glass vials were broken and the resulting material was tested visually and by touching. With the use of DVB-80 as a cross-linker, it was impossible to obtain a strong and tough polymer. The resulting polymer would be a plastic, yellowish, cracked and fragile material, not suitable for resisting the required stirring conditions. Thus, adequate divinylbenzene polymers could not be achieved using the different tested solvents (acetonitrile or methanol), even changing the toluene percentage (from 0% to 90% toluene). As for EGDMA, in every case the polymers were stronger than those obtained with DVB, showing more resistance to compressive strength and even more when using methanol (combined with toluene) instead of acetonitrile. As for the color, EGDMA polymers were white and DVB yellowish. The mixture of methanol and toluene (90:10) with EGDMA proved to be useful, resulting in a hard and white polymer. Consequently, this combination was used to prepare the magnetic molecularly imprinted stir-bar (MMISB).

### 3.2. Analytical Method and Application of MMISB in Apple Samples

The efficiency of the obtained polymer was evaluated with a very simple extraction protocol for the isolation of PAT from apple samples prior to analysis. The protocol consisted simply of pouring in, stirring and pouring out solvents from a beaker. To carry out the extraction process, it was only necessary to have a magnetic stirrer and an external magnet. Satisfactory recoveries of PAT were obtained using this molecularly imprinted glass-covered stir-bar, with 60%–70% of recovery in a minimum loading time of 45 min under stirring. These results were in the range of the recoveries obtained by Wang et al. using a graphene-based magnetic material to extract PAT from apple juice [8]. In this context, Lucci et al. applied commercial MIP-SPE columns for detection of patulin in apple products, obtaining recoveries of >77% [12]. Figure 3 shows a chromatogram of a blank apple sample (*a*) and a sample spiked with patulin (*b*) at 50 ng·g$^{-1}$, both extracted with the MMISB protocol and analyzed by HPLC-MS/MS. The magnetic NIP stir-bar showed a mean recovery difference of 20% less.

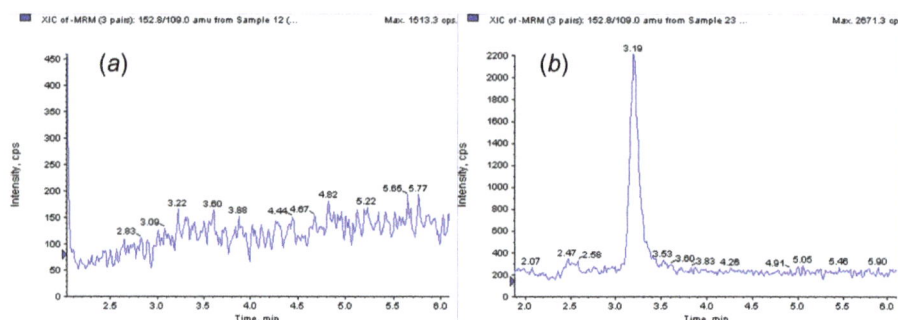

**Figure 3.** Reconstructed LC-MS/MS chromatograms of a blank apple sample (**a**) and the same sample spiked with patulin at 50 ng·g$^{-1}$ (**b**), both extracted using the MMIP stir-bar.

On the other hand, the deterioration of the polymeric coat of the stir-bar was imperceptible after several hours of use, resisting the whole set of experiments, thus indicating that the adhesion of the polymer was strong and stable. The glass layer as a cohesion agent between the bar and the polymeric coat can be assumed. Imprinted polymers, especially the ones cross-linked with DVB, are stable over a long time and can be reused [17]. The analytical limits were calculated from the signal-to-noise ratio. Using this method, the limits of detection (LOD) and quantification (LOQ) correspond to the concentrations of the analyte that would yield a signal equal to three and 10 times the noise level, respectively [18]. The calculated LOD (S/N = 3) was 10 ng·g$^{-1}$ and the LOQ (S/N = 10) was 50 ng·g$^{-1}$.

### 4. Conclusions

The proposed magnetic extraction has demonstrated its usefulness for the isolation of PAT in apple. The stir-bars are easy to use and the extraction protocol was reduced to simply add and remove solvents from a beaker, with the aid of a magnetic stirrer and an external magnet. The main drawback of the proposed methodology lies in its limits, which could be improved using more sensitive instruments or further optimizing the whole method. Additionally, the design approach applied to obtain the imprinted stir-bar was easy and fast, readily reproducible in other laboratories and for different analytes.

**Acknowledgments:** This research was supported by the project EM 2012/153 from Consellería de Cultura, Educación e Ordenacion Universitaria, Xunta de Galicia.

**Author Contributions:** P.R. and M.D.-B. conceived and designed the experiments and wrote the paper; R.B. performed the experiments and contributed to the writing; C.F. and A.C. lead the work and revised the paper.

**Conflicts of Interest:** The authors declare no conflict of interest.

## References

1. Desmarchelier, A.; Mujahid, C.; Racault, L.; Perring, L.; Lancova, K. Analysis of Patulin in Pear- and Apple-Based Foodstuffs by Liquid Chromatography Electrospray Ionization Tandem Mass Spectrometry. *J. Agric. Food Chem.* **2011**, *59*, 7659–7665. [CrossRef] [PubMed]

2. Silva, S.J.N.D.; Schuch, P.Z.; Bernardi, C.R.; Vainstein, M.H.; Jablonski, A.; Bender, R.J. Patulin in food: State-of-the-art and analytical trends. *Rev. Bras. Frutic.* **2007**, *29*, 406–413.

3. EC Commission Regulation (EC) No. 165/2010 of 26 February 2010 amending Regulation (EC) No. 1881/2006 Setting the Maximum Levels for Certain Contaminants in Foodstuffs as Regards Aflatoxins. *Off. J. Eur. Union* **2010**, *50*, 8–11.

4. Brause, A.R.; Trucksess, M.W.; Thomas, F.S.; Page, S.W. Determination of patulin in apple juice by liquid chromatography: Collaborative study. *J. AOAC Int.* **1996**, *79*, 451–455. [PubMed]

5. Barreira, M.J.; Alvito, P.C.; Almeida, C.M. Occurrence of patulin in apple-based-foods in Portugal. *Food Chem.* **2010**, *121*, 653–658. [CrossRef]

6. Regal, P.; Diaz-Bao, M.; Barreiro, R.; Cepeda, A.; Fente, C. Application of molecularly imprinted polymers in food analysis: Clean-up and chromatographic improvements. *Cent. Eur. J. Chem.* **2012**, *10*, 766–784. [CrossRef]

7. Díaz-Bao, M.; Regal, P.; Barreiro, R.; Fente, C.A.; Cepeda, A. A facile method for the fabrication of magnetic molecularly imprinted stir-bars: A practical example with aflatoxins in baby foods. *J. Chromatogr. A* **2016**, *1471*, 51–59. [CrossRef] [PubMed]

8. Wang, Y.; Wen, Y.; Ling, Y. Graphene oxide-based magnetic solid phase extraction combined with high performance liquid chromatography for determination of Patulin in Apple Juice. *Food Anal. Method* **2017**, *10*, 210–218. [CrossRef]

9. Khorrami, A.; Taherkhani, M. Synthesis and Evaluation of a Molecularly Imprinted Polymer for Pre-concentration of Patulin from Apple Juice. *Chromatographia* **2011**, *73*, 151–156. [CrossRef]

10. Turiel, E.; Martín-Esteban, A. Molecularly imprinted stir bars for selective extraction of thiabendazole in citrus samples. *J. Sep. Sci.* **2012**, *35*, 2962–2969. [CrossRef] [PubMed]

11. Díaz-Bao, M.; Regal, P.; Barreiro, R.; Miranda, J.; Cepeda, A. Magnetic molecularly imprinted stirring bar for isolation of patulin using grafting technique. *Int. Electron. Conf. Synth. Org. Chem.* **2015**, *19*, d002. [CrossRef]

12. Lucci, P.; Moret, S.; Bettin, S.; Conte, L. Selective solid-phase extraction using a molecularly imprinted polymer for the analysis of patulin in apple-based foods. *J. Sep. Sci.* **2017**, *40*, 458–465. [CrossRef] [PubMed]

13. Cao, J.; Kong, W.; Zhou, S.; Yin, L.; Wan, L.; Yang, M. Molecularly imprinted polymer-based solid phase clean-up for analysis of ochratoxin A in beer, red wine, and grape juice. *J. Sep. Sci.* **2013**, *36*, 1291–1297. [CrossRef] [PubMed]

14. Razali, M.; Kim, J.F.; Attfield, M.; Budd, P.M.; Drioli, E.; Lee, Y.M.; Szekely, G. Sustainable wastewater treatment and recycling in membrane manufacturing. *Green Chem.* **2015**, *17*, 5196–5205. [CrossRef]

15. Vasapollo, G.; Del Sole, R.; Mergola, L.; Lazzoi, M.R.; Scardino, A.; Scorrano, S.; Mele, G. Molecularly imprinted polymers: Present and future prospective. *Int. J. Mol. Sci.* **2011**, *12*, 5908–5945. [CrossRef] [PubMed]

16. Chen, L.; Wang, X.; Lu, W.; Wu, X.; Li, J. Molecular imprinting: Perspectives and applications. *Chem. Soc. Rev.* **2016**, *45*, 2137–2211. [CrossRef] [PubMed]

17. Kupai, J.; Razali, M.; Buyuktiryaki, S.; Kecili, R.; Szekely, G. Long-term stability and reusability of molecularly imprinted polymers. *Polym. Chem.* **2017**, *8*, 666–673. [CrossRef]

18. Şengül, Ü. Comparing determination methods of detection and quantification limits for aflatoxin analysis in hazelnut. *J. Food Drug Anal.* **2016**, *24*, 56–62. [CrossRef]

*separations*

MDPI

*Article*

# A Sensitive LC-MS Method for Anthocyanins and Comparison of Byproducts and Equivalent Wine Content

Evangelos D. Trikas [1,2], Rigini M. Papi [2], Dimitrios A. Kyriakidis [2] and George A. Zachariadis [1,*]

[1]   Laboratory of Analytical Chemistry, Department of Chemistry, Aristotle University, Thessaloniki 54124, Greece; trikas.vag@hotmail.com

[2]   Laboratory of Biochemistry, Department of Chemistry, Aristotle University, Thessaloniki 54124, Greece; rigini@chem.auth.gr (R.M.P.); kyr@chem.auth.gr (D.A.K.)

\*   Correspondence: zacharia@chem.auth.gr; Tel.: +30-231-099-7707

Academic Editor: Frank L. Dorman
Received: 27 January 2016; Accepted: 18 May 2016; Published: 3 June 2016

**Abstract:** Anthocyanins are a group of phenolic compounds with great importance, not only because they play a crucial role in a wine's quality, but also due to the fact that they can have beneficial effects on human health. In this work, a method was developed for the detection and identification of these compounds in solid wastes of the wine-making industry (red grape skins and pomace), using liquid-liquid extraction (LLE) prior to the liquid chromatography-mass spectrometry technique (LC-MS). The complete process was investigated and optimized, starting from the extraction conditions (extraction solution selection, dried matter-to-solvent volume ratio, water bath extraction duration, and necessary consecutive extraction rounds) and continuing to the mobile phase selection. The extraction solution chosen was a methanol/phosphoric acid solution (95/5, $v/v$), while three rounds of consecutive extraction were necessary in order to extract the maximum amount of anthocyanins from the byproducts. During the LC-MS analysis, acetonitrile was selected as the organic solvent since, compared with methanol, not only did it exhibit increased elution strength, but it also produced significantly narrower peaks. To enable accurate identification of the analytes and optimization of the developed method, kuromanin chloride and myrtillin chloride were used as standards. Furthermore, the wine variety (Syrah) from which the specific byproducts were produced was analyzed for its anthocyanin content, leading to interesting conclusions about which anthocyanins are transferred from grapes to wine during the vinification procedure, and to what extent. The results of this study showed that the total concentration of anthocyanins estimated in wine byproducts exceeded almost 12 times the equivalent concentration in Syrah wine, while the four categories of detected anthocyanins, simple glucosides, acetyl glucosides, cinnamoyl glucosides, and pyroanthocyanins, were present in different ratios among the two samples, ranging from 18.20 to 1, to 5.83 to 1. These results not only confirmed the potential value of these byproducts, but also indicated the complexity of the anthocyanins' transfer mechanism between a wine and its byproducts.

**Keywords:** anthocyanins; wine; wine byproducts; wastes utilization; liquid chromatography; mass spectrometry

---

## 1. Introduction

Nowadays, the number of people who are interested in their physical health and hence opt for a healthier way of living is increasing, and one of the main ways to achieve this is through their nutrition. Phytochemicals, which can be found in vegetables, nuts, fruits, juices, wines and related matrices, have beneficial effects on human health. Various compound groups belong to this category, including flavonols, flavanols, and anthocyanins.

The chemical structure of anthocyanins consists of two or three moieties, an aglycon moiety called anthocyanidin, the sugar/sugars moiety, and possibly acylating groups. Structural differences among anthocyanins arise from the number of hydroxyl and methoxyl groups, the sugar moieties, and the esterification type of the molecule. These differences lead to the large amount of diverse anthocyanins (more than 600). In spite of this variety, the vast majority of anthocyanins are derived from just six anthocyanidins which are more common than the others. These are, namely, cyanidin, delphinidin, malvidin, peonidin, petunidin and pelargonidin (Figure 1) (Table 1) [1–4].

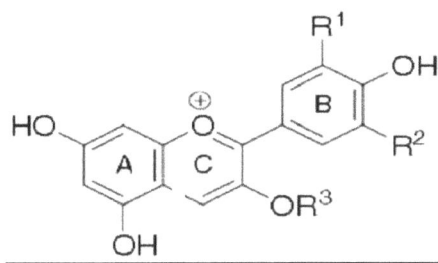

**Figure 1.** Structure of anthocyanins $R^3$ = sugar, and anthocyanidins $R^3$ = H; $R^1$ = OH / OCH$_3$ / H; $R^2$ = OH / OCH$_3$ / H; $R^3$ = H/Glc (Glycine).

**Table 1.** More common anthocyanins and anthocyanidins.

| Name | Abrev. | $R^1$ | $R^2$ | $R^3$ |
|---|---|---|---|---|
| Delphinidin | Dp | OH | OH | H |
| Delphinidin-3-O-glucoside | Dp-3-O-glu | OH | OH | Glc |
| Petunidin | Pt | OH | OCH$_3$ | H |
| Petunidin-3-O-glucoside | Pt-3-O-glu | OH | OCH$_3$ | Glc |
| Malvidin | Mv | OCH$_3$ | OCH$_3$ | H |
| Malvidin-3-O-glucoside | Mv-3-O-glu | OCH$_3$ | OCH$_3$ | Glc |
| Cyanidin | Cn | OH | H | H |
| Cyanidin-3-O-glucoside | Cn-3-O-glu | OH | H | Glc |
| Peonidin | Pn | OCH$_3$ | H | H |
| Peonidin-3-O-glucoside | Pn-3-O-glu | OCH$_3$ | H | Glc |
| Pelargonidin | Pg | H | H | H |
| Pelargonidin-3-O-glucoside | Pg-3-O-glu | H | H | Glc |

One of the most interesting and distinguishing characteristics of the anthocyanins group is that in different pH conditions, their structure changes, resulting in a change of their color as well. This characteristic property is used to measure the amount of the total monomeric anthocyanin content (TAC) in samples, in an easy, quick, and cost-effective way. This is known as the "pH Differential Method" and was introduced by Sondheimer and Kertesz in 1948. Nowadays, the two pH values used for the above purpose are 1.0 and 4.5, respectively, and in order to avoid any potential interference with the calculation of the TAC, the absorbance is measured both at 520 nm and at 700 nm.

However, in order to evaluate the concentration of individual anthocyanins, other selective analytical techniques must be employed. In the past, capillary electrophoresis and various classic chromatographic methods such as paper and thin-layer chromatography were used [5–11]. Nowadays, the separation technique which is commonly applied is liquid chromatography using mainly C18 reversed-phase columns. This technique combined with UV-Vis (PDA (Photodiode array)) and/or mass spectrometry detectors has been used for the determination of anthocyanin content in various matrices [12–28].

One of the most common sources of anthocyanins in the human diet is red wine. Therefore, the determination of these compounds is of great importance, since, in conjunction with other phenolic compounds [29], inorganic content (such as calcium) [30], added sulfites [31], and some off-flavor substances [32,33], they are known to play a crucial role in defining the sensorial characteristics of wines.

Two papers that study the effect of the presence of lees on the phenolic compounds of wine and apply the HPLC-UV-Vis technique are those published by Fernandez *et al.* [12] and by Mazauric and Salmon [13]. The same technique was also used in other studies [14–16] for the determination of anthocyanins in wines, grape extracts, and red and black currant extracts. HPLC-DAD-MS has been widely used for the analysis of anthocyanins in various samples, such as purple corn cob [17–22], while the LC-nano-DESI-MS has been used in red wine analysis [23]. Last but not least, the use of LC-MS/MS has been applied in order to analyze wine grape skins [24], berry skins [22,25], human saliva [26], elderberry extracts [27] and red wine vinegar [28].

In this paper, a sensitive method for the quantitative extraction and determination of anthocyanins from winery byproducts is presented. The specific method investigates and optimizes not only the extraction procedure, but also the liquid chromatography-mass spectrometry (LC-PDA-MS) determination parameters. The purpose of this research was to investigate to what extent the anthocyanins are distributed between the wine (Syrah) and the byproducts that are produced during its vinification procedure, and if the estimated ratio of total anthocyanins between byproducts and wine is the same for every anthocyanin group or not. It is the first time that a comparative study about the concentration of anthocyanins' between a specific wine variety (Syrah) and its byproducts is made and the obtained results could point to a direction towards the possible benefits from the utilization of these "wastes".

## 2. Materials and Methods

### 2.1. Instrumentation

The operating conditions of LC-PDA-MS system and the applied elution program are given in detail in Table 2. A Shimadzu LCMS-2010 EV (Shimadzu, Kyoto, Japan), using electrospray ionization (ESI) process and a quadrupole mass analyzer, equipped with an SPD-M20A PDA (Shimadzu Corporation, Kyoto, Japan) detector was used for the chromatographic separation and identification of anthocyanins. The ion transmission to the quadrupole analyzer was enhanced using a Q-array and octapole configuration while ions were detected with a secondary electron multiplier detector (Shimadzu, Kyoto, Japan). A gradient elution program was employed, using water/formic acid (99/1, $v/v$) and acetonitrile/formic acid (99/1, $v/v$) as elution solvents. The flow rate was 0.5 mL· min$^{-1}$ with a 45 min gradient elution program as follows: 0 min, 10% B; 0–5 min, 10%–22% B; 5–14 min, 22%–28% B; 14–35 min, 28%–42% B; 35–40 min, 42%–100% B; 40–45 min, 100%–10% B.

**Table 2.** Operating conditions of liquid chromatography-mass spectrometry (LC-PDA-MS) instrument. ESI: electrospray ionization.

| Parameter | Value |
| --- | --- |
| Ionization process | ESI |
| Ionization mode | Positive |
| Secondary electron multiplier detector voltage | 1.8 kV |
| Capillary voltage | 4.0 kV |
| Mass scanning range | 200 m/z–1200 m/z |
| Solvent A (volume/volume) | Water/Formic Acid (99/1) |
| Solvent B (volume/volume) | Acetonitrile/Formic Acid (99/1) |
| Mobile phase flow-rate | 0.5 mL/min |
| Injection volume | 20 μL |
| Chromatographic Column | Dionex, RP-C18, 150 mm × 4.6 mm × 5 μm |
| Elution program type | Gradient |

## 2.2. Reagents and Solutions

All reagents used were of analytical-reagent grade and deionized water was used for preparation of all aqueous solutions. Methanol (MeOH, Chem-Lab, Zedelgem, Belgium), ethanol (EtOH, Chem-Lab), acetic acid (CH$_3$COOH, Panreac, Barcelona, Spain), formic acid (HCOOH, Panreac), and ortho-phosphoric acid (H$_3$PO$_4$ 85%, Panreac) were used for the preparation of the extraction solutions tested during the optimization of the extraction procedure. Two buffer solutions were prepared for the application of the pH Differential Method, using potassium chloride (KCl, J.T. Baker, Deventer, The Netherlands), hydrochloric acid (HCl 37%, Carlo Erba Reagents, Milan, Italy), acetic acid (CH$_3$COOH, Panreac), and sodium acetate (CH$_3$COONa, J.T. Baker). Acetonitrile (ACN, LC-MS grade, Sigma-Aldrich, St. Louis, MO, USA), methanol (MeOH, LC-MS grade, Sigma-Aldrich), and water (H$_2$O, Thermo Fischer Scientific, Waltham, MA, USA), were used as elution solvents during liquid chromatography.

Methanol solutions of kuromanin chloride (cyanidin-3-O-glucoside, C$_{21}$H$_{21}$O$_{11}$Cl, $\geq$96%, Extrasynthese, Lyon, France), Myrtillin chloride (delphinidin-3-O-glucoside, C$_{21}$H$_{21}$O$_{12}$Cl, Extrasynthese), and Oenin chloride (malvidin-3-O-glucoside chloride, C$_{23}$H$_{25}$O$_{12}$Cl, Extrasynthese) were prepared by dissolving 1 mg of the standard in 5 mL of methanol. In this way three standard solutions containing 200 mg·L$^{-1}$ of cyanidin-3-O-glucoside chloride, 200 mg·L$^{-1}$ of delphinidin-3-O-glucoside chloride, and 200 mg·L$^{-1}$ of malvidin-3-O-glucoside chloride, respectively, were prepared. Solutions at eight different concentration levels were prepared by diluting the above stock solutions appropriately to yield concentrations of 0.01 mg·L$^{-1}$, 0.10 mg·L$^{-1}$, 1.00 mg·L$^{-1}$, 2.50 mg·L$^{-1}$, 10.0 mg·L$^{-1}$, 25.0 mg·L$^{-1}$, 50.0 mg·L$^{-1}$, and 100 mg·L$^{-1}$. The wine variety Syrah and the byproducts produced during its vinification procedure were obtained from the 'Ktima Gerovassiliou' winery, Thessaloniki, Greece.

## 2.3. Extraction Procedure

The solid wastes were stored in a freezer ($-20$ °C). Amount of these wastes were freeze-dried and the resulted powder was collected and stored in the fridge (5 °C). The analogy between the amount before the freeze-drying and the resulting powder was 3 to 1.

After that, a certain quantity of the powder was mixed with the extraction solution. The mixture was vortexed, transferred to an ultrasound bath, and after that to a water bath at 45 °C for 2 h. The next step was the centrifugation of the mixture at 4000 rpm for 10 min and the collection of the supernatant.

The final step, in order to evaluate the effectiveness of the extraction procedure was the application of the pH Differential Method in the collected supernatant fluid. In order to do so, two buffers had previously been prepared, one for pH = 1, HCl-KCl, and one for pH = 4.5, CH$_3$COOH-CH$_3$COONa.

## 3. Results and Discussion

### 3.1. Optimization of the Extraction Procedure

Four parameters were investigated in order to optimize the extraction procedure of anthocyanins from these byproducts. Those were the selection of the extraction solution, the determination of the optimum analogy between dried matter and the extraction solution, the water bath extraction duration, and the determination of consecutive extraction rounds needed in order to retrieve the maximum amount of anthocyanins from the examined wastes. During the optimization studies of the extraction procedure, the pH differential method was applied for the evaluation of results.

### 3.1.1. Extraction Solution Selection

Anthocyanins are extracted using organic solvents, and the most commonly used are methanol, ethanol, and acetone. In many cases acidified solutions are used because in this way it becomes easier for the anthocyanins to pass through the cellular membranes and be obtained by the extraction solution. Weak organic acids, or low concentrations of strong acids such as HCl, should be used.

In this work, two organic solvents were investigated, methanol and ethanol, in various combinations with acids and water. In total, 14 extraction solutions were tested (Figure 2), using formic, acetic, and phosphoric acid, each one at two different concentration levels, and water. The results showed that methanol performs better than ethanol as a solvent for the extraction of anthocyanins from the specific substrate, as has also been reported in the paper of Metivier *et al.* [34]. More specifically, the methanol/phosphoric acid solution (95/5, *v/v*) was chosen to be used as the extraction solution in this work since it exhibits the best results compared to all other solutions tested.

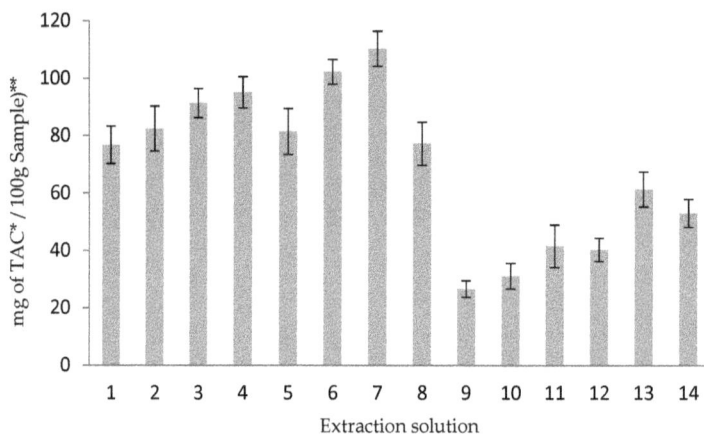

**Figure 2.** Extraction solutions tested for the extraction of anthocyanins, methanol/water 50/50 (1); methanol (2); methanol/acetic acid 99.5/0.5 (3); methanol/acetic acid 95/5 (4); methanol/formic acid 95/5 (5); methanol/phosphoric acid 98/2 (6); methanol/phosphoric acid 95/5 (7); ethanol/water 50/50 (8); ethanol (9); ethanol/acetic acid 99.5/0.5 (10); ethanol/acetic acid 95/5 (11); ethanol/formic acid 95/5 (12); ethanol/phosphoric acid 98/2 (13); ethanol/phosphoric acid 95/5 (14). The results are expressed as total monomeric anthocyanin content (TAC) per 100 g of dried wastes. * Total Anthocyanin Content; ** Average of $n = 5$ measurements and $\pm$ one standard deviation (SD).

### 3.1.2. Dried Matter–to–Solvent Volume Ratio Optimization

During this stage, four dried matter-to-solvent volume ratios were tested: 1/10, 1/20, 1/30, 1/40 (g/mL) (Table 3). Statistical analysis of the results was employed for the selection of the optimum dried matter-to-solvent volume ratio, specifically the "Student's *t*-test" at a 95% confidence level. It was found that the difference between the 1/20 and 1/10 ratios is statistically significant, while among the 1/20, 1/30, and 1/40 ratios there is no statistically significant difference. Consequently, the 1/20 ratio was selected, since the maximum amount of anthocyanins is retrieved from the examined wine byproducts with the least use of the extraction solvent.

**Table 3.** Dry sample mass–to–solvent volume ratio in correlation with total anthocyanins concentration per 100 g of sample.

| Dry Matter Mass (g) | Solvent Volume (mL) | Dry Matter Mass (g)/Solvent Volume (mL) | mg TAC */100 g sample ** |
|---|---|---|---|
| 0.5 | 5.0 | 1/10 | 99.9 ± 10.1 |
| 0.5 | 10.0 | 1/20 | 117 ± 10.9 |
| 0.5 | 15.0 | 1/30 | 116 ± 10.8 |
| 0.5 | 20.0 | 1/40 | 119 ± 11.3 |

* Total Anthocyanin Content; ** All results are mean values of $n = 5$ measurements and uncertainty is expressed as the standard deviation.

### 3.1.3. Optimization of Water Bath Extraction Duration

Four different dried matter–extraction solution mixtures were prepared and they remained in the water bath at 45 °C for one, two, three and four hours, respectively (Table 4). Statistical analysis of the results was employed for the selection of the optimum water bath extraction duration. The "Student's *t*-test" at a 95% confidence level was used again. It was found that the difference between one and two hours is statistically significant, while among two, three, and four hours there is no statistically significant difference. Therefore, two hours was selected as the optimum water bath extraction duration, since the maximum amount of anthocyanins is retrieved in the minimum amount of time.

**Table 4.** Water bath extraction duration, in correlation with anthocyanins concentration to supernatant.

| Duration Time (h) in Water Bath | mg TAC */100 g Sample ** |
|:---:|:---:|
| 1 | 93.4 ± 9.41 |
| 2 | 129 ± 11.2 |
| 3 | 128 ± 11.1 |
| 4 | 130 ± 11.6 |

\* Total Anthocyanin Content; ** All results are mean values of $n = 5$ measurements and uncertainty is expressed as the standard deviation.

### 3.1.4. Consecutive Extraction Rounds Optimization

As illustrated in Table 5, 81% of anthocyanins that can be retrieved from these wastes have been obtained after one extraction round only. However, in order to retrieve the maximum possible amount of anthocyanins from the byproducts, two additional extraction rounds are necessary.

**Table 5.** Consecutive extraction rounds in correlation with anthocyanins concentration to supernatant.

| Successive Extraction Number | mg TAC */100 g Sample ** |
|:---:|:---:|
| 1 | 130 ± 14 |
| 2 | 27.6 ± 3.3 |
| 3 | 2.92 ± 0.33 |
| 4 | 0.08 ± 0.02 |

\* Total Anthocyanin Content; ** All results are mean values of $n = 5$ measurements and uncertainty is expressed as the standard deviation.

### 3.2. Mobile Phase Selection

Using the same elution program, two organic elution solvents were tested as Solvent B (Table 2). The first was acetonitrile and the second was methanol (Figure 3). The two compounds are delphinidin-3-O-glucoside and cyanidin-3-O-glucoside (peaks 1 and 2). After the completion of the experiments, acetonitrile was selected as the optimum solvent because the chromatograms obtained with the use of that specific solvent outclassed the ones with methanol in two parameters.

First of all, the peaks' widths were much narrower compared with the methanol-eluted peaks, improving the resolution between these peaks eluting in close elution times. Secondly, using the same elution program, the peaks were eluted much earlier, thus reducing the duration of the analysis. The elution time could also be reduced through the use of methanol as the organic solvent but this would require an increase in the ratio between the organic solvent and the water used, producing an increased volume of organic wastes.

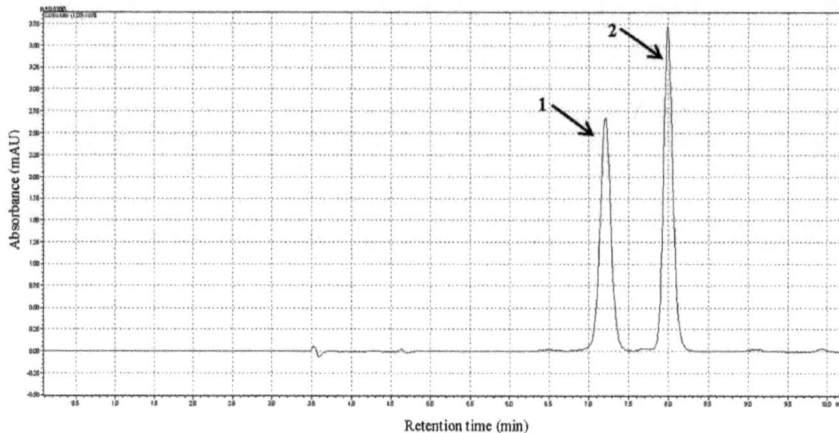

(a)

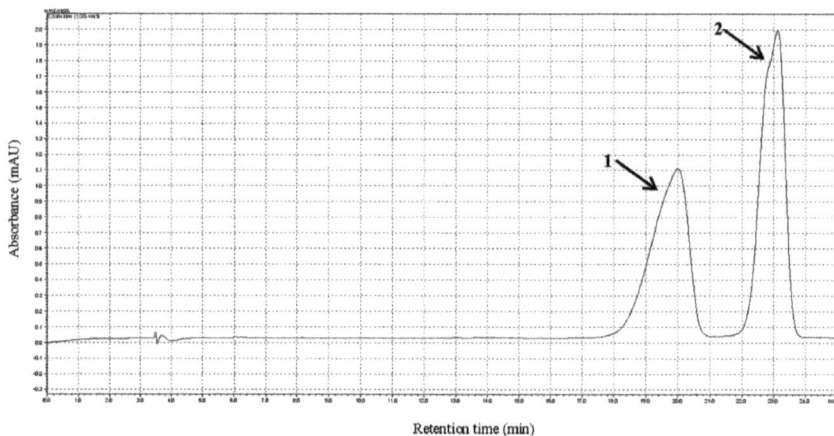

(b)

**Figure 3.** HPLC-PDA * chromatograms obtained using (**a**) acetonitrile as Solvent B and (**b**) methanol as Solvent B (See Table 2). Peaks 1 and 2 are delphinidin-3-O-glucoside and cyanidin-3-O-glucoside, respectively. * High-performance liquid chromatography-Photodiode array detector.

### 3.3. Figures of Merit of the Proposed Method

The performance characteristics of the proposed method were derived from calibration using a series of standard samples at different concentration levels of oenin chloride (malvidin-3-O-glucoside chloride) solutions. The results, using linear regression analysis on peak area signals, are listed in Table 6, together with the repeatability, the limit of detection (LOD) and the limit of quantification (LOQ) of the method. The repeatability was calculated from $n = 5$ measurements at a 10.0 mg·$L^{-1}$ concentration level. The detection and quantification limits were determined based on the standard deviation of the standard solution at the lowest concentration level. In particular, the limit of detection (LOD) was calculated as three times and the limit of quantitation (LOQ) as 10 times the above standard deviation divided by the slope of the calibration curve. The LOD and LOQ thus were computed as 9.0 ng·$mL^{-1}$ and 27 ng·$mL^{-1}$, respectively. The dynamic range extends between 0.03 and 10.0 mg·$L^{-1}$.

**Table 6.** Analytical characteristics of the proposed LC-MS method with ESI. RSD: Relative Standard Deviation.

| Slope, S (Peak Area/(mg· L$^{-1}$)) | Correlation Coefficient, r | Instrumental RSD * | LOD (ng· mL$^{-1}$) | Dynamic Range (mg· L$^{-1}$) |
|---|---|---|---|---|
| $1.34 \times 10^5$ | 0.9996 | 5.8% | 9.0 | 0.03–10.0 |

* Values were calculated from $n = 5$ repetitive measurements of the standard.

*3.4. Identification and Quantification of Anthocyanins in Wine Variety (Syrah) and Its Byproducts*

The proposed method was applied to two real samples, to the wine variety Syrah, and to the byproducts that were produced during the vinification procedure of the specific wine, and there were four main remarks after the analysis.

In total, 23 anthocyanins were identified. All of them were present in the wine, while 21 of them were detected in the wine byproducts (Figures 4 and 5). The two anthocyanins that were detected in the wine, even in small amounts, but not in the wine byproducts were the cyanidin-3-O-glucoside and the delphinidin-glucoside-pyruvate derivative.

During the preliminary studies the samples' mass spectra were acquired using both the scan and SIM (Selected-ion monitoring) mode for specific anthocyanins. However, during the last steps of the determination procedure, all of the detected anthocyanins were quantified using the SIM mode.

The analyzed samples, both the byproducts' extract and the wine, apart from anthocyanins, were also very rich in many other phenolic compounds that could co-elute with some of the compounds of interest. By combining the use of the SIM mode at specific $m/z$ (mass-to-charge ratio) values with the use of standard compounds and elution order data, there were no interferences problems that occurred during the quantification procedure due to matrix's complexity.

In both samples the dominant anthocyanins were the malvidin-based ones, malvidin-3-O-glucoside, Mv-3-(6″-acetylglucoside), and Mv-3-(6″-p-coumaroylglucoside). However, the analogy in which those three compounds were present in each sample was quite different. In wine byproducts the most abundant compound was the Mv-3-(6″-acetylglucoside), and the concentration ratio of Mv-3-O-glucoside/Mv-3-(6″-acetylglucoside)/Mv-3-(6″-p-coumaroylglucoside) equals 1.67/1.73/1.00, while in wine the most abundant compound was the Mv-3-O-glucoside, and the respective ratio equals 3.48/2.79/1.00.

Furthermore, after the analysis of the samples, it was found that the four categories of the detected anthocyanins, simple glucosides, acetyl glucosides, cinnamoyl glucosides, and pyroanthocyanins, were not present in a standard ratio among the two samples, wine byproducts and wine (Table 7). The simple and the acetyl glucosides exhibited a ratio of almost 10 to 1, while cinnamoyl glucosides had a ratio of 18.20 to 1, and pyroanthocyanins had a ratio of 5.83 to 1.

**Table 7.** Anthocyanin groups concentrations and ratio between the two analyzed samples. DW: Dried Wastes.

| Anthocyanins Group / Samples | Concentration in Wine Byproducts (mg 1000 g$^{-1}$ of DW) * | Concentration in Wine (Syrah) (mg· L$^{-1}$ of Wine) * | Ratio Between Samples |
|---|---|---|---|
| Simple glucosides | 4367 | 457.7 | 9.54 |
| Acetyl glucosides | 5224 | 421.3 | 12.40 |
| Cinnamoyl glucosides | 2779 | 152.7 | 18.20 |
| Pyroanthocyanins | 73.3 | 12.6 | 5.82 |
| Total anthocyanins | 12,443.3 | 1044.3 | 11.92 |

* Values were calculated from $n = 5$ repetitive measurements of the standard.

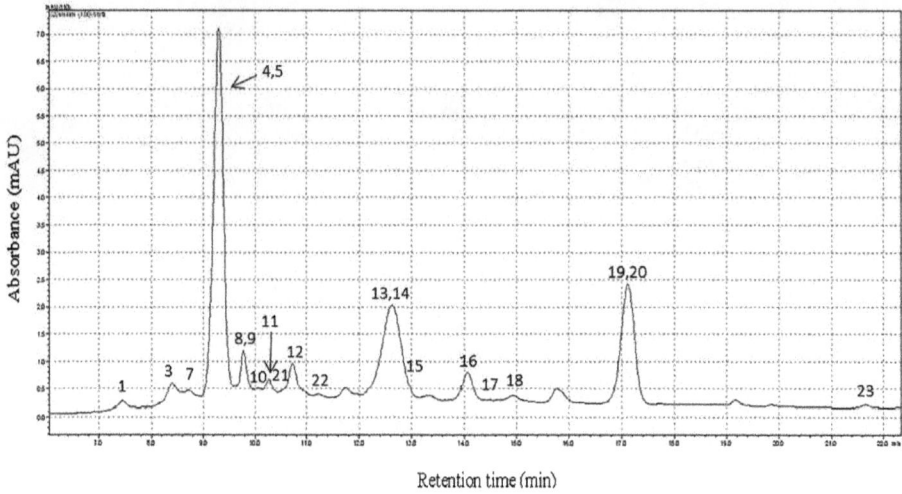

**Figure 4.** HPLC-PDA chromatographic profile of the anthocyanins determined in Syrah grape pomace sample. Peak numbers correspond to those compounds mentioned in Table 8.

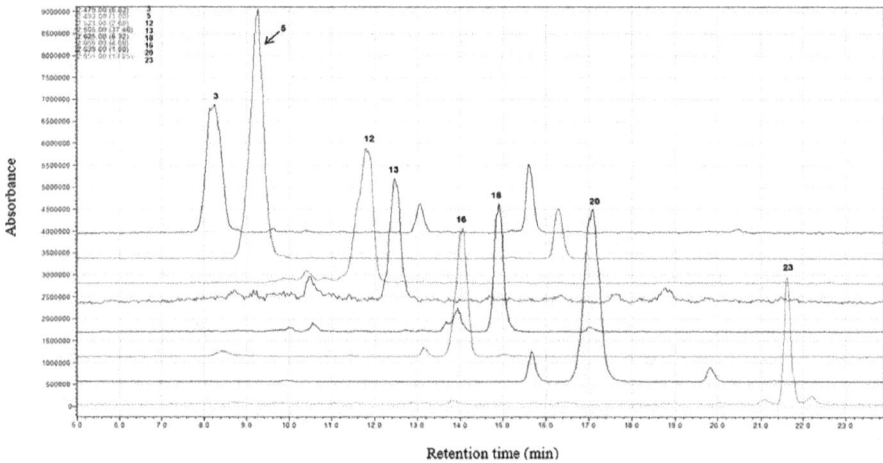

**Figure 5.** LC-MS chromatogram of selected anthocyanins, determined in Syrah grape pomace sample, at SIM mode for characteristics $m/z$ ($M^+$). The numbers above the peaks corresponds to anthocyanins as listed in Table 8. The five-digit numbers at the upper left corner of the figure correspond to the characteristic $m/z$ ($M^+$) of detected compounds, and the number in the parenthesis indicates a multiplying factor of each peak in the chromatogram.

All the above clearly show that the transfer mechanism of anthocyanins between a wine (Syrah) and its byproducts is not a simple procedure that functions and performs in the same way for every anthocyanin, indicating that it is probably affected by the chemical structure of every compound.

The total concentration of anthocyanins in wine byproducts was estimated to be almost 12 times higher than the equivalent concentration in Syrah wine (Table 8). This is extremely important since it indicates the potential that these "wastes" could have as a raw material for the recovery of that kind of compound, thus revealing their true value.

**Table 8.** Anthocyanin compounds identified by LC-PDA-MS in the analyzed samples. $t_r$: retention time.

| Peak No. | Identity | $t_r$ (min) | $m/z$ ($M^+$) | Concentration (mg $1000 \ g^{-1}$ of DW) * | Concentration (mg· $L^{-1}$ of Wine) * |
|---|---|---|---|---|---|
| **Simple glucosides** | | | | | |
| 1 | Delphinidin-3-O-glucoside | 7.14 | 465 | 4.0 ± 0.2 | 9.6 ± 0.7 |
| 2 | Cyanidin-3-O-glucoside | 7.95 | 449 | - | 2.9 ± 0.1 |
| 3 | Petunidin-3-O-glucoside | 8.25 | 479 | 89.0 ± 5.7 | 76.8 ± 5.0 |
| 4 | Peonidin-3-O-glucoside | 9.06 | 463 | 51.7 ± 3.7 | 26.4 ± 1.7 |
| 5 | Malvidin-3-O-glucoside | 9.27 | 493 | 4222 ± 178 | 342.0 ± 14.0 |
| **Acetyl glucosides** | | | | | |
| 6 | Dp-gls-pyruvate derivative | 7.53 | 533 | - | 6.0 ± 0.5 |
| 7 | Pt-gls-pyruvate derivative | 8.67 | 547 | 1.1 ± 0.1 | 2.4 ± 0.1 |
| 8 | Dp-3-(6″-acetylglucoside) | 9.72 | 507 | 24.9 ± 1.5 | 7.6 ± 0.5 |
| 9 | Mv-3-gls-pyruvate (Vitisin A) | 9.84 | 561 | 54.0 ± 3.2 | 10.7 ± 0.9 |
| 10 | Vitisin B (Mv derivative) | 10.20 | 517 | 1.0 ± 0.1 | 8.5 ± 0.7 |
| 11 | Cn-3-(6″-acetylglucoside) | 10.47 | 491 | 332.1 ± 12.5 | 3.3 ± 0.2 |
| 12 | Pt-3-(6″-acetylglucoside) | 10.92 | 521 | 423.3 ± 27.2 | 79.2 ± 5.0 |
| 13 | Pn-3-(6″-acetylglucoside) | 12.48 | 505 | 6.4 ± 0.5 | 29.9 ± 1.9 |
| 14 | Mv-3-(6″-acetylglucoside) | 12.57 | 535 | 4381 ± 306 | 273.7 ± 16.1 |
| **Cinnamoyl glucosides** | | | | | |
| 15 | Dp-3-(6″-p-coumaroylglucoside) | 12.93 | 611 | 7.8 ± 0.8 | 5.7 ± 0.4 |
| 16 | Mv-3-(6″-caffeoylglucoside) | 14.07 | 655 | 95.3 ± 6.2 | 2.2 ± 0.2 |
| 17 | Cn-3-(6″-p-coumaroylglucoside) | 14.67 | 595 | 1.3 ± 0.1 | 3.4 ± 0.3 |
| 18 | Pt-3-(6″-p-coumaroylglucoside) | 14.91 | 625 | 55.7 ± 3.1 | 15.7 ± 0.9 |
| 19 | Pn-3-(6″-p-coumaroylglucoside) | 17.01 | 609 | 84.7 ± 6.1 | 27.5 ± 1.7 |
| 20 | Mv-3-(6″-p-coumaroylglucoside) | 17.16 | 639 | 2534 ± 142 | 98.3 ± 6.9 |
| **Pyroanthocyanins** | | | | | |
| 21 | Mv-3-(6″-acetylglucoside) pyruvate | 10.68 | 603 | 10.0 ± 0.8 | 7.6 ± 0.5 |
| 22 | Mv-3-glucoside-ethyl-catechin | 11.19 | 809 | 8.5 ± 0.6 | 2.1 ± 0.2 |
| 23 | Mv-3-(6″-acetylglucoside)-4-vinylphenol | 21.66 | 651 | 54.9 ± 4.0 | 2.9 ± 0.2 |
| | Total (g/kg) | | | 12.44 | 1.04 |

\* All results are mean values of $n = 5$ measurements and uncertainty is expressed as standard deviation.

## 4. Conclusions

A sensitive method for the determination of anthocyanin compounds in winery byproducts was developed. The complete extraction procedure, along with some of the LC-PDA-MS analysis parameters, was investigated and optimized. The influence of five parameters was studied during the whole procedure, four during the extraction and one affecting the analysis of the supernatant. In the first category, the selection of the extraction solution, the dried matter–to–solvent volume ratio, the water bath extraction duration, and the necessary consecutive extraction rounds are included, while the second category covers the mobile phase selection. None of these parameters proved to be irrelevant with the final outcome of the analysis, and all of them could play a more or less crucial role in the obtained results. The last step of this work was the application of the method to real samples. A wine variety (Syrah) and the byproducts produced after the vinification procedure of the specific wine were analyzed for their anthocyanin content. This was the first time that an approach such as this was followed in order to evaluate which anthocyanins transferred from grapes to wine during the fermentation process, and to what degree. The results revealed that the transfer mechanism of anthocyanins is quite complex and affects every compound differently. Moreover, the anthocyanin content in the specific byproducts was found to be 12 times higher compared to the content in the wine (Syrah). Considering the fact that these compounds can have beneficial effects on human health, these results confirm the potential value of these byproducts and indicate a direction that should be followed towards their utilization.

**Acknowledgments:** This work was supported by "11SYN_2_1992" action "COOPERATION 2011" of EYDE-ETAK funded by the Operational Program "Competitiveness and Entrepreneurship" (EPAN-II).

**Author Contributions:** Evangelos Trikas, Rigini Papi, Dimitrios Kyriakidis and George Zachariadis conceived and designed the experiments; Evangelos Trikas performed the experiments as part of his Ph.D. thesis, analyzed the data, and wrote the paper.

**Conflicts of Interest:** The authors declare no conflict of interest.

## Abbreviations

The following abbreviations are used in this manuscript:

| | |
|---|---|
| Mv | Malvidin |
| LC | liquid chromatography |
| PDA | photodiode array |
| MS | mass spectrometry |

## References

1. Bueno, J.; Ramos-Escudero, F.; Saez-Plaza, P.; Munoz, A.; Navas, M.J.; Asuero, A. Analysis and antioxidant capacity of anthocyanin pigments. Part I: General considerations concerning polyphenols and flavonoids. *Crit. Rev. Anal. Chem.* **2012**, *42*, 102–105. [CrossRef]
2. Bueno, J.; Saez-Plaza, P.; Ramos-Escudero, F.; Jimenez, A.; Fett, R.; Asuero, A. Analysis and antioxidant capacity of anthocyanin pigments. Part II: Chemical structure, color, and intake of anthocyanins. *Crit. Rev. Anal. Chem.* **2012**, *42*, 126–151. [CrossRef]
3. Navas, M.; Jimenez, A.; Bueno, J.; Saez-Plaza, P.; Asuero, A. Analysis and Antioxidant Capacity of Anthocyanin Pigments. Part III: An Introduction to Sample Preparation and Extraction. *Crit. Rev. Anal. Chem.* **2012**, *42*, 284–312. [CrossRef]
4. Navas, M.; Jimenez, A.; Bueno, J.; Saez-Plaza, P.; Asuero, A. Analysis and antioxidant capacity of anthocyanin pigments. Part IV: Extraction of anthocyanins. *Crit. Rev. Anal. Chem.* **2012**, *42*, 313–342. [CrossRef]
5. Corradini, E.; Foglia, P.; Giansanti, P.; Gubbiotti, R.; Samperi, R.; Lagana, A. Flavonoids: Chemical properties and analytical methodologies of identification and quantitation in foods and plants. *Nat. Prod. Res.* **2011**, *25*, 469–495. [CrossRef] [PubMed]
6. Ignat, I.; Volf, I.; Popa, V. A critical review of methods of characterization of polyphenol compounds in fruit and vegetables. *Food Chem.* **2011**, *126*, 1821–1835. [CrossRef] [PubMed]
7. Cote, J.; Caillet, S.; Doyon, G.; Sylvain, J.; Lacroix, M. Analyzing cranberry bioactive compounds. *Crit. Rev. Food Sci. Nutr.* **2010**, *50*, 872–888. [CrossRef] [PubMed]
8. Unger, M. Capillary electrophoresis of natural products: current applications and recent advances. *Planta Med.* **2009**, *75*, 735–745. [CrossRef] [PubMed]
9. Welch, C.; Wu, Q.; Simon, J. Recent advances in anthocyanin analysis and characterization. *Curr. Anal. Chem.* **2008**, *4*, 75–101. [CrossRef] [PubMed]
10. Mazza, G.; Cacae, J.; Kay, C. Methods of analysis for anthocyanins in plant and biological fluids. *J. AOAC Int.* **2004**, *87*, 129–145. [PubMed]
11. Kong, J.; Chia, L.; Goh, N.; Chia, T.; Brouillard, R. Analysis and biological activities of anthocyanins. *Phytochemistry* **2003**, *64*, 923–933. [CrossRef]
12. Fernandez, O.; Martinez, O.; Hernandez, Z.; Guadalupe, Z.; Ayestaran, B. Effect of the presence of lysated lees on polysaccharides, color and main phenolic compounds of red wine during barrel ageing. *Food Res. Int.* **2011**, *44*, 84–91. [CrossRef]
13. Mazauric, J.; Salmon, J. Interactions between yeast lees and wine polyphenols during simulation of wine aging. II. Analysis of desorbed polyphenol compounds from yeast lees. *J. Agric. Food Chem.* **2006**, *54*, 3876–3881. [CrossRef] [PubMed]

14. Sanchez-Moreno, C.; Cao, G.; Ou, B.; Prior, R. Anthocyanin and proanthocyanidin content in selected white and red wines. Oxygen radical absorbance capacity comparison with nontraditional wines obtained from highbush blueberry. *J. Agric. Food Chem.* **2003**, *51*, 4889–4896. [CrossRef] [PubMed]

15. Lapornik, B.; Prosek, M.; Wondra, A. Comparison of extracts prepared from plant by-products using different solvents and extraction time. *J. Food Eng.* **2005**, *7*, 214–222. [CrossRef]

16. Revilla, E.; Ryan, J.; Ortega, G. Comparison of several procedures used for the extraction of anthocyanins from red grapes. *J. Agric. Food Chem.* **1998**, *46*, 4592–4597. [CrossRef]

17. Pascual-Teresa, S.; Santos, B.C.; Rivas-Gonzalo, C. LC-MS analysis of anthocyanins from purple corn cob. *J. Sci. Food Agric.* **2002**, *82*, 1003–1006. [CrossRef]

18. Arribas, M.; Gomez, C.; Alvarez, P. Evolution of red wine anthocyanins during malolactic fermentation, postfermentative treatments and ageing with lees. *Food Chem.* **2008**, *109*, 149–158. [CrossRef] [PubMed]

19. Corrales, M.; Garcia, A.; Butz, P.; Tauscher, B. Extraction of anthocyanins from grape skins assisted by high hydrostatic pressure. *J. Food Eng.* **2009**, *90*, 415–421. [CrossRef]

20. Ju, Z.; Howard, L. Effects of solvent and temperature on pressurized liquid extraction of anthocyanins and total phenolics from dried red grape skin. *J. Agric. Food Chem.* **2003**, *51*, 5207–5213. [CrossRef] [PubMed]

21. Corrales, M.; Toepfl, S.; Butz, P.; Knorr, D.; Tauscher, B. Extraction of anthocyanins from grape by-products assisted by ultrasonics, high hydrostatic pressure or pulsed electric fields: A comparison. *Innov. Food Sci. Emerg. Technol.* **2008**, *9*, 85–91. [CrossRef]

22. Revilla, I.; Perez, M.S.; Gonzalez, S.M.; Beltran, S. Identification of anthocyanin derivatives in grape skin extracts and red wines by liquid chromatography with diode array and mass spectrometric detection. *J. Chromatogr. A* **1999**, *847*, 83–90. [CrossRef]

23. Hartmanova, L.; Ranc, V.; Papouskova, B.; Bednar, P.; Havlicek, V.; Lemr, K. Fast profiling of anthocyanins in wine by desorption nano-electrospray ionization mass spectrometry. *J. Chromatogr. A* **2010**, *1217*, 4223–4228. [CrossRef] [PubMed]

24. Li, Z.; Pan, Q.; Cui, X.; Duan, C. Optimization on anthocyanins extraction from wine grape skins using orthogonal test design. *Food Sci. Biotechnol.* **2010**, *19*, 1047–1053. [CrossRef]

25. Liang, Z.; Wu, B.; Fan, P.; Yang, C.; Duan, W.; Zheng, X.; Liu, C.; Li, S. Anthocyanin composition and content in grape berry skin in *Vitis* germplasm. *Food Chem.* **2008**, *111*, 837–844. [CrossRef]

26. Ling, Y.; Ren, C.; Mallery, S.; Ugalde, C.; Pei, P.; Saradhi, U.; Stoner, G.; Chan, K.; Liu, Z. A rapid and sensitive LC-MS/MS method for quantification of four anthocyanins and its application in a clinical pharmacology study of a bioadhesive black raspberry gel. *J. Chromatogr. B* **2009**, *877*, 4027–4034. [CrossRef] [PubMed]

27. Duymus, G.; Goger, F.; Baser, C. *In vitro* antioxidant properties and anthocyanin compositions of elderberry extracts. *Food Chem.* **2014**, *155*, 112–119. [CrossRef] [PubMed]

28. Cerezo, A.; Cuevas, E.; Winterhalter, P.; Garcia, M.; Troncoso, A. Anthocyanin composition in *Cabernet Sauvignon* red wine vinegar obtained by submerged acetification. *Food Res. Int.* **2010**, *43*, 1577–1584. [CrossRef]

29. Papadoyannis, I.; Samanidou, V.; Antoniou, C. Gradient RP-HPLC Determination of Free Phenolic Acids in Wines and Wine Vinegar Samples after SPE, with Photodiode Array Identification. *J. Liq. Chromatogr. Relat. Tech.* **2001**, *24*, 2161–2176.

30. Themelis, D.; Tzanavaras, P.; Anthemidis, A.; Stratis, J. Direct, selective flow injection spectrophotometric determination of calcium in wines using methylthymol blue and an on-line cascade dilution system. *Anal. Chim. Acta* **1999**, *402*, 259–266. [CrossRef]

31. Tzanavaras, P.D.; Thiakouli, E.; Themelis, D. Hybrid sequential injection–flow injection manifold for the spectrophotometric determination of total sulfite in wines using o-phthalaldehyde and gas-diffusion. *Talanta* **2009**, *77*, 1614–1619. [CrossRef] [PubMed]

32. Gomez-Ariza, L.; Garcia-Barrera, T.; Lorenzo, F. Dynamic headspace coupled to perevaporation for the analysis of anisoles in wine by gas chromatography-ion-trap tandem mass spectrometry. *J. Chromatogr. A* **2004**, *1056*, 243–247. [CrossRef] [PubMed]

33. Gomez-Ariza, L.; Garcia-Barrera, T.; Lorenzo, F.; Beltran, R. Use of multiple headspace solid-phase microextraction and pervaporation for the determination of off-flavours in wine. *J. Chromatogr. A* **2006**, *1112*, 133–140. [CrossRef] [PubMed]
34. Metivier, R.; Francis, F.; Clydesdale, F. Solvent extraction of anthocyanins from wine pomace. *J. Food Sci.* **1980**, *45*, 1099–1100. [CrossRef]

*Article*

# Development of an Automated Method for Selected Aromas of Red Wines from Cold-Hardy Grapes Using Solid-Phase Microextraction and Gas Chromatography-Mass Spectrometry-Olfactometry

**Lingshuang Cai [1,2], Somchai Rice [1,3], Jacek A. Koziel [1,3,4,*] and Murlidhar Dharmadhikari [4]**

[1]  Department of Agricultural and Biosystems Engineering, Iowa State University, Ames, IA 50011, USA;
    Lingshuang.Cai@dupont.com (L.C.); somchai@iastate.edu (S.R.)
[2]  DuPont Crop Protection, Stine-Haskell Research Center, Newark, DE 19711, USA
[3]  Interdepartmental Toxicology Graduate Program, Iowa State University, Ames, IA 50011, USA
[4]  Department of Food Science and Human Nutrition, Iowa State University, Ames, IA 50011, USA;
    murli@iastate.edu
*   Correspondence: koziel@iastate.edu; Tel.: +1-515-294-4206

Received: 1 June 2017; Accepted: 30 June 2017; Published: 5 July 2017

**Abstract:** The aroma profile of red wine is complex and research focusing on aroma compounds and their links to viticultural and enological practices is needed. Current research is limited to wines made from cold-hardy cultivars (interspecific hybrids of vinifera and native N. American grapes). The objective of this research was to develop a fully automated solid phase microextraction (SPME) method, using tandem gas chromatography-mass spectrometry (GC-MS)-olfactometry for the simultaneous chemical and sensory analysis of volatile/semi-volatile compounds and aroma in cold-hardy red wines. Specifically, the effects of SPME coating selection, extraction time, extraction temperature, incubation time, sample volume, desorption time, and salt addition were studied. The developed method was used to determine the aroma profiles of seven selected red wines originating from four different cold-hardy grape cultivars. Thirty-six aroma compounds were identified from Maréchal Foch, St. Croix, Frontenac, Vincent, and a Maréchal Foch/Frontenac blend. Among these 36 aroma compounds, isoamyl alcohol, ethyl caproate, benzeneethanol, ethyl decanoate, and ethyl caproate are the top five most abundant aroma compounds. Olfactometry helps to identify compounds not identified by MS. The presented method can be useful for grape growers and wine makers for the screening of aroma compounds in a wide variety of wines and can be used to balance desired wine aroma characteristics.

**Keywords:** cold-hardy grapes; wine; gas chromatography-mass spectrometry (GC-MS); olfactometry; solid phase microextraction (SPME); aroma

## 1. Introduction

In order to understand the aroma of wine and make a marketable product, it is necessary to separate, identify, and quantify the chemical compounds that impart these aromas. Aromas to note are the primary (varietal aroma), secondary (aromas due to fermentation from yeasts), and bouquet (aromas due to aging and storage) aromas. The pleasant volatiles in wines are due to the presence of higher aliphatic alcohols, ethyl esters, and acetates [1,2]. Wine aroma increased in complexity after malo-lactic fermentation (MLF), which produces changes in the carbonyl compounds [3]. Wines that have undergone MLF can be associated with herbaceous aromas from aliphatic aldehydes [4,5] or buttery aromas from diacetyl [6]. Wine can also have off odors due to volatile sulfur compounds, described as garlic, onion, or cabbage [7]. Vinylphenols have been described as phenolic, medicinal, smoky, spicy, and clove-like [8,9]. An experienced palate can often distinguish the "foxy" characteristic of the main North American vines (*Vitis labrusca* and *Vitis rotundifolia*) from vinifera vines caused by methyl anthranilate [9,10].

Cold climate grapes are newer, and the aroma profiles are less characterized than Vinifera varieties. The identification and quantification of the most aromatic compounds can help the industry maximize the aroma quality in these wines. The varietal flavor profile was used to demonstrate good examples of wine production and the best grape growing and innovative vinification techniques. The first step in the aroma analysis of wines is to extract volatile organic compounds (VOCs). Solid phase microextraction (SPME) coupled with gas chromatography-mass spectrometry (GC-MS) is useful in extracting and pre-concentrating VOCs in wine. The inception of GC-olfactometry (GC-O) in 1964 allowed researchers to link an aroma descriptor to these separated compounds [11]. Although many detectors have been used to identify and quantify aroma compounds from wine, MS is the most widely used [12]. A review and summary of many experimental parameters are available elsewhere [12,13].

In this research, an automated headspace SPME-GCMS-O method was developed and the aroma profiles of seven cold-hardy wine samples were investigated. The chemicals in selected cold-hardy wines were isolated and tentatively identified by matching the mass spectral and aroma character. The grape and wine industry has expanded exponentially in cold climates. Therefore, there is a need to research aroma compounds and their links to grape growing and wine making practices in cold climates. Such information can be used for the monitoring of fruit maturity, developing the best viticultural and wine making practices, and the development of appropriate wine styles specific to cold climates. The flavor and aroma profiling of cold-hardy wine enables the development of high quality and unique wines. Therefore, a method was developed to evaluate the full chemical and sensory aroma profile of wine from cold climate grapes.

## 2. Materials and Methods

### 2.1. Samples, Internal Standard, SPME

Seven different red wines were obtained from various wineries in Iowa. Varieties included two Maréchal Foch from separate wineries, two Frontenac from separate wineries, a St. Croix, a Vincent, and a Maréchal Foch/Frontenac blend. Wines were not stored after initial opening for analysis. 3-nonanone (99%), CAS 925-78-0 (Sigma-Aldrich, St. Louis, MO, USA), was used as an internal standard (IS) for the semi-quantification of aroma compounds. 3-nonanone was chosen because the compound is odor-active and not present in these wine samples. The final concentration of IS in wine (0.206 mg/L) was achieved by adding 10 μL of IS in ethanol (82.5 mg/L) to each 4 mL of wine. Each wine sample bottle was opened immediately before each analysis, and triplicate runs were performed for each experiment (*n* = 3). SPME fibers with seven different coatings were purchased from Sigma-Aldrich (St. Louis, MO, USA). These coatings included: 50/30 μm Divinylbenzene (DVB)/Carboxen (CAR)/Polydimethylsiloxane (PDMS), 100 μm PDMS, 7 μm PDMS, 85 μm Polyacrylate (PA), 65 μm PDMS/DVB, 70 μm Carbowax (CW)/DVB, and 85 μm CAR/PDMS. All SPME fibers were 1 cm in length. Details of SPME fiber cores, coatings, and the internal structure can be found elsewhere [14].

The optimized method used for the automated analysis of wine aroma used a 1 cm 50/30 μm DVB/CAR/PDMS SPME fiber, 10 min extraction time at 50 °C, and 2 min desorption in the heated GC inlet. A 10 min incubation in the heated agitator was used to equilibrate VOCs in the headspace of 4 mL of the wine sample in a 10 mL vial, facilitated by the addition of 2 g of sodium chloride.

### 2.2. GC-MS-Olfactometry System

The analysis was performed on a standard 6890N GC/5973Network Platform (Agilent Technologies, Santa Clara, CA, USA) and a CTC CombiPal™ autosampler equipped with a heated agitator (Trajan Scientific, Pflugerville, TX, USA). A constant agitation speed of 500 rpm was used throughout this research, so that extraction would only depend on the SPME fiber geometry and diffusion coefficients of the aroma compounds. The instrument was modified after marketing with a Dean's switch for heartcutting, the ability for cryogenic focusing, FID, and the olfactometry port. A detailed schematic of the instrument can be found elsewhere [15]. The GC contains two columns connected in series. The first non-polar column was BPX-5 stationary phase with the following dimensions: 43.5 m length × 0.53 mm ID × 1.0 μm film thickness (SGE Analytical Science, by Trajan, Austin, TX, USA). The second cross-linked polar column was BP-20 (Wax) with the following dimensions: 25 m length × 0.53 mm ID × 1.0 μm film thickness (SGE Analytical Science, by Trajan, Austin, TX, USA). A constant pressure of 5.8 psi was provided at the midpoint between the first and second column using MultiTrax™ V.6.00 (Microanalytics, a Volatile Analysis company, Round Rock, TX, USA) system automation and MSD ChemStation™ E.01.01.335 data acquisition software (Agilent Technologies, Santa Clara, CA, USA). Additional analysis to obtain the total compound chromatogram (TCC) was done using a MassHunter Workstation (Agilent, Santa Clara, CA, USA). Flow from the second analytical column was directed to the single quadrupole mass selective detector and the olfactometry port by fixed restrictor tubing in an open-split interface.

For this research, a full heartcut was utilized from 0.05 to 35.00 min. In other words, the sample flow was first directed through the non-polar column, and then the second polar column, yielding results similar to a mid-polarity GC separation on a long column. Therefore, retention indices were not used for identification in this research, due to the configuration of the two GC capillary columns connected in series.

The following instrument parameters were used: GC inlet temperature, 260 °C; FID, 280 °C; column, 40 °C initial, 3.0 min hold, 7 °C per min ramp, 220 °C final, 11.29 min hold; carrier gas, UHP helium (99.999%) with an inline filter trap. The mass detector was operated in electron ionization (EI) mode with an ionization energy of 70 eV. The mass detector ion source and quadrupole were held at 230 °C and 150 °C, respectively. Full spectrum scans were collected with the mass filter set from $m/z$ 33 to $m/z$ 450. The MS was auto-tuned daily before analysis. The use of a full scan for data acquisition allowed for library search techniques using NIST05 and Wiley 6th edition mass spectral databases.

Olfactometry data was generated using AromaTrax™ V.6.61 software (Microanalytics, a Volatile Analysis company, Grant, AL, USA). Recorded parameters included an aroma descriptor and the perceived intensity. The editable descriptor panel is shown in Figure 1. The area under the peak of each aroma note in the aromagram is calculated as width × intensity × 100, where the width is the length of time that the aroma persisted in minutes. The sum of the areas under the peaks in the aromagram is the total odor, a dimensionless value used to analyze the total aroma detected by the human nose.

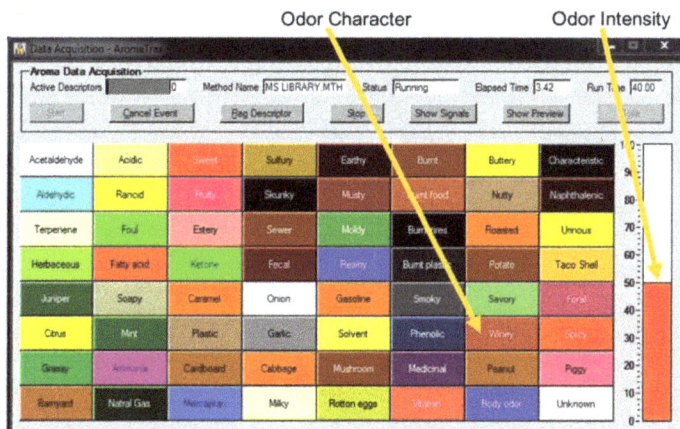

**Figure 1.** Aroma descriptor panel used to characterize volatiles and perceived intensities.

## 3. Results

### 3.1. SPME Optimization

#### 3.1.1. SPME Coating Selection

Seven commercially available SPME coatings were selected for optimization in the extraction of aroma compounds from Iowa red wines (Table 1). The response of the mass spectrometer to volatiles detected throughout the run was used to determine the extraction efficiency of the SPME coating. It was shown that the use of a 50/30 μm DVB/CAR/PDMS SPME coating was most appropriate for the rest of the experiments. In Table 1, a coating with a number lower than 100 indicates a lower extraction efficiency for that analyte when compared to 50/30 μm DVB/CAR/PDMS. These analytes spanned the entire chromatographic run, representing the range of analytes in wine aroma.

**Table 1.** Optimization of solid phase microextraction (SPME) extraction conditions—Fiber selection.

| RT (min) | Compound | 50/30 μm DVB/CAR/PDMS | 100 μm PDMS | 7 μm PDMS | 85 μm PA | 65 μm PDMS/DVB | 70 μm CW/DVB | 85 μm CAR/PDMS |
|---|---|---|---|---|---|---|---|---|
| 4.77 | Ethyl isobutyrate | 100 | 58 | 0 | 13 | 75 | 29 | 104 |
| 5.53 | Isobutyl alcohol | 100 | 42 | 1 | 95 | 81 | 97 | 83 |
| 8.10 | Isoamyl alcohol | 100 | 35 | 0 | 73 | 75 | 66 | 92 |
| 11.00 | Ethyl lactate | 100 | 17 | 0 | 71 | 60 | 79 | 132 |
| 11.75 | Ethyl caproate | 100 | 24 | 0 | 7 | 53 | 17 | 109 |
| 12.92 | Acetic acid | 100 | 5 | 3 | 124 | 38 | 242 | 158 |
| 16.38 | Ethyl caprylate | 100 | 67 | 1 | 21 | 98 | 45 | 64 |
| 18.44 | Vitispirane | 100 | 61 | 0 | 16 | 90 | 38 | 99 |
| 18.34 | Diethyl succinate | 100 | 52 | 0 | 45 | 105 | 56 | 63 |
| 20.42 | Ethyl decanoate | 100 | 114 | 12 | 56 | 121 | 72 | 53 |
| 21.13 | Benzenethanol | 100 | 23 | 0 | 74 | 86 | 67 | 82 |
| 24.03 | Ethyl myristate | 100 | 94 | 32 | 59 | 92 | 66 | 41 |
| 30.37 | Ethyl palmitate | 100 | 187 | 132 | 135 | 194 | 155 | 17 |

Bolded numbers indicate when a SPME fiber coating extracted more mass than the 50/30 μm DVB/CAR/PDMS coating. DVB: Divinylbenzene; PA: Polyacrylate; PDMS: Polydimethylsiloxane; CW: Carbowax; CAR: Carboxen.

#### 3.1.2. Extraction Time

Different extraction times were tested using the autosampler. These times were 10 s, 30 s, 1 min, 3 min, 5 min, 10 min, 15 min, 20 min, 30 min, and 60 min. Plots of the mass extracted versus the extraction time varied in shape. The profiles were typically linear or logarithmic, with the exception of acetic acid (Figure 2). Equilibrium was reached for most compounds (i.e., the logarithmic curve had started to flatten out), and was not excessively long, with a figure of about 10 min. Additional SPME fiber sorption capacity limitations were noticed after 10 min, for example, ethyl isobutyrate, ethyl lactate, and acetic acid. An extraction time of 10 min was chosen to avoid these possible interactions due to competitive adsorption and apparent analyte displacement after 10 min.

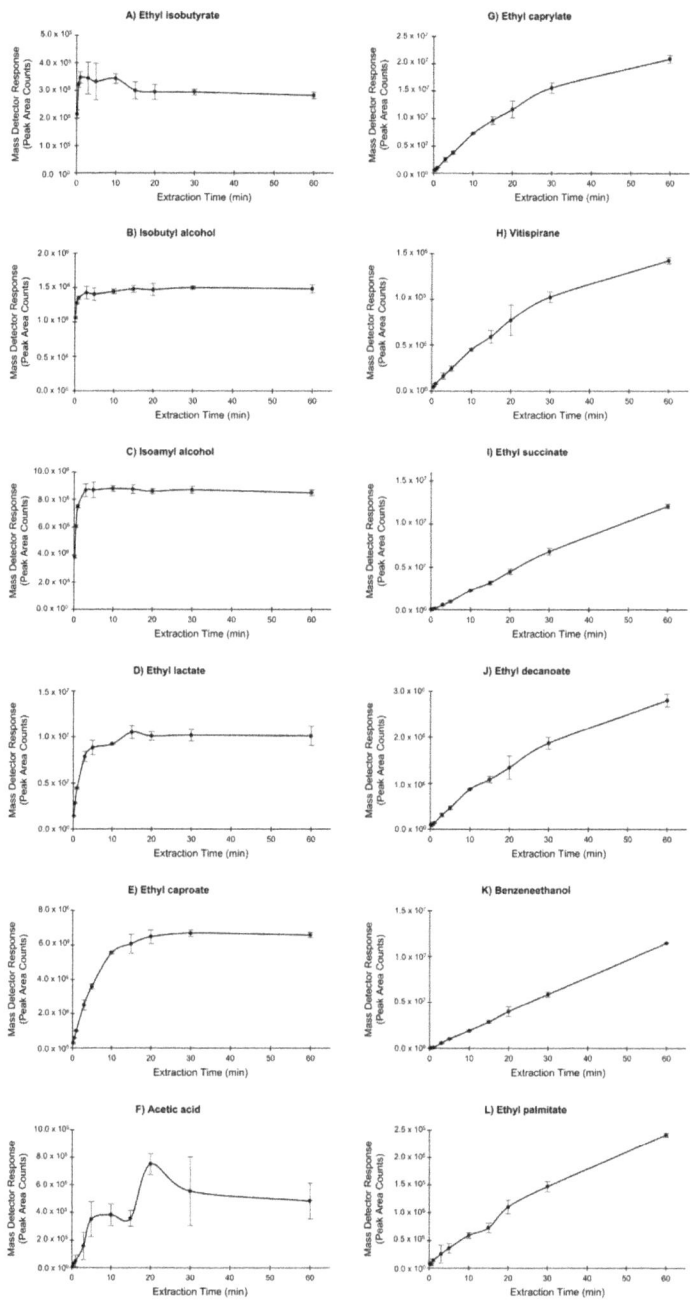

**Figure 2.** Extraction time profile of selected volatiles extracted with HS-SPME in an Iowa Maréchal Foch. Conditions: 5 min pre-sampling desorption of 50/30 μm DVB/CAR/PDMS SPME fiber; 4 mL wine sample in a 10 mL threaded glass amber vial with PTFE/silicone septa; 10 min incubation at 35 °C; 10 s, 30 s, 1 min, 3 min, 5 min, 10 min, 20 min, 30 min, and 60 min extraction time at 35 °C; agitation speed 500 rpm; 5 min desorption time into GC inlet. 10 min extraction time was chosen for the method.

### 3.1.3. Extraction Temperature

Extraction times can be shortened by the efficient use of a higher extraction temperature, as seen in Figure 3. Masses extracted of ethyl caprylate, ethyl succinate, ethyl decanoate, benzeneethanol, and ethyl palmitate increased with a higher temperature. Higher temperatures can also decrease the amount of analyte extracted, as observed in ethyl isobutyrate, isoamyl alcohol, and ethyl caproate. Extraction temperatures of 35, 40, 50, 60, 70, and 80 °C were investigated. To efficiently extract the range of volatiles and semi-volatiles, the optimal temperature chosen for this experiment was 50 °C.

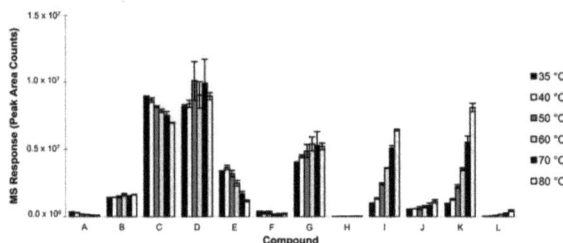

**Figure 3.** Effects of extraction temperature on selected volatiles extracted with HS-SPME in an Iowa Maréchal Foch. Compound letters correspond to compounds found in Figure 2 (i.e., A—ethyl isobutyrate, B—isobutyl alcohol, L—ethyl palmitate). Conditions: 5 min pre-sampling desorption of 50/30 μm DVB/CAR/PDMS SPME fiber; 4 mL wine sample in a 10 mL threaded glass amber vial with PTFE/silicone septa; 10 min incubation at 35, 40, 50, 60, 70, 80 °C; 10 min extraction time at 35, 40, 50, 60, 70, 80 °C; agitation speed 500 rpm; 5 min desorption time into GC inlet. 50 °C was chosen for the method.

### 3.1.4. Incubation Time

An extended incubation time allows for volatiles to equilibrate in the headspace before sampling. This was useful in the extraction of less volatile compounds in the wine sample, such as ethyl caprylate or ethyl decanoate (Figure 4). The incubation time did not have a large effect on the extraction efficiency of more volatile compounds such as ethyl isobutyrate, isobutyl alcohol, and isoamyl alcohol, likely due to their abundance in the headspace of the sample. Incubation times of 0, 5 min, 10 min, 15 min, and 20 min were investigated to establish the equilibrium of analytes in the headspace. The incubation time of 10 min was chosen, as determined by the extracted mass of semi-volatiles.

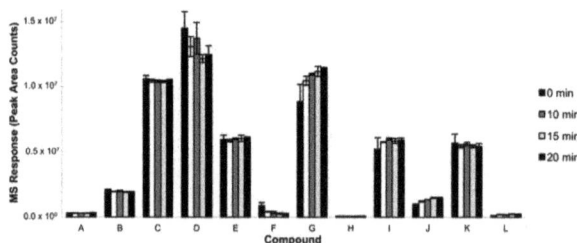

**Figure 4.** Effects of incubation time on selected volatiles extracted using HS-SPME in an Iowa Maréchal Foch. Compound letters correspond to compounds found in Figure 2 (i.e., A—ethyl isobutyrate, B—isobutyl alcohol, L—ethyl palmitate). Conditions: 5 min pre-sampling desorption of 50/30 μm DVB/CAR/PDMS SPME fiber; 4 mL wine sample in a 10 mL threaded glass amber vial with PTFE/silicone septa; 0, 5 min, 10 min, 15 min, and 20 min incubation at 50 °C; 10 min extraction time at 50 °C; agitation speed 500 rpm; 5 min desorption time into GC inlet. 10 min incubation time was chosen for the method.

### 3.1.5. Sample Volume

A 10 mL glass amber vial was used with the autosampler. The sample volume was investigated to maximize the mass extracted from the headspace by SPME. A higher volume of wine would yield a greater mass of volatiles in the headspace, up to the equilibrium. This was observed as an increase in the mass extracted was directly proportional to an increase in the sample volume in ethyl caproate, ethyl caprylate, ethyl succinate, ethyl decanoate, benzeneethanol, and ethyl palmitate (Figure 5). From this experiment, a 4 mL sample volume in a 10 mL vial was chosen for the method.

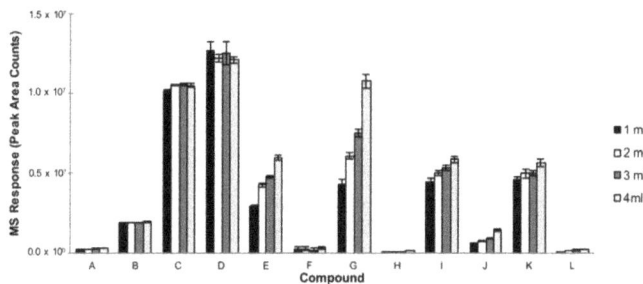

**Figure 5.** Effects of sample volume on selected volatiles extracted using HS-SPME in an Iowa Maréchal Foch. Compound letters correspond to compounds found in Figure 2 (i.e., A—ethyl isobutyrate, B—isobutyl alcohol, L—ethyl palmitate). Conditions: 5 min pre-sampling desorption of 50/30 µm DVB/CAR/PDMS SPME fiber; 1, 2, 3, and 4 mL wine sample in a 10 mL threaded glass amber vial with PTFE/silicone septa; 10 min incubation at 50 °C; 10 min extraction time at 50 °C; agitation speed 500 rpm; 5 min desorption time into GC inlet. 4 mL sample volume was chosen for the method.

### 3.1.6. Desorption Time

Optimizing the desorption time maximizes the transfer of analytes into the instrument for analysis. Desorption times of 30 s, 60 s, 120 s, 180 s, 240 s, and 300 s were used to determine the minimum time needed to desorb analytes from the SPME fiber (Figure 6). A desorption time of 120 s was chosen in a 260 °C injector. The inlet pressure was constant and determined by the pressure needed to maintain balance with the midpoint pressure.

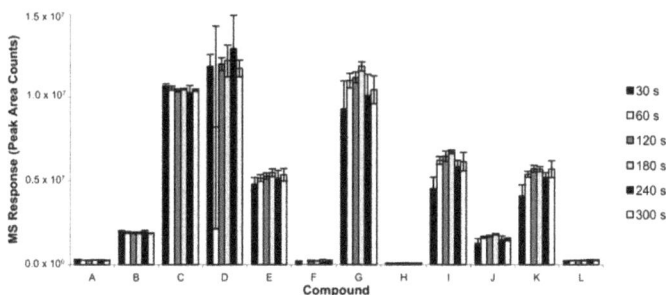

**Figure 6.** Effects of fiber desorption time on selected volatiles extracted in an Iowa Maréchal Foch. Compound letters correspond to compounds found in Figure 2 (i.e., A—ethyl isobutyrate, B—isobutyl alcohol, L—ethyl palmitate). Conditions: 30 s, 60 s, 120 s, 180 s, 240 s, 300 s pre-sampling desorption of 50/30 µm DVB/CAR/PDMS SPME fiber; 4 mL wine sample in a 10 mL threaded glass amber vial with PTFE/silicone septa; 10 min incubation at 50 °C; 10 min extraction time at 50 °C; agitation speed 500 rpm; 30 s, 60 s, 120 s, 180 s, 240 s, 300 s desorption time into GC inlet. 120 s thermal desorption time was chosen for the method.

### 3.1.7. Salt Addition

The addition of sodium chloride was used to adjust the ionic strength of the wine sample. This salting-out effect can help drive analytes to the SPME coating with increasing amounts of salt. The addition of 0.5, 1.0, 1.5, 2, and 2.5 g of salt was investigated to maximize the extraction efficiency (Figure 7). For this experiment, a 2.0 g addition of sodium chloride was chosen for the method.

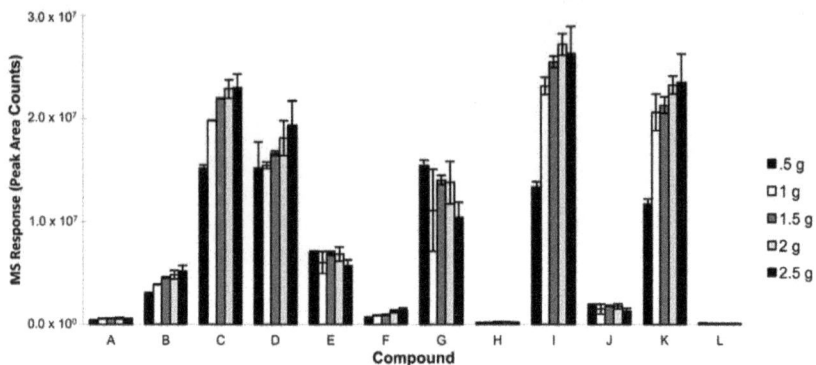

**Figure 7.** Effects of salt addition on selected volatiles extracted in an Iowa Maréchal Foch. Compound letters correspond to compounds found in Figure 2 (i.e., A—ethyl isobutyrate, B—isobutyl alcohol, L—ethyl palmitate). Conditions: 30, 60, 120, 180, 240, 300 s pre-sampling desorption of 50/30 μm DVB/CAR/PDMS SPME fiber; 4 mL wine sample in a 10 mL threaded glass amber vial with PTFE/silicone septa; 10 min incubation at 50 °C; 10 min extraction time at 50 °C; agitation speed 500 rpm; 30, 60, 120, 180, 240, 300 s desorption time into GC inlet. 120 s thermal desorption time was chosen for the method.

## 3.2. Analysis of Wine Samples

### GC-MS-Olfactometry

Tentative identifications of thirty-six compounds by the mass spectral match of their most significant red wine aroma compounds with the ratio of their compound peak area to the internal standard peak area are listed in Table 2. The molecular weight of the compounds identified by mass spectral match ranged from 60 to 284 amu. The relative quantity of each compound to the corresponding ones in the other wine samples was calculated by the peak area to internal standard ratio. The aroma profile of the seven Iowa red wines varied considerably between samples. These results can reflect the influence of the climate, grape variety, vintage, and different viticultural and enological practices in seven different Iowa red wines. Further research is warranted to link the variables to wine aroma.

**Table 2.** Tentative identification by mass spectral match of the most significant red wine aroma wine aroma compounds, listed as the volatile compound peak area: internal standard peak area ratio.

| # | RT (min) | LRI [A] | Stationary Phase 1–10 | Compound | MW | B Foch | B St. Croix | B Frontenac | B Foch/Frontenac | C Vincent | D Frontenac | D Foch |
|---|---|---|---|---|---|---|---|---|---|---|---|---|
| 1 | 3.58 | 949 963 955 558 619 575 976 | 2 2 4 5 8 8 9 | 2,3-Butanedione | 86 | 1.11 | 5.18 | 0.85 | 0.32 | 0.94 | 1.39 | 1.05 |
| 2 | 3.7 | 880 710 | 2 8 | Acetal | 118 | 1.24 | 0.96 | 8.57 | 23.1 | 6.75 | 4.9 | 8.08 |
| 3 | 4.08 | 774 | 3 | Ethyl propanoate | 102 | 1.73 | 1.05 | 7.26 | 19.6 | 3.42 | 6.85 | 5.38 |
| 4 | 4.72 | 956 746 | 2 8 | Ethyl isobutyrate | 116 | 5.94 | 1.67 | 3.99 | 3.54 | 6.36 | 2.3 | 1 |
| 5 | 5.49 | 1110 1054 1083 1093 609 616 | 2 2 4 4 6 8 | Isobutyl alcohol | 74 | 14.6 | 23.8 | 21.9 | 27.6 | 86.1 | 21.4 | 28.7 |
| 6 | 6.07 | 863 788 | 3 5 | Ethyl butanoate | 116 | 1.55 | 3.43 | 3.5 | 2.2 | 2.63 | 1.05 | 3.59 |
| 7 | 6.66 | 1138 1149 634 653 | 4 4 5 6 | n-Butanol | 74 | 0.9 | 1.85 | 1.44 | 4.07 | 1.76 | 4.32 | 1.68 |
| 8 | 7.22 | 1060 856 840 | 2 6 8 | Ethyl isovalerate | 130 | 4.29 | 0.41 | 2 | 1.76 | 0.9 | 0.38 | 0.29 |
| 9 | 8.05 | 1184 719 | 2 8 | Isoamyl alcohol | 88 | 132 | 355 | 275 | 477 | 475 | 221 | 377 |
| 10 | 8.12 | 1110 860 | 2 8 | Isoamyl acetate | 130 | 32.2 | 3.89 | 3.87 | 4.02 | 3.89 | 3.22 | 0 |
| 11 | 10.03 | 1250 | 2 | Styrene | 104 | nd | 16.9 | 3.22 | 0.05 | 0.03 | nd | 0.68 |

**Table 2.** *Cont.*

| # | RT (min) | LRI [A] | Stationary Phase 1–10 | Compound | MW | B Foch | B St. Croix | B Frontenac | B Foch/Frontenac | C Vincent | D Frontenac | D Foch |
|---|---|---|---|---|---|---|---|---|---|---|---|---|
| 12 | 10.91 | 1312 1341 803 | 2 4 8 | Ethyl lactate | 118 | 13.7 | 34.1 | 39.9 | 34.3 | 30 | 67.1 | 19.8 |
| 13 | 11.15 | 1230 1060 1232 1238 985 981 996 996 | 2 3 4 4 5 5 6 6 | Ethyl caproate | 144 | 26.4 | 71 | 40.4 | 33 | 34.7 | 21.5 | 64.3 |
| 14 | 11.27 | 1316 1330 1332 1352 847 848 848 862 1354 | 2 2 2 4 5 5 5 6 9 | n-Hexanol | 102 | 8.64 | 0 | 0 | 19.6 | 12.6 | 7.82 | 51.7 |
| 15 | 12.81 | 1400 1401 791 1435 1442 1459 621 723 | 2 2 3 4 4 4 5 8 | Acetic acid | 60 | 51.4 | 6.61 | 58.2 | 10.6 | 11.6 | 13.2 | 18.7 |
| 16 | 13.25 | 1437 1438 1450 1449 1447 | 2 2 2 2 4 | Furfural | 96 | nd | nd | nd | nd | nd | nd | nd |

Table 2. *Cont.*

| # | RT (min) | LRI [A] | Stationary Phase 1–10 | Compound | MW | [B] Foch | [B] St. Croix | [B] Frontenac | [B] Foch/Frontenac | [C] Vincent | [D] Frontenac | [D] Foch |
|---|---|---|---|---|---|---|---|---|---|---|---|---|
| | | 1456 | 4 | | | | | | | | | |
| | | 1466 | 4 | | | | | | | | | |
| | | 1465 | 4 | | | | | | | | | |
| | | 802 | 5 | | | | | | | | | |
| | | 829 | 5 | | | | | | | | | |
| | | 800 | 5 | | | | | | | | | |
| | | 800 | 5 | | | | | | | | | |
| | | 836 | 6 | | | | | | | | | |
| | | 868 | 6 | | | | | | | | | |
| | | 830 | 6 | | | | | | | | | |
| | | 815 | 8 | | | | | | | | | |
| 17 | 13.57 | | | 3-Nonanone (IS) | 142 | 100 | 100 | 100 | 100 | 100 | 100 | 100 |
| 18 | 14.37 | 1375 | 1 | Methyl octanoate | 158 | 0.39 | 1.01 | 0.32 | 0.23 | 0.35 | 0.18 | 0.33 |
| | | 1392 | 4 | | | | | | | | | |
| | | 1109 | 10 | | | | | | | | | |
| 19 | 14.48 | 1692 | 2 | 1,3-Butanediol | 90 | 1.89 | 4.03 | 11 | 3.38 | 1.48 | 8.15 | 4.43 |
| | | 941 | 8 | | | | | | | | | |
| 20 | 14.81 | 1518 | 1 | Benzaldehyde | 106 | 0.18 | 0.93 | 0.95 | 1.14 | 0.73 | 0.12 | 55.2 |
| | | 1516 | 1 | | | | | | | | | |
| | | 1509 | 2 | | | | | | | | | |
| | | 1454 | 2 | | | | | | | | | |
| | | 1520 | 2 | | | | | | | | | |
| | | 1482 | 2 | | | | | | | | | |
| | | 1502 | 3 | | | | | | | | | |
| | | 1086 | 4 | | | | | | | | | |
| | | 1515 | 4 | | | | | | | | | |
| | | 1496 | 4 | | | | | | | | | |
| | | 1513 | 4 | | | | | | | | | |
| | | 1538 | 4 | | | | | | | | | |
| | | 1530 | 4 | | | | | | | | | |
| | | 1522 | 4 | | | | | | | | | |
| | | 1516 | 4 | | | | | | | | | |

**Table 2.** *Cont.*

| # | RT (min) | LRI [A] | Stationary Phase 1–10 | Compound | MW | B Foch | B St. Croix | B Frontenac | B Foch/Frontenac | C Vincent | D Frontenac | D Foch |
|---|---|---|---|---|---|---|---|---|---|---|---|---|
| | | 926 | 5 | | | | | | | | | |
| | | 926 | 5 | | | | | | | | | |
| | | 926 | 5 | | | | | | | | | |
| | | 924 | 6 | | | | | | | | | |
| | | 960 | 6 | | | | | | | | | |
| | | 962 | 6 | | | | | | | | | |
| | | 957 | 6 | | | | | | | | | |
| | | 961 | 7 | | | | | | | | | |
| | | 944 | 7 | | | | | | | | | |
| | | 938 | 8 | | | | | | | | | |
| | | 947 | 8 | | | | | | | | | |
| | | 947 | 9 | | | | | | | | | |
| | | 1540 | | | | | | | | | | |
| 21 | 15.02 | 1501 | 2 | Isobutyric acid | 88 | 2.45 | 0.28 | 1.38 | 0.78 | 2.46 | 0.45 | 0.32 |
| | | 935 | 3 | | | | | | | | | |
| 22 | 15.15 | | | 2,3-Butanediol | 90 | 1.41 | 1.56 | 2.31 | 2.03 | 1.5 | 2.62 | 2.35 |
| 23 | 15.72 | 1258 | 3 | Ethyl caprylate | 172 | 164 | 398 | 209 | 116 | 187 | 181 | 205 |
| | | 1429 | 4 | | | | | | | | | |
| | | 1466 | 4 | | | | | | | | | |
| | | 1196 | 6 | | | | | | | | | |
| | | 1193 | 6 | | | | | | | | | |
| | | 1195 | 6 | | | | | | | | | |
| 24 | 16.83 | 1631 | 2 | Isovaleric acid | 102 | 8.97 | 2.58 | 7.79 | 5.1 | 3.73 | 3.54 | 3.55 |
| | | 834 | 6 | | | | | | | | | |
| 25 | 17.66 | 1276 | 7 | Vitispirane | 192 | 0 | 3.02 | 0.31 | 4.5 | 15.7 | 3.39 | 0 |
| 26 | 17.79 | 1278 | 5 | Ethyl nonanoate | 186 | 0.83 | nd | 1.68 | 0 | 0 | 0 | 0.78 |
| 27 | 17.99 | 1642 | 2 | Diethyl succinate | 174 | 3.62 | 11.5 | 18.9 | 43.8 | 12.9 | 5.77 | 122 |
| | | 1153 | 8 | | | | | | | | | |
| 28 | 19.74 | 1390 | 6 | Ethyl decanoate | 200 | 54.9 | 99.5 | 83.9 | 29.8 | 33.3 | 98.5 | 34.4 |
| | | 1391 | 6 | | | | | | | | | |
| | | 1394 | 6 | | | | | | | | | |

Note: # is the chromatographic peak number. RT is the retention time in minutes. A is the GC capillary column linear retention index from the LRI & Odour Database [16]. GC stationary phases are: 1—CP-Wax, 2—CW-20M, 3—DB-1701, 4—DB-Wax, 5—DB1, 6—DB5, 7—HP-1, 8—OV-101, 9—SP-Wax, 10—SPB-1. MW is the molecular weight. B, C, and D stand for the three different Iowa wineries. IS = internal standard.

**Table 2.** *Cont.*

| # | RT (min) | LRI [A] | Stationary Phase 1–10 | Compound | MW | [B] Foch | [B] St. Croix | [B] Frontenac | [B] Foch/Frontenac | [C] Vincent | [D] Frontenac | [D] Foch |
|---|---|---|---|---|---|---|---|---|---|---|---|---|
| 29 | 20.27 | 1788<br>1785<br>1233 | 2<br>2<br>8 | Phenethyl acetate | 164 | 0.95 | 1.56 | 1.26 | 1.41 | 0.95 | 1.47 | 1.2 |
| 30 | 20.98 | 1903 | 4 | Benzeneethanol | 122 | 41.3 | 146 | 56 | 232 | 111 | 53.7 | 134 |
| 31 | 23.06 | 2007<br>2007<br>2100<br>2013<br>2075<br>1183<br>1256 | 2<br>2<br>2<br>4<br>6<br>7 | Octanoic acid | 144 | 1.39 | 1.07 | 6.63 | 0.09 | 1.99 | 2.22 | 0.7 |
| 32 | 23.36 | 1595 | 6 | Ethyl laurate | 228 | 3.3 | 2.63 | 6.73 | 1.62 | 1.28 | 8.24 | 2.72 |
| 33 | 24.76 | | | 4-Ethylphenol | 122 | nd | 7.49 | 2.32 | nd | nd | 3.4 | 2.33 |
| 34 | 24.81 | | | 8-Pentadecanone | 226 | 1.06 | 7.49 | 2.31 | 2.53 | 2.33 | 3.4 | 2.32 |
| 35 | 26.57 | | | Glycerol | 92 | nd | nd | 26.8 | nd | nd | nd | nd |
| 36 | 26.63 | 1793 | 6 | Ethyl myristate | 256 | 0.39 | 0.53 | 0.48 | 0.38 | 0.57 | 0.84 | 0.99 |
| 37 | 29.63 | 1993<br>1985 | 6<br>7 | Ethyl palmitate | 284 | 2.35 | 2.1 | 5.75 | 1.92 | 2.81 | 4.12 | 5.14 |

Note: # is the chromatographic peak number. RT is the retention time in minutes. A is the GC capillary column linear retention index from the LRI & Odour Database [16]. GC stationary phases are: 1—CP-Wax, 2—CW-20M, 3—DB-1701, 4—DB-Wax, 5—DB1, 6—DB5, 7—HP-1, 8—OV-101, 9—SP-Wax, 10—SPB-1. MW is the molecular weight. B, C, and D stand for the three different Iowa wineries. IS = internal standard.

Simultaneous olfactometry, when used with GC-MS, can verify compounds by aroma character. An example of this is highlighted in Figure 8. The chromatographic peak at 19 min was not identified by mass spectral comparison, but it was recorded with an aroma character of burnt food (aromagram peak number 27). An open source aroma database search narrows the possible identification of this compound from hundreds to six: octanol, indole, 3-methyl-1-butanol, ethyldimethylpyrazine, dimethyl sulfone, or furfuryl alcohol [17]. The use of olfactometry, by means of the human nose as a detector, is a very valuable tool for the identification of unknown compounds. The total aroma values, calculated as the sum of the area under the aromagram peaks, are compared in Figure 9 and no significant differences are exhibited between the seven Iowa red wines.

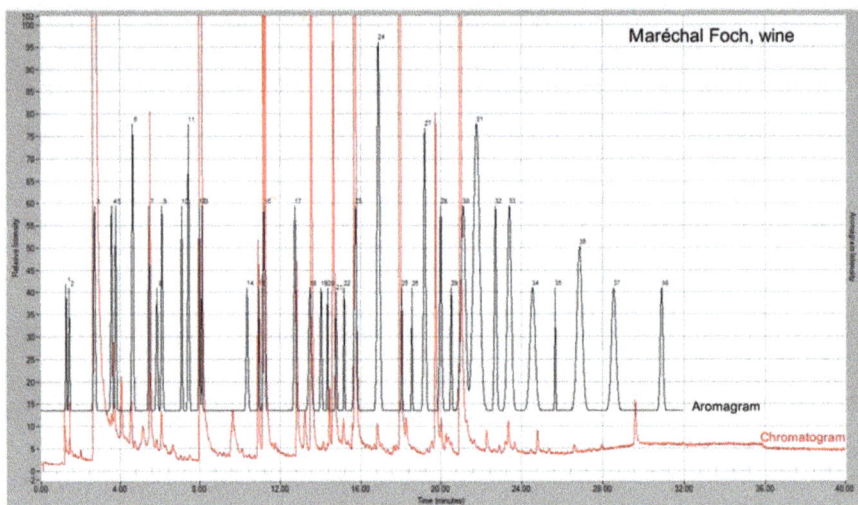

**Figure 8.** Overlay of total ion chromatogram and aromagram of an Iowa Maréchal Foch wine using the optimized method. Intense aromas (observed as increased peak height, black signal) and responsible chemical compounds (TIC, red signal) were aromagram peak: (#6) sweet, fruity—ethyl isobutyrate (4.71 min); (#11) fruity—ethyl isovalerate (7.21 min); (#24) rancid, sweaty, body odor, burnt—isovaleric acid (16.83 min) (#27) burnt, burnt food—unknown compound (19 min); (#31) sweet, fruity, winey—ethyl laurate (23.26 min). Conditions: 2 min pre-sampling desorption of 50/30 μm DVB/CAR/PDMS SPME fiber; 4 mL wine sample in a 10 mL threaded glass amber vial with PTFE/silicone septa; 10 min incubation at 50 °C; 10 min extraction time at 50 °C; agitation speed 500 rpm; 2 min desorption time into GC inlet.

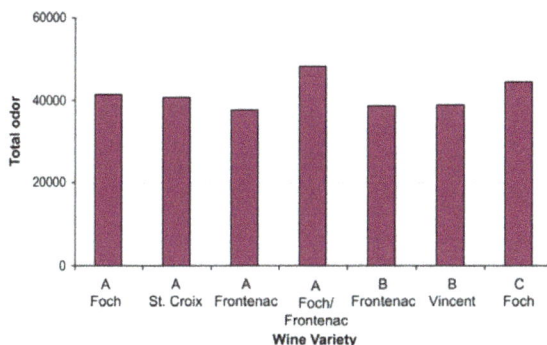

**Figure 9.** Overlay of total ion chromatogram and aromagram of an Iowa Maréchal Foch wine using the optimized method. Intense aromas (observed as increased peak height, black signal) and responsible chemical compounds (TIC, red signal) were aromagram peaks: (#6) sweet, fruity—ethyl isobutyrate (4.71 min); (#11) fruity—ethyl isovalerate (7.21 min); (#24) rancid, sweaty, body odor, burnt—isovaleric acid (16.83 min) (#27) burnt, burnt food—unknown compound (19 min); (#31) sweet, fruity, winey—ethyl laurate (23.26 min). Conditions: 2 min pre-sampling desorption of 50/30 μm DVB/CAR/PDMS SPME fiber; 4 mL wine sample in a 10 mL threaded glass amber vial with PTFE/silicone septa; 10 min incubation at 50 °C; 10 min extraction time at 50 °C; agitation speed 500 rpm; 2 min desorption time into GC inlet.

## 4. Discussion

The mixed adsorbent beds of the 50/30 μm DVB/CAR/PDMS coating were best suited for the extraction of wine volatiles, and extracted a greater mass of analytes than the other six fibers. The dual layer Car/PDMS/DVB fiber has been used to overcome the lack of selectivity toward some of the compounds in the single and double-phase fibers [18] and is consistent with previous work by Howard et al. [19]. Even though extraction equilibrium was not reached in some analytes (i.e., the profile is linear in shape at 10 min), precision was assured by using the autosampler to control the mass transfer conditions. The addition of salt can improve the extraction efficiency up to a point, where the target analytes may interact with the salt ions in solution. These interactions will then reduce the extraction efficiency. This phenomenon has been shown to be related to the pKa of the analyte [20]. The total wine aroma is a balance between the heavier aroma of the alcohols, esters, acids, and the unpleasant rancid odors of the aliphatic acids and carbonyls which can be formed during the fermentation process. It should be noted that these 36 compounds were detected in the presence of a highly volatile organic solvent. In Figure 10, ethanol is present at 2.8–3.0 min and is the most abundant compound in the headspace of wine, as expected.

Four esters, including ethyl caproate, ethyl isobutyrate, ethyl isovalerate, and isoamyl acetate, were detected in seven Iowa red wines. In only one previous study, nonanal, (E,Z)-2,6-nonadienal, β-damascenone, ethyl caprylate, and isoamyl acetate had the highest OAVs in Frontenac, Marquette, Maréchal Foch, Sabrevois, and St. Croix wines, using SPME-GCMS(TOF) [21]. Ethyl caproate was previously reported in an analysis of Frontenac and Marquette juice from Quebec using SPME-GCMS [22]. Five compounds in selected Iowa red wines (i.e., isoamyl alcohol, ethyl caproate, benzeneethanol, ethyl decanoate, and ethyl caproate) were also found in Cabernet Sauvignon and Merlot, where ethyl caproate and ethyl caprylate were reported as the most abundant ethyl esters in these vinifera varieties [23]. These compounds have been attributed to yeast metabolism and do not impart any varietal characteristics to wine [24].

A principal components analysis followed by hierarchical clustering analysis is shown in Figure 11. St. Croix (from winery B) and Frontenac (from wineries B and D) are distinguishable by the grape variety and from the other five wine samples. Frontenac from winery B was more significantly associated with (31) octanoic acid than Frontenac from winery D. Maréchal Foch wines (from wineries B and D) were not similar to each other or when blended with Frontenac (from winery B). Maréchal Foch from winery B was associated (8) with ethyl isovalerate, Maréchal Foch from winery D was associated with (20) benzaldehyde, and Maréchal Foch/Frontenac blend from winery B was associated with (16) furfural. Vincent wine from winery C and the Maréchal Foch/Frontenac blend from winery B were similar in wine aroma. It cannot be determined if the difference in aroma is due to the variety or winemaking practices for the Maréchal Foch, Maréchal Foch/Frontenac blend, or Vincent wines.

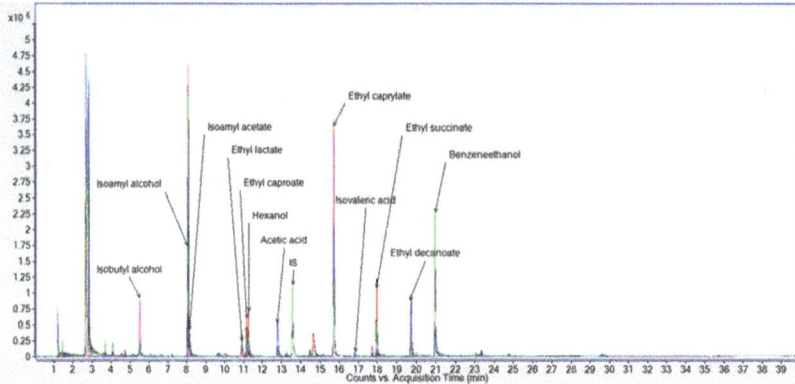

**Figure 10.** Overlay of total compound chromatograms of seven selected Iowa red wines using the optimized method. Wines included: two bottles of Maréchal Foch, one bottle of St. Croix, two bottles of Frontenac, one bottle of Vincent, and one bottle of Maréchal Foch/Frontenac blend. Conditions: 2 min pre-sampling desorption of 50/30 µm DVB/CAR/PDMS SPME fiber; 4 mL wine sample in a 10 mL threaded glass amber vial with PTFE/silicone septa; 10 min incubation at 50 °C; 10 min extraction time at 50 °C; agitation speed 500 rpm; 2 min desorption time into GC inlet.

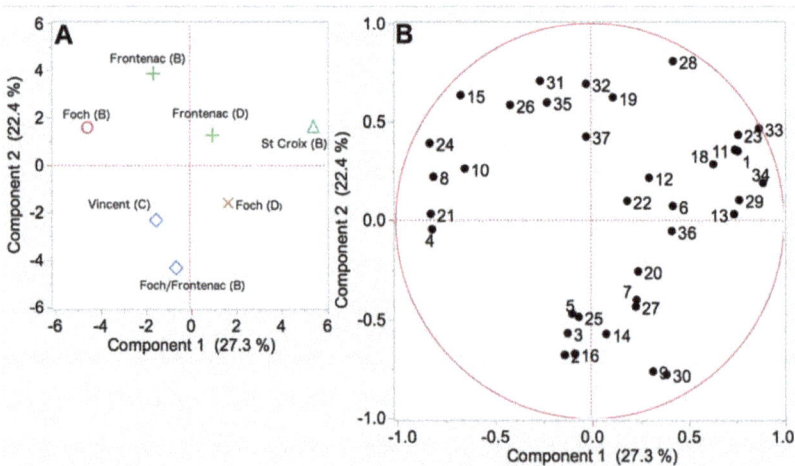

**Figure 11.** A principal components analysis of wine aroma of seven wines from three different Iowa wineries (**A**) and variables (**B**). Numbered variables refer to peak numbers in Table 2. Internal standard (17) 3-nonanone is not included in (**B**) because there was no variation between the samples.

An automated headspace SPME method, coupled with GCMS-olfactometry was developed to characterize wine aroma. This method was applied to characterize 36 aroma compounds present in seven Iowa red wines from three different wineries. The interactions between the experimental factors were not considered in this research. A multivariate experimental design aimed to determine the main factors followed by a response surface methodology [25] could yield better results. Although a distinct varietal aroma character was not 'pinpointed' for these Iowa red wines, there were notable differences in the aroma profile by grape variety and by the winery. Linking aroma compounds to grape variety and winemaking practices continues to be important in producing high quality Iowa wines.

**Acknowledgments:** The authors acknowledge funding provided through the Iowa Department of Agricultural and Land Stewardship (07-1273; Project title: Development of a rapid screening method for chemical and aroma analyses of Iowa wines: transresveratrol and aroma-enhancing compounds).

**Author Contributions:** L.C., J.A.K., and M.D. conceived and designed the experiments; L.C. performed the experiments; L.C. and S.R. analyzed the data; J.A.K. and M.D. contributed reagents/materials/analysis tools; L.C., S.R., J.A.K., and D.M. wrote the paper.

**Conflicts of Interest:** The authors declare no conflict of interest. The funding sponsors had no role in the design of the study; in the collection, analyses, or interpretation of data; in the writing of the manuscript, and in the decision to publish the results.

# References

1. Schreier, P.; Jennings, W.G. Flavor composition of wines: A review. *CRC Crit. Rev. Food Sci. Nutr.* **1979**, *12*, 59–111. [CrossRef] [PubMed]
2. Rapp, A. Wine aroma substances from gas chromatographic analysis. In *Modern Methods of Plant Analysis*; Linskens, H.F., Jackson, J.F., Eds.; Springer: Berlin, Germany, 1988; Volume 6, pp. 29–66. [CrossRef]
3. Sauvageot, F.; Vivier, P. Effects of malolactic fermentation on sensory properties of four burgundy wines. *Am. J. Enol. Vitic.* **1997**, *48*, 187–192.
4. De Revel, G.; Bertrand, A. Dicarbonyl compounds and their reduction products in wine. Identification of wine aldehydes. In *Trends in Flavour Research*; Maarse, H., van der Heij, D.G., Eds.; Elsevier Science B.V.: Amsterdam, The Netherlands, 1994; Volume 35, p. 353. ISBN 0444815872.
5. Allen, M. What level of methoxypyrazines is desired in red wines? The flavour perspective of the classic red wines of Bordeaux. *Aust. Grapegrow. Winemak.* **1995**, *381*, 7–9.
6. Davis, C.R.; Wibowo, D.; Eschenbruch, R.; Lee, T.H.; Fleet, G.H. Practical implications of malolactic fermentation: A review. *Am. J. Enol. Vitic.* **1985**, *36*, 290–301.
7. Rapp, A.; Güntert, M.; Almy, J. Identification and significance of several sulfur-containing compounds in wine. *Am. J. Enol. Vitic.* **1985**, *36*, 219–221.
8. Montedoro, G.; Bertuccioli, M. The flavour of wines, vermouth and fortified wines. In *Food flavours, Part B, The Flavour of Beverages*; Morton, I.D., MacLeod, A.J., Eds.; Elsevier Science Ltd.: Amsterdam, The Netherlands, 1986; p. 171. ISBN 9780444425997.
9. Rapp, A.; Versini, G. Methylanthranilate ("foxy taint") concentrations of hybrid and Vitis vinifera wines. *Vitis* **1996**, *35*, 215–216.
10. Margalit, Y. *Concepts in Wine Chemistry*, 3rd ed.; The Wine Appreciation Guild: San Francisco, CA, USA, 2012; p. 203. ISBN 1-935879-94-4.
11. Fuller, G.H.; Steltenkamp, R.; Tisserand, G.A. The gas chromatograph with human sensor: Perfumer model. *Ann. N. Y. Acad. Sci.* **1964**, *116*, 711–724. [CrossRef] [PubMed]
12. Robinson, A.L.; Boss, P.K.; Solomon, P.S.; Trengrove, R.D.; Heymann, H.; Ebeler, S.E. Origins of grape and wine aroma. Part 2. Chemical and sensory analysis. *Am. J. Enol. Vitic.* **2014**, *65*, 25–42. [CrossRef]
13. Panighel, A.; Flamini, R. Solid phase extraction and solid phase microextraction in grape and wine volatile compounds analysis. *Sample Prep.* **2015**, *2*, 55–65. [CrossRef]
14. Pawliszyn, J. *Handbook of Solid Phase Microextraction*; Chemical Industry Press: Beijing, China, 2009; pp. 90–103. ISBN 978-7-122-04701-4.

15. Zhang, S.; Koziel, J.A.; Cai, L.; Hoff, S.J.; Heathcote, K.Y.; Chen, L.; Jacobson, L.D.; Akdeniz, N.; Hetchler, B.P.; Parker, D.B.; et al. Odor and odorous chemical emissions from animal buildings: Part 5. Simultaneous chemical and sensory analysis with gas chromatography-mass spectrometry-olfactometry. *Trans. ASABE* **2015**, *58*, 1349–1359. [CrossRef]

16. LRI & Odour Database. Available online: http://www.webcitation.org/6rDRATLKY (accessed on 14 June 2017).

17. Flavornet and Human Odor Space. Available online: http://www.webcitation.org/6qr4xTprN (accessed on 30 May 2017).

18. Ferreira, A.C.S.; de Pinho, P.G. Analytical method for determination of some aroma compounds on white wines by solid phase microextraction and gas chromatography. *J. Food Sci.* **2003**, *68*, 2817–2820. [CrossRef]

19. Howard, K.L.; Mike, J.H.; Riesen, R. Validation of a solid-phase microextraction method for headspace analysis of wine aroma compounds. *Am. J. Enol. Vitic.* **2005**, *56*, 37–45.

20. Hall, B.J.; Brodbelt, J.S. Determination of barbiturates by solid-phase microextraction (SPME) and ion trap gas chromatography-mass spectrometry. *J. Chromatogr. A* **1997**, *777*, 275–282. [CrossRef]

21. Slegers, A.; Angers, P.; Ouellet, E.; Truchon, T.; Pedneault, K. Volatile compounds from grape skin, juice, and wine from five interspecific hybrid grape cultivars grown in Quebec (Canada) for wine production. *Molecules* **2015**, *20*, 10980–11016. [CrossRef] [PubMed]

22. Pedneault, K.; Martine, D.; Angers, P. Flavor of cold-hardy grapes: Impact of berry maturity and environmental conditions. *J. Agric. Food Chem.* **2013**, *61*, 10418–10438. [CrossRef] [PubMed]

23. Cheng, G.; Liu, Y.; Yue, T.X.; Zhang, Z.W. Comparison between aroma compounds in wines from four Vitis vinifera grape varieties grown in different shoot positions. *Food Sci. Technol. Camp.* **2015**, *35*, 237–246. [CrossRef]

24. Jackson, R.S. *Wine Science Principles and Applications*, 3rd ed.; Elsevier: Amsterdam, The Netherlands, 2008; p. 662. ISBN 978-0-12-373646-8.

25. Gunst, R.F. Response surface methodology: Process and product optimization using designed experiments. *Technometrics* **1996**, *38*, 284–286. [CrossRef]

MDPI AG

St. Alban-Anlage 66

4052 Basel, Switzerland

Tel. +41 61 683 77 34

Fax +41 61 302 89 18

http://www.mdpi.com

*Separations* Editorial Office

E-mail: separations@mdpi.com

http://www.mdpi.com/journal/separations